A DRY OASIS

A DRY OASIS

INSTITUTIONAL ADAPTATION TO CLIMATE ON THE CANADIAN PLAINS

edited by
GREGORY P. MARCHILDON

University of Regina

CPRC PRESS

Printed and bound in Canada at Friesens.
This book is printed on 100% post-consumer recycled paper.

Edited by: Gregory P. Marchildon
Cover and text design by Duncan Campbel, CPRC.
Editor for the Press: David McLennan, CPRC.

Library and Archives Canada Cataloguing in Publication

A dry oasis : institutional adaptation to climate on the Canadian plains / edited by Gregory P. Marchildon.

Includes index.
Special issue of Prairie forum, vol. 34, No. 1.
ISBN 978-0-88977-217-5

1. Pallisers Triangle—History. 2. Pallisers Triangle—Environmental conditions. 3. South Saskatchewan River Watershed (Alta. and Sask.). 4. Climatic changes—Environmental aspects—Pallisers Triangle. 5. Climatic changes—Environmental aspects—South Saskatchewan River Watershed (Alta. and Sask.). 6. Climatic changes—Prairie Provinces—Case studies. 7. Climatic changes—Prairie Provinces—Forecasting. 8. Climatic changes—Government policy—Canada.
I. Marchildon, Gregory P., 1956–
II. University of Regina. Canadian Plains Research Center

GE160.C3D79 2009 363.738'740971242 C2009-905813-8

Canadian Plains Research Center
University of Regina
Regina, Saskatchewan
Canada, S4S 0A2
TEL: (306) 585-4758 FAX: (306) 585-4699
E-MAIL: canadian.plains@uregina.ca
WEB: www.cprc.uregina.ca
www.cprcpress.com

We acknowledge the financial support of the Government of Canada through the Book Publishing Industry Development Program (BPIDP) for our publishing activities. We also acknowledge the support of the Canada Council for the Arts for our publishing program.

Canadian Heritage Patrimoine canadien

Canada Council for the Arts Conseil des Arts du Canada

CONTENTS

PART III: CASE STUDIES OF VULNERABILITY

INTRODUCTION

INSTITUTIONAL ADAPTATION TO CLIMATE CHANGE IN THE CANADIAN PLAINS

GREGORY P. MARCHILDON[1]

For the past two decades, the United Nations' Intergovernmental Panel on Climate Change (IPCC) has issued increasingly dire reports concerning the impact of global warming. In its fourth report issued in 2007, the IPCC concluded that world temperatures are likely to increase between 1.10 and 6.4°C over this century. For decades, scientists have shown that a significant increase in greenhouse gases will lead to a more drought-prone climate in the Great Plains of North America even if there are some extremely wet years.[2] Current global climate models not only project a drying of the summer continental interior, but a 66% probability of an increase in the area affected by drought in the 21st century as compared to the 20th century.[3]

The implications of these predicted changes are dire for the Canadian Plains, where even a small change in temperature can have a dramatic impact on the availability of water. During his mid-19th century survey, Captain John Palliser viewed the Canadian Plains as an extension of the Great American Desert. A large portion of the Canadian Plains—called the Palliser Triangle—was viewed as unsuitable for agriculture given the lack of precipitation and soil moisture.[4]

Nestled largely (but not exclusively) within the Palliser Triangle, the South Saskatchewan River Basin (SSRB) is the major watershed in the southern part of the Prairie Provinces. With its limited precipitation, a sizeable population within the Palliser Triangle are highly dependent—directly and indirectly—on the rivers of the SSRB as the main source of water for drinking, irrigated agriculture, ranching, industry and recreation. The SSRB stretches from the Rocky Mountains across southern Alberta and Saskatchewan. From west to east, it covers an area of roughly 170,000 square kilometres and contains a population

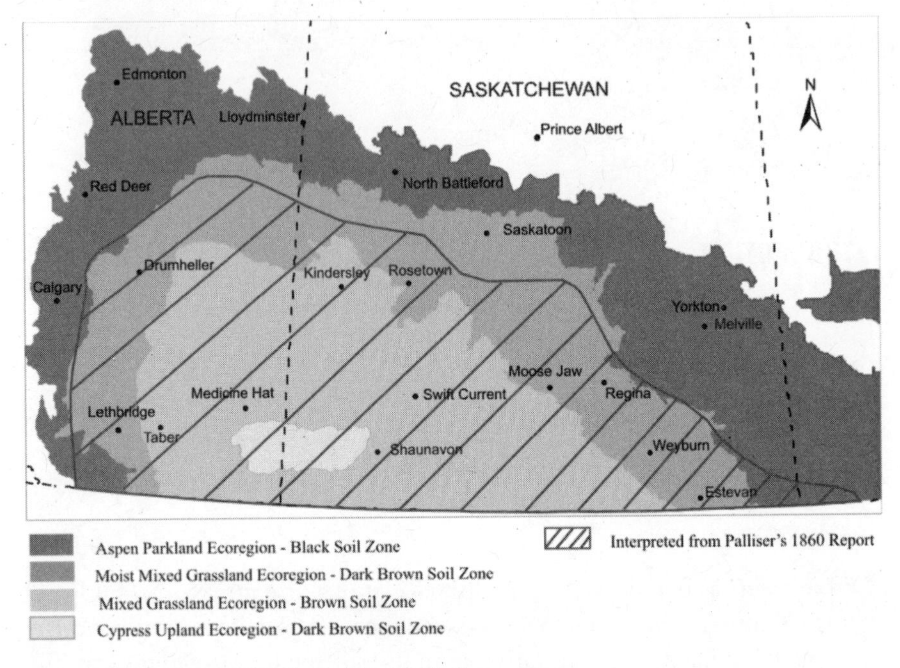

Figure 1. The Palliser Triangle and the Prairie Ecoregions.

of 2.2 million people who rely on the major river systems within the basin.[5] Although 80% of the SSRB's population lives within major urban centres such as Calgary and Saskatoon, there are 225 rural communities (townspeople as well as farmers and ranchers) with small, dispersed populations.[6]

Most of the land within the basin is devoted to dryland cultivation, particularly commercial cereal crops such as wheat and canola which are produced on medium to large-scale farms. Significant portions of the basin are pasture lands devoted to livestock production because of the dry nature of the SSRB. There is large-scale irrigation concentrated in parts of the SSRB such as the region around Lethbridge, Alberta. The largest dammed reservoir is Lake Diefenbaker, water from which is used for drinking (45% of the population of Saskatchewan depends on water from this source), agricultural irrigation, recreation, and to generate hydro-electricity.[7]

The purpose of this volume is to better understand how individuals living in the Prairie Provinces—responding through their respective communities and governments—can best adjust to the impact of global warming in the 21st century. Institutional adaptations are organized and collective responses to climate change. For example, in response to the Dust Bowl of the 1930s, while some families abandoned their farms and fled to more promising regions, other farm families stayed and adjusted their agricultural methods, aided by government assistance and new organizations such as the Government of

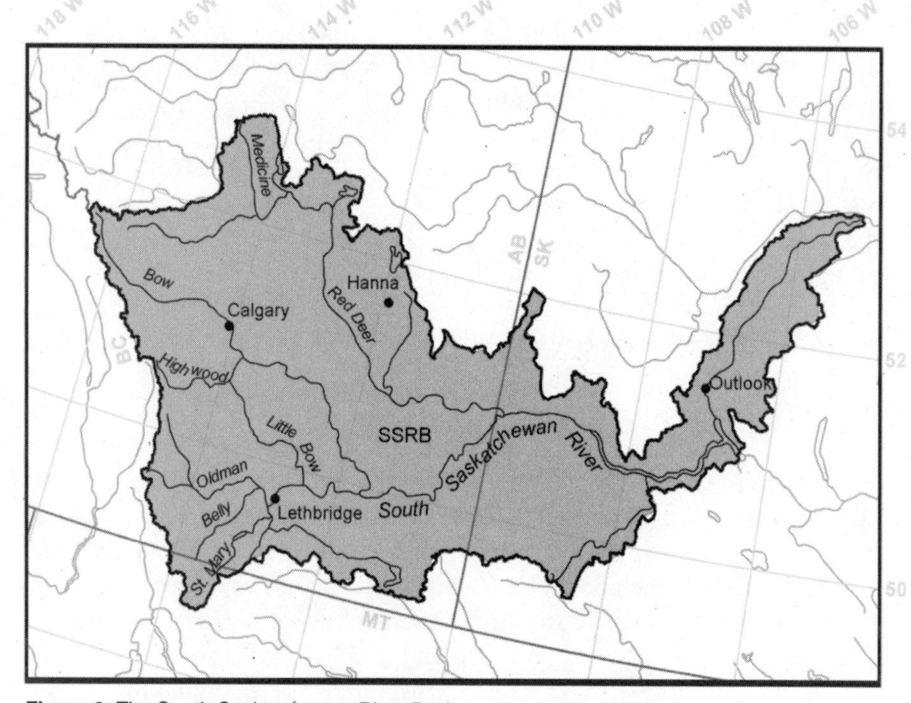

Figure 2. The South Saskatchewan River Basin.

Alberta's Special Areas Administration and the Government of Canada's Prairie Farm Rehabilitation Administration (PFRA).[8] In this way, they were able to feed and clothe their children, seed their fields, pay for their inputs, provide additional grazing land for their cattle, improve their dryland farming practices and water their livestock. From this history, we know that institutions, including the legal and policy framework in which they operate, will be critical in reducing vulnerability and facilitating adaptation to climate change in the 21st century.

The IPCC defines vulnerability as the degree to which a given community is susceptible to, or unable to cope with, the adverse effects of climate change, including any increase in climate variability and extremes.[9] As shown in Figure 3, vulnerability is a function of exposure (including sensitivity as a result of land use patterns) and the adaptive capacity of a given community to absorb, cope with, manage or recover from the stress of climate change. The most vulnerable communities are those that are the most exposed and the least able to adapt. An example of the overlap between exposure and adaptive capacity is land tenure. The decision to use land for grain farming or ranching is a factor in determining the degree of exposure, and the ability to make changes in land tenure can be part of the adaptive capacity of a community.[10]

Figure 4 illustrates the linkages between past and present vulnerability to future climate change vulnerability. In particular, the future vulnerability of the Canadian Plains will be determined by long-term climate change that may well exacerbate the recurring cycle of prolonged droughts the region has historically experienced, including prolonged droughts in the mid-19th century and in the 1920s and 1930s. It will also be shaped by existing land tenure and human habitation patterns that, beyond climatic factors, influence the degree of exposure to future water shortages and droughts. Finally, future vulnerability is determined by future institutional adaptations that are, in turn, built upon past and current laws, organizations and public and community-based institutions, which include the PFRA (now a branch of Agriculture and Agri-Food Canada), provincial watershed authorities, as well as local government and civil society initiatives. While this model of vulnerability is not deterministic, past and present structures and behaviours play a large role in determining future adaptive structures and behaviours.

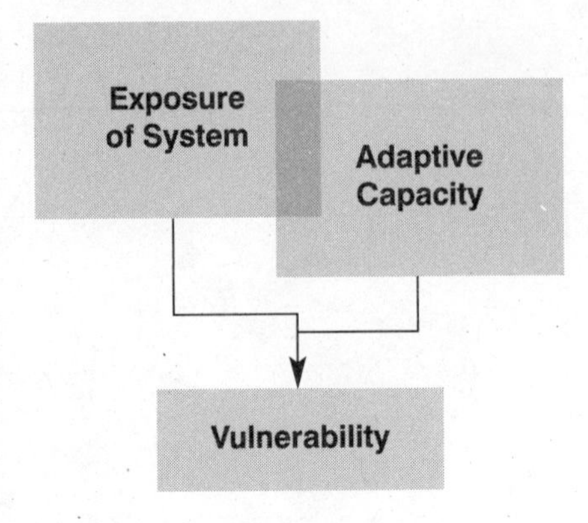

Figure 3. The Concept of Vulnerability.

Almost all of the authors of the chapters in this book have been closely associated with a five-year comparative project on Institutional Adaptation to Climate Change covering the SSRB in Canada and a similarly vulnerable region in Chile known as the Elqui River Basin.[11] To assist their work, the authors have used a common conceptual framework concerning vulnerability to climate change, more specifically addressing vulnerability in light of severe water shortages produced or exacerbated by climate change.

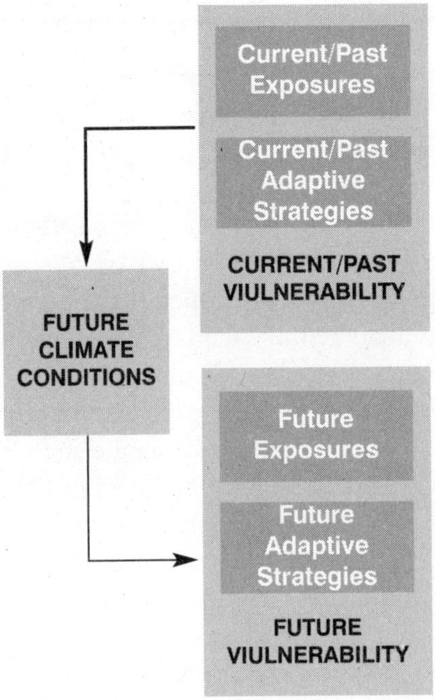

Figure 4. A Model of Vulnerability.

The first four chapters approach

the question of institutional adaptation to climate change from broad historical and spatial perspectives. James Daschuk examines the susceptibility of the northern grasslands from the arrival of the first Indigenous societies until the introduction of Euro-Canadian agriculture. Jeremy Pittman, David Sauchyn and I investigate the changing pattern and geographical extent of aridity reflecting periods of prolonged drought from the agricultural settlement of the Prairie West until the end of the 20th century. Margot Hurlbert guides us through the different provincial models of water governance in Western Canada and highlights the criteria against which water governance in the 21st century can be evaluated. Lorena Patiño and David Gauthier propose a new approach to geographic information system (GIS) mapping of climate change involving extensive public participation.

The next four chapters focus on describing and analysing the SSRB. Brenda Toth and her colleagues provide an overview of the South Saskatchewan River Basin's natural characteristics including its geography (ecozones), hydrology (river flows, and the quantity and quality of surface and ground water), and climate in terms of prolonged drought—a marked feature of the SSRB. Joel Bruneau and his colleagues examine the human characteristics of the SSRB, discussing the ways in which water has been managed, diverted and consumed in this dry region of Canada. Based upon the best scientific evidence concerning the impact of climate change in the region, Suzan Lapp and her co-authors hypothesize what will likely be the climate and water supply in the SSRB in the future. Margo Hurlbert with her co-authors analyse the impact of water governance on rural communities within the SSRB using interviews with stakeholders and water institutions to answer the question of how these same communities might cope with future climate change.

The final group of articles are all case studies of vulnerability in specific communities in the SSRB. Johanna Wandel, Gwen Young, and Barry Smit study recent drought in the Special Areas of southeast Alberta in order to examine the degree of exposure and the adaptive strategies of communities and organizations in what has historically been the driest sub-region of the Palliser Triangle. Alejandro Rojas and his colleagues use the conflict among First Nations, environmentalists, government, farmers and ranchers over the construction and impact of the Oldman River Dam to determine what lessons can be used in future water conflicts. Harry Diaz and his co-authors assess two small communities in southwestern Saskatchewan—Cabri and Stewart Valley—and their response to the drought of 2001–02 to determine the elements of effective adaptation to future climate change. Based on his detailed study of the Blood Tribe reserve community in southwest Alberta, Lorenzo Magzul explores the vulnerabilities of marginalized Indigenous communities in coping with major climatic change in the future.

All of these studies emphasize the tenuous connection between nature and society. As the consequences of climate change become more pronounced, the necessity of understanding the various impacts of our 'built' environment on the natural environment of the Canadian Plains become more urgent.

Endnotes

1. The editor would like to thank the journal's two anonymous referees for so carefully reviewing the articles in this special issue as well as Angela Scott of the Johnson-Shoyama Graduate School of Public Policy at the University of Regina for her expert editing assistance.
2. G.D.V. Williams, R.A. Fautley, K.H. Jones, R.B. Stewart, and E.E. Wheaton, "Estimating Effects of Climate Change on Agriculture in Saskatchewan, Canada," in M.L. Perry, T.R. Carter, and N.T. Konijn (eds.), *The Impact of Climatic Variations on Agriculture: Volume 1, Assessment in Cool Temperate and Cold Regions* (London: Kluwer Academic Publishers, 1988).
3. Prairie Adaptation Research Collaborative, *Climate Change Impacts on Canada's Prairie Provinces: A Summary of our State of Knowledge* (Regina: PARC, 2008). Intergovernmental Panel on Climate Change, *Climate Change 2007: The Physical Science Basis: Summary for Policy Makers.* (Geneva: IPCC Secretariat, United Nations, 2007).
4. Irene M. Spry, *The Palliser Expedition: An Account of John Palliser's British North American Expedition, 1857–1860* (Toronto: Macmillan of Canada, 1963). Irene M. Spry (ed.), *The Papers of the Palliser Expedition, 1857–1960* (Toronto: Champlain Society, 1968). Donald Lemmen and Lisa Dale-Burnett, "The Palliser Triangle," in Kai-iu Fung, ed., *Atlas of Saskatchewan: Second Edition* (Saskatoon: University of Saskatchewan, 1999), 41.
5. Prairie Farm and Rehabilitation Administration, PFRA Hydrology Report #104, addendum 8 (Ottawa: PFRA, 2001). Statistics Canada, "2006 Census of Agriculture," online Dec. 18, 2007: http://www.statcan.ca/english/agcensus2006/index.htm.
6. D. Sobool and S. Kulshreshtha, "Socio-Economic Database: South Saskatchewan River Basin (Saskatchewan and Alberta), unpublished report, Department of Agricultural Economics, University of Saskatchewan, 2003.
7. H. Diaz and D. Gauthier, "Institutional Capacity for Agriculture in the South Saskatchewan River Basin," in E. Wall, B. Smit and J. Wandel (eds), *Farming in Climage Change: Agriculture and Adaptation in Canada* (Vancouver: UBC Press, 2007).
8. G.P. Marchildon, "Institutional Adaptation to Drought and the Special Areas of Alberta, 1909–1939," *Prairie Forum* 32, 2 (2007): 251–71. G.P. Marchildon, "The Prairie Farm Rehabilitation Administration: Climate Crisis and Federal-Provincial Relations during the Great Depression," *Canadian Historical Review* 90 (2009): 275-301.

9. Intergovernmental Panel on Climate Change, *Climate Change 2001: Impacts, Adaptation, and Vulnerability Technical Summary.* A Report of Working Group II of the IPCC (New York: WMO and UNEP, 2001).
10. B. Smit and J. Wandel, "Adaptation, Adaptive Capacity and Vulnerability," *Global Environmental Change* 16 (2006): 282–92.
11. This project was generously supported by the Social Sciences and Humanities Research Council of Canada (MCRI project # 412–2003–1001) from 2003 until 2009.

PART 1: THE GREAT PLAINS

A DRY OASIS: THE CANADIAN PLAINS IN LATE PREHISTORY

JAMES DASCHUK

Abstract

The susceptibility of the Canadian plains to drought is widely considered the greatest threat to the economic and social stability of the region. Rather than simply a condition of the physical environment, vulnerability to drought is a cultural phenomenon with its origins in the spread of the modern world system to the northern Great Plains. This point is illustrated by the relationship of indigenous societies to the region in the late prehistoric period. For hundreds of years after the climatic downturn of the 13th century, the grasslands served as a refuge for many groups experiencing climatically driven hardship in other regions, even though the country was more prone to desiccation than at any time in the recent past. Two key subsistence strategies that were used in the region were pedestrian bison hunting and the management of water through the proscription of beaver hunting. As market forces took hold, these ancient and effective practices were lost, contributing to climatic vulnerability, violence and ultimately to the end of the bison economy.

Sommaire

La susceptibilité des Plaines canadiennes à la sècheresse est considérée par plusieurs comme étant la plus grande menace à la stabilité sociale et économique de la région. Plutôt qu'une condition de l'environnement physique, la vulnérabilité à la sècheresse est un phénomène culturel qui prend son origine dans l'extension du système du monde moderne jusqu'au nord des Grandes Plaines. Ceci est illustré par la relation qu'avaient les sociétés indigènes avec la région pendant la période préhistorique récente. Les Prairies ont servi de refuge à plusieurs groupes éprouvés par les conditions climatiques dans d'autres régions pendant des centaines d'années après le refroidissement climatique du treizième siècle, et ce, malgré que le pays était plus enclin au dessèchement qu'à tout autre temps récemment. Deux techniques de subsistance clé utilisées à l'époque dans la région furent la chasse au bison à pied et la gestion de l'eau par l'interdiction de chasser le castor. Lorsque les forces du marché prirent racines dans la région, ces pratiques anciennes et efficaces furent délaissées, contribuant à la vulnérabilité climatique, la violence, et à l'écroulement de l'économie fondée sur le bison.

Introduction

The susceptibility of the northern Great Plains to drought is considered the greatest threat to the stability and prosperity of the region. As other papers in this volume show, vulnerability to desiccation is not simply a natural phenomenon, the consequence of a shortage of precipitation; rather, the limitation of human populations in areas of variable or low rainfall is a measure of a societal weakness to adapt to the range of conditions imposed by the environment. The primacy of market forces in our collective relationship with the land and climate of the northern Great Plains has been the Achilles heel of the region since the establishment of agrarian capitalism more than a century ago. Even before the recasting of the Canadian prairies as an agricultural region, the growing influence of the fur trade changed the relationship between the indigenous population and their environment. As communities looked to global markets for the produce of the region, they turned their backs on the need to maintain a sustainable relationship with the conditions imposed by the region. Prior to the extension of the modern world system to the northern Great Plains, indigenous societies were far better adapted to the environmental regime imposed by the climatic variability than those established with the coming of Europeans. The role of the northern Plains as a refuge for communities displaced by climate change during the late prehistoric period should be a lesson to those of us considering our own adaptations to the region in this time of environmental uncertainty resulting from climate change in our own time (please refer to the map at the end of this chapter for entities discussed in this paper).

Before market forces took hold of North America, climate was the dominant consideration in group decision making among indigenous societies. On the northern Great Plains, the shift away from climate toward trade with Europeans may have occurred as late as 1675.[1] The quality of prehistoric adaptation to climate is illustrated by the fact that, from the 13th century on, the region served as a refuge—an oasis for environmentally dislocated people from other regions. This was at a time when the northern Plains were considerably more prone to desiccation than it has been in the recent past.

Linking Archaeological Evidence to Historical Communities

Although human communities have inhabited the northern Great Plains for more than 11,000 years, the connection between prehistoric societies, identified through the analysis of archaeological remains, and historic and contemporary First Nations, is neither readily apparent nor free from dispute. Uncovering the link between prehistoric and contemporary societies in western North America is critical to understanding the impact of climate change on the development of aboriginal communities.

The connection between prehistoric and existing populations, however, remains a subject of debate. J.R. Vickers attributed the difficulty of relating prehistoric and contemporary populations to both limited data and the inconsistency of its interpretation in the field of archaeology. He added, "none of the current models [of cultural change in the prehistoric] can be demonstrated to be other than guesses."[2] On the Canadian Plains, there is little consensus surrounding the archaeological past of existing First Nations populations. One thing was certain; the yearly cycle of humans following bison herds from the grasslands in summer to sheltered areas in the parklands or wooded areas from the fall to the spring was essentially an unchanging and unchangeable phenomenon:

> This then must serve to approximate the human-land relationships for the past two thousand years. No matter what the coming and going of archaeological phases or historic tribes, all must have constructed their adaptation to the seasonal availability of resources. Deviation from the seasonal round outlined here would seem impossible for a hunting people on the Northwestern Plains. Climatic change may have advanced or retarded the schedule of resource exploitation slightly, but the basic adaptation must have remained constant.[3]

Vickers' assertion reveals several truths about human adaptation to the climate and environment of the northern Great Plains. First, human communities relied primarily on bison for their subsistence. Other resources were undoubtedly used as a supplement, but the herds remained the staple food of the plains for millennia. Second, archaeological groups who participated in the seemingly endless cycle of bison predation changed over time. Although changes in material cultures did not always signal changes in the ethnic identity of their makers, it is certain different ethnic groups occupied different regions of the plains over time. Third, climatic change over thousands of years would have altered the annual movement of the herds and the human populations that hunted them, though, as Vickers stressed, "the basic adaptation must have remained constant."[4] While the plains bison hunt served as a stable food supply for millennia, the reliability of the herds as a source of food acted as a cultural magnet during the five centuries of climatic instability that preceded the arrival of European influence to the northern Great Plains.

The Atlantic Climate Episode and its Effect on North America

Across the northern hemisphere, the climate from the 9th to the 13th centuries was marked by consistently warm temperatures.[5] In the 1960s, two pioneers of climate history, Reid Bryson and Wayne Wendland, designated the period from a.d. 850–900 to 1200 the "Neo-Atlantic Climatic Episode."[6] During this time, global climatic patterns—the central tendencies of weather conditions or 'mean' temperatures—underwent a long-term period of warm temperatures and adequate precipitation.[7] The period has also been variously described as the 'Climatic Optimum' and in Europe as the 'Medieval Warm Period.' Bryson and Wendland described the general effect of the Neo-Atlantic Climate Episode on North America:

> (O)pen water appeared in the Canadian Arctic Archipelago, the tree-line re-advanced into the tundra, summer rains extended farther into the southwest and corn-farming became practical across the Great Plains. Glaciers disappeared from the U.S. Rockies. These evidences suggest weaker westerlies and more meridional [southerly] circulation. The boreal forest probably expanded both south and north, while prairies shifted west at the expense of the steppe.[8]

The centuries of benign weather associated with the Neo-Atlantic fuelled an unprecedented period of cultural and technological expansion through much of North America. In the Arctic, long periods of open water made it possible for village-dwelling whale hunters known as the Thule to expand eastward from Alaska to Greenland between a.d. 900 and 1250.[9] At the same time, the Norse moved westward from Iceland, growing cereals and raising cattle along the coast of Greenland. At the turn of the first Millennium, they briefly maintained an outpost on the Island of Newfoundland.[10]

In the Northeast, the Neo-Atlantic brought nothing less than a revolution in subsistence practices. The adoption of maize agriculture set the stage for the present as "either specific cultures or the general patterns of historically documented groups" appeared with the diffusion of the horticultural triumvirate of corn, beans and squash.[11] The sudden appearance of the Iroquoian-speaking Owasco culture south of Lake Ontario around a.d. 900 signalled the spread of maize-centred agriculture to the area.[12] North of Lake

Ontario, the Huron Confederacy grew crops within walking distance of the Canadian Shield, an indication that the acidic and rocky soils of the coniferous forest were a greater barrier to the spread of cultivation than was climate. By a.d. 1000, maize and the technology associated with it were spread across the woodlands from the Atlantic Coast, through the American Midwest to the Great Plains.[13] Improvements in storage techniques and the development of maize varieties[14] contributed to the new mode of production, but the centuries of favourable weather that came with the climatic optimum have been accepted as the key factor in the northward spread of corn cultivation for decades.

The people responsible for what was essentially a 'Neolithic' revolution in eastern North America during this remarkable time are known archaeologically as the Late Woodland cultures.[15] Their culture was centred in a region known as the American Bottoms, a low-lying region encompassing the confluence of the Mississippi and Missouri Rivers and Ohio Rivers. Maize had been grown in the area since the time of Christ, but the crop did not become a major economic factor until a.d. 750.[16] The heart of the Mississippian culture, the city of Cahokia, was nothing less than a metropolis, with a population of between 10 and 20 thousand people in a.d. 1100.[17] With its large population and unparalleled size of its earthen mounds, Cahokia represents the apex of social organization and social stratification in pre-Columbian North America.[18] Most researchers agree Mississippian societies such as Cahokia were 'chiefdoms,' although some have argued the sophisticated relationship between Cahokia and its hinterland would best be described as a "state."[19] Whatever the most appropriate designation, the city reached its demographic and cultural zenith during the halcyon days of the late Neo-Atlantic.

As should be the case with such profound societal changes over such a vast geographical area, the full nature of the diffusion of maize-centred agriculture was complex, and is still not fully understood. What is certain is that Late Woodland agriculture spread over a massive area in a remarkably brief period characterized by warm northern hemispheric temperatures. Communities that adopted cultivation may not have been simple offshoots of Cahokian culture, but the energy from corn and her sister crops fuelled an unprecedented cultural florescence across eastern North America.

In addition to effecting the Iroquoian and Algonkian speakers of Lake Ontario, the St. Lawrence Lowlands, and the eastern seaboard of the United States, corn and the crops associated with it were instrumental to qualitative cultural changes in the Great Lakes and Midwest. Along the Ohio River, the agricultural societies such as the Fort Ancient and Monongahela complexes[20] that emerged around the turn of the first millennium a.d. borrowed heavily from the societies centred at Cahokia.[21] Farming spread up the Missouri and Mississippi Rivers into Iowa and South Dakota where it was adopted by Siouan-speaking peoples;[22] among them were the Great Oasis Complex and the Mill Creek cultural system.[23] The archaeological assemblages of these communities bore strong affinities to the highly urbanized Cahokia Culture, indicating close cultural ties or possibly migration to northern Iowa. In Wisconsin, Iowa and southern Minnesota, the Oneota culture emerged contemporaneously with the ascendancy of Cahokia, a sign that the diffusion of cultivation was not the simple spread of the technology from a metropolis to a hinterland.[24] As was the case in the diffusion of cultivation to other groups, the long warming trend associated with the Neo-Atlantic episode was a critical factor in the spread of the new subsistence economy.

As temperatures improved, communities in the deciduous forest of northern Minnesota added maize-based technology to their economy based on the combination hunting and the harvesting of wild rice. These people, known archaeologically as Blackduck, emerged from their heartland around a.d. 800 and swiftly expanded eastward along both shores of Lake Superior and into the Interlake region of Manitoba, and possibly even further north.[25] As they broadened their range, the makers of Blackduck ceramics co-existed with longstanding inhabitants of the shield, the Laurel people. The northward expansion of both the Blackduck and Laurel complexes has been attributed to the improved environmental conditions experienced during "this benign climatic episode."[26] As people shifted northward, so too did the environment they exploited. During the Neo-Atlantic, the tree line may have advanced as far as 280 km north of its present location west of Hudson Bay.[27]

Blackduck and Laurel remained distinct until a.d. 1000 when a hybrid group known by their material culture as the Rainy River Composite emerged in the Boundary Waters region. To the north, on the Canadian Shield the Laurel and Blackduck traditions came together in another hybrid configuration known as Selkirk. Although not an exact science, linguistic analysis indicates these changes had profound implications for the future occupation of central Canada:

> The result of using a selected Algonkian word list indicates a Cree-Ojibwa split between a.d. 987 and a.d. 1079. This range closely approximates the proposed emergence of the Selkirk (representing the northern "Cree" people) and the Rainy River (representing the more southern "Ojibwa" people) composites.[28]

During the Neo-Atlantic, corn-based cultivation spread westward along floodplains of river valleys to the margins of the Great Plains where the new technology supplemented the hunting and collecting economy focussed on bison predation. There, the climatic optimum came with an increase in precipitation, "(m)esic conditions enabled the spread of corn gardening and promoted the increase in the overall biomass which enhanced the productivity of hunter-gatherers."[29] The ancestors of the Hidatsa and the Mandan continued to hunt bison, but by a.d. 1000 they had established semi-sedentary horticultural communities along the Missouri River in North and South Dakota.[30] Among the oldest of these communities were those attributed to the Awatixa branch of the Hidatsa, located on the Knife and Heart Rivers, only 200 km south of the Manitoba, Saskatchewan and North Dakota borders. The doyen of Middle Missouri archaeology, W.R. Wood, stressed the connection between the favourable climate and the establishment of corn horticulture in the region: "It is surely no accident that the initial variant makes its appearance on the Prairie-Plains border and High Plains at the time it does . . . This was a warm period when more moisture was available than previously on the High Plains."[31] The villages of the Middle Missouri underwent cycles of expansion and contraction, but the persistence of their life way for 800 years stands as a testament to their ability to maintain a successful coping range with regard to variability in climate.[32]

The Neolithic revolution that engulfed the eastern half of North America abruptly halted its westward march at the Middle Missouri villages of North Dakota. Dale Walde argued climate did not limit the spread of maize cultivation on the plains; rather, the hunting and gathering communities beyond the margins of cultivation were "existing tribally organized semi-sedentary communal bison-hunting peoples" powerful enough to withstand the encroachment of horticulturalists.[33] Hunters on the Great Plains did not adopt cultivation because their needs were more than adequately served from their existing economy based on large-scale bison hunting. Walde stressed that the inhabitants of the plains were organized in a far more complex manner than has previously been recognized. Communal bison hunting through the use of jumps or pounds exploited large concentrations of bison and created great quantities of food, described as an "almost industrial level of resource exploitation."[34]

To secure food in this way required the participation of large populations to drive herds over many kilometres, and then to dispatch and process prey. High concentrations of labour were rewarded with large surpluses of food, enough for future consumption, trade, and, eventually, heightened levels of social organization as stockpiled meat provided people with time to devote to pursuits beyond simple survival. Walde asserted these communities were not nomadic hunting bands, but maintained "semi-sedentary" communities occupied for six months or more at a time.[35] The move toward increased social complexity among plains bison hunters may have begun as long as 2,000 years ago, when the people who occupied the entire northern plains came under the influence of the mound-building Woodland people who maintained villages before the general diffusion of maize-based cultivation.

While the bison hunters of the grasslands did not undergo a wholesale change in subsistence practices during the Neo-Atlantic, it was a time when hunting flourished:

> Perhaps even more important than the expansion of gardening practices was the overall build-up of the regional biomass. Grasslands and bison flourished. Availability of bison seems to have been the persistent limiting factor for human population growth and stability throughout the entire prehistory in the area.[36]

On the Northern Great Plains, the Neo-Atlantic brought centuries of resource and cultural stability. Although the region remained subject to drought,[37] the bison served as a tried and true subsistence base. Archaeological evidence indicates that cultural change during the period was gradual and without intrusion from other regions.[38] At the beginning of the Neo-Atlantic, the inhabitants of the southern plains of Saskatchewan and Alberta were represented by only two distinct archaeological assemblages, the Besant and Avonlea phases.[39] According to Walde, both were influenced by eastern woodland communities in their early development and practiced large-scale communal bison hunting.[40]

Besant first appeared on the plains about 200 b.c. associated with a group known as the Sonota, mound-building bison hunters on the forest plains ecotone of Minnesota. Avonlea first appeared in the dry country of southern Alberta and Saskatchewan about a.d. 100. The makers of Avonlea should not be thought of as a 'people' or as a specific ethnic group. The geographical and temporal spread of Avonlea styles across the plains indicates the people responsible for it were "from a variety of ethnic backgrounds."[41]

One major Avonlea innovation was the development of bow and arrow technology, in contrast to the atlatls (throwing sticks) used by the Besant. The spread of the new military technology may have been what Walde described as "a signal of mutual support" wherein local communities came together across the plains to prevent Besant encroachment on their hunting grounds.[42] Over time, Avonlea technology expanded eastward to the margins of the woodlands of Minnesota and South Dakota. There, they traded with groups such as the Truman Mound Builders whose ceramics are present in Avonlea sites on the northern plains.[43] As large-scale communal bison hunters, both Avonlea and Besant communities had abundant dried meat and hides that they traded for pottery and agricultural produce of the villages to the east. Evidence of the exchange between hunters and farmers has been found in the Moose Bay Burial Mound near Crooked Lake, Saskatchewan. While well beyond the region where cultivation was practiced, the discovery of a bone squash knife indicates that, while people of the Qu'Appelle Valley did not cultivate the soil, they were in contact with people who did.[44] A recent analysis of food residue of ceramics from beyond the range of known horticulture indicated that populations previously thought to be solely dependent on the herds were consuming maize, an indication of interregional trade.[45]

The centuries of good hunting on the plains contributed to the emergence of a new culture, Old Woman's, in southern Alberta about a.d. 1000. This group replaced the Avonlea tradition in the region, but the change was so gradual that scholars agree the appearance of Old Woman's represented an evolutionary change, an in situ development rather than the replacement of one population with another.[46] Old Woman's is acknowledged to be the prehistoric manifestation of Blackfoot culture.[47] On the Canadian plains, the Neo-Atlantic brought centuries of cultural stability and, in the case of Old Woman's, regional development. Climatic reconstructions of Alberta have shown the period between a.d. 900 and a.d. 1350–1450 was marked by greater drought severity than the period that followed it.[48] This factor indicates human communities in the region were well adapted to dry conditions, living within their means in an arid environment.

The untroubled years of the Neo-Atlantic, when climatic conditions were analogous to those of the 1960s,[49] of course, did not last. Although climate varied from the end of the Neo-Atlantic to the modern period, conditions marked by drought or cold, or both, continued in the northern Great Plains until at least the mid-19th century. As conditions deteriorated, indigenous populations across the continent were confronted with hard choices merely to survive.

The Pacific Climate Episode: The Exodus from the Eastern Woodlands and the Plains as a Refuge, a.d. 1250–1550

The end of the unprecedented period of cultural florescence in eastern North America came swiftly. Global conditions changed so rapidly they have been attributed to a single cataclysmic event, a massive volcanic eruption in a.d. 1259.[50] The magnitude of the blast is hard to fathom. In ice cores, it produced three times the concentration of sulphates found in layers associated with 1816, when the explosion of Mount Tambora resulted in the global phenomenon known as "the year without summer."[51] Recent scientific scholarship has stressed that volcanism is a key factor in abrupt large-scale climate change.[52]

In Europe, written accounts documented the monumental change over the coming decades. By 1309, Christmas feasts took place on the frozen Thames River in the city of London.[53] Along with cold, Europe was drenched with torrential rains. In 1315, unceasing rains brought chaos to Europe, thousands died in floods and from subsequent crop failures. In Greenland, the Norse settlements began their painful slide into oblivion by the end of the 13th century.[54] In the Canadian Arctic, the colder temperatures brought longer periods of sea ice making whaling unsustainable, forcing the population to diversify their subsistence base into a pattern that evolved into that of the historic Inuit.[55]

On the North American continent, the changes in climate marked the beginning of the long-term climatic regime known as the Pacific Climate Episode. Colder temperatures brought reduced moisture to much of western North America because the expansion of the circumpolar vortex reduced the amount of tropical air entering North America during the summer.[56] Bryson and Murray explained, "expansion of the westerlies [was]-in effect the expansion of the Arctic."[57] In the 13th century, the boreal forest began its 600-year retreat south as trees gave way to tundra.[58] During cold periods, such as those experienced after a.d. 1200, the increasing importance of westerlies as the prevailing winds would have blocked moist air from entering the plains, "enlarging and intensifying the dry shadow of the Rockies." In the American southwest, drought and the conflict that came in its wake were central aspects of the depopulation and subsequent abandonment of the Pueblo communities of the Four Corners region by 1300.[59]

In eastern North America, the deteriorating climate shook the farming cultures that had spread across the eastern half of the continent. By the turn of the 14th century, the complex trade and cultural network controlled by Cahokia was breaking up.[60] Within 100 years, the Mississippian occupation of the American Bottoms region was over. Cahokia was abandoned. By 1450, the mid-continental U.S. centred on the confluence of the Ohio and Mississippi Rivers was so depopulated it has been described as the "Vacant Quarter."[61] Violence undoubtedly accompanied the failure of Cahokia and its satellite communities, as the masses who had toiled so long in the construction of mound complexes lost faith in the elite who were perceived to have been responsible for large-scale failure of harvests.

In the Northeast, Iroquoian farmers experienced long-term decline of their harvest as people sought safety behind the defensive palisades of fewer, but larger communities in what descended into a 200-year cycle of intervillage warfare.[62] The fighting stopped only in the early 16th century when the feuding communities came together with the formation of the League of the Iroquois.

The farming villages northwest of the crumbling city of Cahokia also suffered severe hardship with the arrival of the Pacific Climate regime. This was particularly the case of the region known as the Prairie Peninsula, a finger of dry climate and corresponding vegetation extending eastward from the Dakotas through Iowa into Illinois and Wisconsin.[63] The Oneota people experienced severe stress by the turn of the 14th century:

> (L)arge settlements had seemingly fragmented into smaller dispersed communities and some relocation of population had occurred. This fractionalization was probably effected by several interrelated factors, including the deterioration of the climate (the Pacific climatic episode) and the decline of the Middle Mississippi influences to the south. Reliability of crops presumably changed as the length of the growing season and the moisture content of the soil both decreased.[64]

Rather than clinging to an unsustainable way of life, the Oneota abandoned their villages and migrated to the encroaching grasslands of the Prairie Peninsula where they hunted the bison who were expanding their range eastward as forest gave way to prairie.[65] As displaced Oneota farmers took up their new life to the south and west, they came into conflict with the Mill Creek villages that were themselves undergoing severe hardship from diminished harvests and the breakdown of the Mississippian trade network.[66]

Because the Mill Creek people occupied the Iowa ecotone,[67] the transitional area between the arid short grass prairie to the west and the wetter tall grass prairie to the east, they too were vulnerable to climate change. Because of their precarious environmental position, and the abandonment of their villages during the Pacific Climate Episode, the Mill Creek sites were the subject of a pioneering study in environmental history entitled *Climates of Hunger.* The authors explained the change that came with the Pacific Climate Episode, "(w)hile western Europe grew damp and gloomy, the farmers of the plains 800 years ago must have seen their corn wither and turn white with drought, their game die or move away."[68] Faunal studies indicated a shift from tree browsing deer to bison which were consumers of grass and more resilient to dry conditions. Most of the sites showed a decline in ceramics as the drought period progressed. As pottery was required for the storage and preparation of corn, the decline in ceramics was an indication less corn was being consumed. Pollen analysis, another indicator, showed a shift from oak, the dominant tree species during the Neo-Atlantic, to willow during the Pacific. Changes were marked in other plant communities harvested by the Mill Creek people, "(t)hey moved from a higher percentages of composites—which include sunflowers and asters—to a large percentage of grasses, which have smaller leaf surfaces and require less moisture; they are favoured by a drier climate."[69]

In the case of the Prairie Peninsula, grasslands advanced far to the east during the Pacific Episode. In essence, the plains came to the people of the region, leading them to abandon their settlements and take up communal bison hunting. Many of these prehistoric environmental refugees from the eastern woodlands moved toward the Great Plains as they adopted the bison hunt as their primary means of subsistence. Oneota communities continued to move west until as late as the mid-15th century, displacing the local small-scale farming villages of the Central Plains Tradition in southern Nebraska and western Iowa.[70] Although drawn west by the reliability of the herds, the sophisticated tribal level of social organization they had developed in the eastern woodlands "facilitated colonial expansion" of Oneota society.[71] The western Oneota have been linked to the historic Siouan-speaking Oto and Kansa populations of the Central plains. The Degiha-speaking branch of Siouan speakers, including the ancestors of the Ponca, the Omaha, the Quapaw, and Osage nations may also have undertaken the exodus to the plains from their homeland on the Ohio River.[72]

Along the Missouri River, climate change during the 13th century was responsible for major retrenchment among horticultural villagers. According to Donald Lehmer, the deterioration of the climate in South Dakota came with the shift to the Pacific Climate Episode, "seems to be directly reflected by the abandonment of most of the Missouri valley and the concentration of the remaining villages in a few small areas."[73]

During the Pacific Episode the most significant "unfavourable" aspect of climate was drought. The severity of drought in the American plains during the 1200s was such that it was designated only one of two "megadroughts" on the American Great Plains during the last 1,000 years.[74] Analysis of core sediments indicates the Devil's Lake Basin in North Dakota dried up three times between a.d. 1300 and a.d. 1535. Sediment analyses provide "solid evidence of negative environmental effects of Pacific climatic episode droughts."[75] Although not fully understood, drought probably forced the abandonment of several Middle Missouri villages in South Dakota as populations moved northward into fewer large and well-fortified settlements.[76] Displaced groups, including members of the Caddo-speaking Plains Village Tradition driven from their homes by the Oneota, occupied the deserted villages of South Dakota during a cultural period known as the Initial Coalescent beginning around a.d. 1300.[77]

The most famous Initial Coalescent village is located at Crow Creek, South Dakota. Its history illustrates the region-wide hardship experienced during the Pacific Climate Episode, even though the locality of

Crow Creek itself "is abundantly blessed by essentially every material characteristic that is likely to have attracted human settlement in the region."[78] The Initial Middle Missouri population who brought maize to the region established the community some time in the 12th century, but abandoned the community during the 13th. When Initial Coalescent people occupied the site around 1300, they built both palisades and dug a moat or ditch as protection from the violence surrounding them. The defensive measures were certainly warranted; however, at Crow Creek they proved inadequate. Although estimates of the date vary, a major violent episode occurred at Crow Creek resulting in the deaths of almost 500 individuals. The scene must have been horrific as victims were "killed and mutilated, that the town was burned down around their bodies, and that they were intentionally buried after an extended period of exposure to carnivores."[79] The attack was probably an invasion, as the site was occupied after the massacre.[80] People along the Missouri were killing each other in large numbers to occupy what few feasible horticultural sites remained in the region. In addition to the violence of their deaths, skeletal remains from the mass grave at Crow Creek were marked by malnutrition, another indication of the plight of horticultural villagers during this difficult time.[81]

The large-scale regional migrations of people who inhabited the prairie-woodland border of the central U.S. indicates that climate change during the Pacific Episode was severe enough to trigger the abandonment of farming over a large area and prompt the adoption of bison hunting in the western plains. The Oneota expansion was not the only case of Siouan speakers abandoning their crops in favour of the herds. On the northern Great Plains, the Mortlach people left their homeland in the eastern woodlands during the 13th century, and soon spread across the grasslands through western Manitoba, Saskatchewan, north-western North Dakota and north-eastern Montana.[82] In Saskatchewan, Mortlach displaced the ancestors of the Blackfoot in what Walde considered another case of a sophisticated society with eastern roots encroaching onto the plains.[83]

Walde has shown there were two distinct manifestations of Mortlach culture in western Canada. The southern group, the Lake Midden subphase, whose population hunted in valleys of the Qu'Appelle and the Missouri Rivers, were able to maintain sedentary communities for as long as eight months at a time. In summer, they followed the herds onto the open plains. A clustering of Mortlach sites around the Big Muddy area of Saskatchewan indicates it was used as a corridor to the villages of the Missouri Basin, supporting the view of the interdependence between nomadic hunters and village dwellers.[84] The connection between the Assiniboine descendents of the Mortlach and the Middle Missouri villages continued into the historic period. It was they who guided the La Vérendrye brothers to the Mandan villages in 1741, and the explorer noted the two groups knew each other's languages.[85]

The dry decades of the Pacific Climate Episode were also marked by the dispersal of villages into groups too small to rely on fortification for their protection. These fragmented communities practiced "a subsistence continuum from foraging, through horticulture to agriculture."[86] Even during optimal conditions, the horticultural villages of the Middle Missouri may have relied on hunted meat for as much as half of their food requirements.[87] By the early 15th century, the collapse of large settlements in the woodlands triggered a wave of immigration to western North Dakota and southern Manitoba. The tradition, known as the Scattered Village Complex, was comprised of Mississippian-influenced woodland people who displaced the small remaining Blackduck hunting and gathering communities in southern Manitoba.[88] This group came west with the technology for horticulture and goods that indicate strong trade ties to the south. In Manitoba, these members of the Scattered Village tradition are known as the Vickers focus.[89] They were almost certainly environmental refugees, displaced when climate-induced war in the Missouri Basin "caused a demographic ripple throughout the northeastern Plains, with some groups drifting northwards following stream valleys, including the Pembina and Souris Rivers."[90] The oldest cluster of sites, in the Tiger Hills south of Brandon, is dated to about a.d. 1400. The Vickers focus people were highly selective in their choice of planting locations, on well-watered soils on southern slopes that maximized sunlight and sheltered plants from the north winds. Gardening was undertaken on a small scale, a sign that crops were now more of a supplement than staple.[91] To the east, along the Red River, a brief, but intensive, period of maize cultivation occurred in the early 1400s. Although their identity remains obscure, the planters at Lockport were also "acutely aware of the length of the frost free period so that sites that were also selected for micro-climatic variations which maximized the length of the growing season."[92] The 15th century horticulturalists who walked to Manitoba were undoubtedly experts in their craft, perhaps heading north to exploit higher precipitation levels than their home regions to the south. Their expertise was no match for the global decline of temperatures in the 1450s as attempts at growing corn were marginally successful at best:

> It may be that an initial, limited horticultural success in the region ended in a series of cold summers that prevented the harvesting and drying of corn, with the last crops being consumed as "green corn" or simply lost due to unseasonal [sic], late killing frosts.[93]

Vickers focus people persisted in the Tiger Hills until mid-century then fled the region so fast that studies have attributed the move to a "sudden, drastic cold spike during the little ice age."[94] The event that triggered what may have been the coldest period of the past 1,000 years was the cataclysmic explosion of Mount Kuwae in the South Pacific in 1453–1454.[95] Tree rings from Europe, Asia and California indicate a severe short-term drop in temperatures.[96] In 1454, Aztec society experienced the famine of One Rabbit, "the most devastating one recorded for pre-Columbian central Mexico, [that] led to disease, massive migration, and widespread death . . ."[97] In central Mexico, three years of "utter" drought began "with an unprecedented freeze and snowfall in the winter of 1453–1454 which caught the inhabitants totally unprepared."[98]

In Manitoba, the Vickers focus people headed west turning up in settlements in the Lauder Sand Hills south of the Virden-Oak Lake area almost a century after their departure from the Tiger Hills. In that time, the size of their settlements was "scaled down," there was a decrease in exotic materials such as catlinite and Knife River flint, but, most importantly, there was no evidence of cultivation at any of the western sites.[99] Although the later Vickers focus sites point to the abandonment of crops, the span of their occupation in southern Manitoba points to the use of microhabitats where subsistence options could be maximized:

> The point of interest is that a successful, likely semi-sedentary village culture that followed a horticultural and foraging lifeway, abandoned their home territory and moved to another area is also marked by a high level of biodiversity.[100]

A mixed subsistence base during a time of uncertainty is a prudent survival strategy. In the case of Vickers focus, it may have been used as a bridge until a more profitable resource could be exploited, and some may have returned to their ancestral communities in the Missouri Basin. Oral tradition among the Awaxawis Hidatsa of the Knife River in North Dakota recount the arrival of their people to the region "after having lost their corn" in the north.[101] Improved conditions in the Middle Missouri after a.d. 1450 are indicated by the resettlement of southern portions of the region that had been abandoned earlier.[102] At the Sanderson Site near Estevan, Vickers focus ceramics were found with "huge amounts of bison bone, suggesting a full adaptation to mass killing of bison" although "traditional foraging strategies were still followed to some degree."[103] With their move to the Saskatchewan plains, the conversion of the Vickers focus people from farmers to large-scale bison hunting was complete.

With the exception of the enigmatic and probably equestrian protohistoric group known as One Gun, Besant, Avonlea, Old Woman's, Mortlach and Vickers focus represent all major prehistoric populations on the Northern Great Plains of Canada.[104] Walde has argued all of these groups were either migrants from the large-scale tribally organized societies in the eastern woodlands, or were profoundly influenced by them either as a consequence of ancient trade relationships or, in the case of Avonlea, as a region-wide response to the threat of encroachment. Despite the horticultural roots of the Mortlach and Vickers focus people, all of the inhabitants of the Canadian plains were large-scale bison hunters by the 16th century. They were not small and mobile nomadic hunting bands who chased herds across the plains and parklands, but were large, highly organized, semi-sedentary communities that were able to manipulate their environments to manage herds and ensure large surpluses of meat for trade. The Hidatsa, who may have been the source of Vickers focus, also hunted the plains just above the 49th parallel and were important suppliers of bison meat to the horticultural communities of the Missouri River. The trade between hunters and cultivators may have existed for centuries; however, as Bamforth noted, colder winters and increased year-to-year variability in temperatures associated with the Little Ice Age probably heightened the interdependency of nomadic hunters and village horticulturalists, as both groups increasingly required supplements of stored food to ensure survival through unpredictable winters.[105] As interaction between grassland hunters and horticultural villages increased, so too might have the intensity of hunting practices after a.d. 1500. The excavation of Mortlach sites indicate bison were exploited to their fullest:

> In particular, bison bones were intensively processed by being broken into tiny pieces and then boiled, the vats of water being heated with red hot rocks. Apparently, the bone fragments where then drained and dried, following which they were used as fuel. As a result, these sites contain massive deposits of small pieces of burned bone.[106]

Climate change beginning in the mid-13th century triggered a large-scale abandonment of cultivation and reorientation to large-scale bison hunting in the grasslands from Texas to the Canadian border. While there was undoubtedly some level of group fragmentation as large woodland centres collapsed, they were largely able to maintain a high level of social cohesion and cultural institutions, even though they became hunter-gatherers, a subsistence pattern normally associated with less complex band-level societies. This was possible because, according to Walde, the large investment in labour required for the mass slaughter of bison in pre-equestrian hunts without firearms was rewarded with huge payoffs in meat. Large surpluses were stored and traded, reinforcing the complex nature of social organization through ceremonial activities and regular and profitable contact and trade with other groups.

This pattern of adaptation on the woodlands-plains frontier provided large populations with stability for centuries. But how could communities, in some cases larger than 1,000 or more, maintain residency patterns so stable they are described as "semi-sedentary"? To move communities of such size with only dogs as beasts of burden would have required a huge effort. Even with a high degree of residential stability, pedestrian communities could only have moved a short distance without great hardship. As such, they developed complex strategies to maximize the existing assets of the environment within their limited range of mobility. The widespread success of these large-scale bison hunting communities was predicated on the regularity and reliability of bison movement and exploitation. Principal kill sites such as pounds and jumps were used repeatedly, a sign the movement of herds was predictable.[107]

Grass fires were employed as a means of directing the local movement of herds and steering them toward kill sites, or as a means to draw bison to areas of new growth following winter and early spring burns, where they could be harvested. Pedestrian hunters had to use care in steering their prey towards the kill sites so their methods did not create widespread stampedes or significantly disrupt the movements of the larger herds.[108] In doing so, bison movements remained relatively predictable from year to year and allowed hunting groups to rely on access to their main source of sustenance and maintain a high degree of residential stability. Throughout the pedestrian period bison herds maintained their seasonal cycle, spending the winters in sheltered areas in the parklands or near valleys and the summers on the open plains.[109] Walde has shown that large communities remained stable over the winter and, in the case of the Midden Lake Mortlach site, for as long as eight months in a stretch.[110] Summer campsites on the open plains could be maintained for as long as three to three and a half months.[111]

There is no doubt the reliability of bison in the food quest drew woodland populations onto the plains. Once in the grasslands, communities were confronted with the only issue more pressing than the need for food. To survive, pedestrian groups on the northern Great Plains had to ensure they had an adequate supply of water. Securing water for large-scale hunting communities such as those described by Walde would have been an imperative, especially in the drought prone centuries of the late prehistoric. The inhabitants of the dry landscape of western Saskatchewan and Alberta developed a water management strategy that buffered them from the effects of even long-term drought. Ecological studies have shown Avonlea, and the Old

Woman's tradition that grew from it, protected their water supply by conserving populations of beaver that maintained water level in the valleys shared by both species.[112] Beaver ponds purposely maintained through non-exploitation served as dependable water sources for groups even during periods of extended drought. According to Grace Morgan, "the bison were "the staff of life" although the beaver "were at the core of a profound ideological framework which prized the role of the beaver in the stabilization of water resources."[113]

During protracted droughts, bison populations probably did not move to their summer range in the open grasslands because of the scarcity of water and tended to remain near water supplies in valley complexes.[114] Bison activity during a drought cycle served to re-enforce the relative stability of communities camped along the tributaries of major waterways because herds remained in what would have been their winter range during normal conditions. Beaver whose dams held back even small amounts of water maintained water levels in the valley complexes, thereby stabilizing the limited resource. During the most lengthy drought periods, when water could not be maintained in tributaries, human, bison and beaver populations would have all sought refuge along main channels of waterways.

The success of this ancient strategy continued to guide the descendents of those who developed it until well into the 19th century. David Smyth has detailed how the Niitsitapi, or Blackfoot people, and their allies, the A'aninin (Gros Ventre) and the Athabascan-speaking T'su Tina (Sarcee), refused to participate in the commercial harvest of beaver until as late as the 1820s when military and economic necessity forced them to enter the trade.[115] Their refusal to harvest beaver was not a consequence of the rarity of the species in their territory. On his journey through southern Alberta in 1792–93, Peter Fidler rebuked his Piikani (Peigan) guides for trapping nothing of value as they passed numerous small lakes teeming with beaver.[116] When the Blackfoot did begin to hunt beaver for commercial purposes in the 1820s, Edmonton House, their primary trading post, briefly became the most profitable in the entire domain of the Hudson's Bay Company.[117]

The availability of wood was also a key determinant of winter settlement patterns.[118] In addition to fuel, wood was the key component of bison pounds, fenced enclosures where animals were driven to slaughter. Because pounds could accommodate between 200 and 300 animals, the amount of wood required for their construction would have been enormous, especially given the relative scarcity of trees. Trees cut by beavers, rather than compromising the wood supply of their human neighbours, were often reused and served as sources of firewood and building materials for bison pounds.[119] The absence of beaver remains in archaeological excavations from the Moose Jaw

Creek area suggest the non-exploitation of what would have been easy prey especially in summer, is evidence that the avoidance of beaver hunting was a traditional ecological practice.[120]

Mortlach people and their Assiniboine descendents who had been on the northern Great Plains for hundreds of years before the arrival of Europeans may also have avoided killing beaver. As early as the 1690s, Hudson's Bay Company differentiated Assiniboine groups in the parklands who trapped beaver and their compatriots on the plains who did not.[121] On his journey to the Mandan villages on the Missouri River in 1738, La Vérendrye was told the plains Assiniboine did not know how to hunt beaver and should be instructed in the practice.[122] Early traders who witnessed the abundance of beaver while they travelled in the grasslands were perplexed at the inability, or unwillingness, of the Indians to harvest them.[123] By the late-18th century, plains groups who refused to commercially harvest beaver began to pay a high price for the maintenance of their tradition. Recent immigrant groups from the woodlands, particularly the Cree, the eastern Assiniboine, the Saulteaux, and even the Iroquois, filled the demand for beaver pelts by expanding into the region as commercial trappers. Beaver hunting secured for them a preferred status among the traders and, in turn, they were able to procure firearms that provided them with a military advantage over those groups who maintained the taboo on beaver harvesting. As incoming groups associated with the fur trade displaced indigenous plains communities, beaver populations were soon reduced. The depletion of beaver populations in the parklands was noted in HBC journals by the mid-1750s, forcing trappers onto the margins of the plains.[124] By the end of the century, beaver were extirpated from other regions, such as the North and Lower Saskatchewan Rivers and the Qu'Appelle Valley.[125] In the 1790s, the trader at Fort George on the North Saskatchewan River near Edmonton complained over-hunting had "entirely ruined" the country.[126] As commercial beaver hunting to meet the European demand for furs replaced traditional conservation practices, the key strategy for aboriginal drought mitigation on the northern plains was lost.

Conclusion

For more than 500 years after a.d. 1200, the Great Plains served as a refuge for groups experiencing climatically driven environmental stress. The dynamics of those migrations, undertaken during a period of protracted hardship, is essential to the understanding of both climatic adaptation and tribal occupation of the northern Great Plains. The people who lived on the Canadian plains in late prehistory were not small nomadic band societies constantly on the move in search of prey; rather, they were large, socially complex tribally-organized societies with institutions more akin to the villages of the eastern woodlands than the band societies of the Canadian Shield. As climate deteriorated in the 13th century, horticultural people whose ceremonial and political lives had grown more complex during centuries of affluence during the Neo-Atlantic climate episode, chose to take their ideology and institutions west in search of the herds. In doing so, they were able to continue their traditions while taking on a new role as large-scale bison hunters. Those already on the grasslands, such as the makers of Avonlea, borrowed from woodland groups they with, creating a region-wide, and probably "multicultural," mutual assistance mechanism to meet the expansion of woodland peoples onto the plains.

Large, sophisticated, and stable groups controlled the plains during the pedestrian period, and they were successful in doing so because they were able to live within sustainable limits, even during the drought prone Pacific Climate Episode. Two basic requirements had to be met for sustained habitation: the need for food and the need for water. Bison proved to be a reliable staple; however, to ensure a constant supply of meat for so many mouths, the resource must have been managed carefully. Fire and non-disruptive hunting allowed semi-sedentary hunters to keep their prey within easy reach, walking distance in fact, for months at a time. With the arrival of horses and chase hunting, this tradition was lost. Large-scale hunting communities were able to secure enough flesh to consume, store and trade. Prehistoric people on the plains also managed water.

By abstaining from killing beaver, grassland communities secured something far more valuable than the flesh and fur of the species. The beaver, and the water their dams retained, were a powerful tool for the people of the dry lands. The species proved to be in a fragile position when the fur trade placed a higher value on the pelt of a dead beaver than the work of a live one. The longstanding inhabitants of the grasslands, the Blackfoot, the A'aninin, and perhaps the plains Assiniboine, understood the role of the species in the plains ecosystem, but they were marginalized as new groups extirpated beaver from the plains by the turn of the 19th century. As the modern world system took hold in the west; ancient practices were set aside in favour of market forces, creating vulnerability, hardship and conflict.

Acknowledgements: I thank the following people for their assistance in the completion of this paper, Alwynne Beaudoin, Berks Browne, Henry Dobyns, Renée Fossett, Scott MacNeil, David R. Miller and Dale Walde.

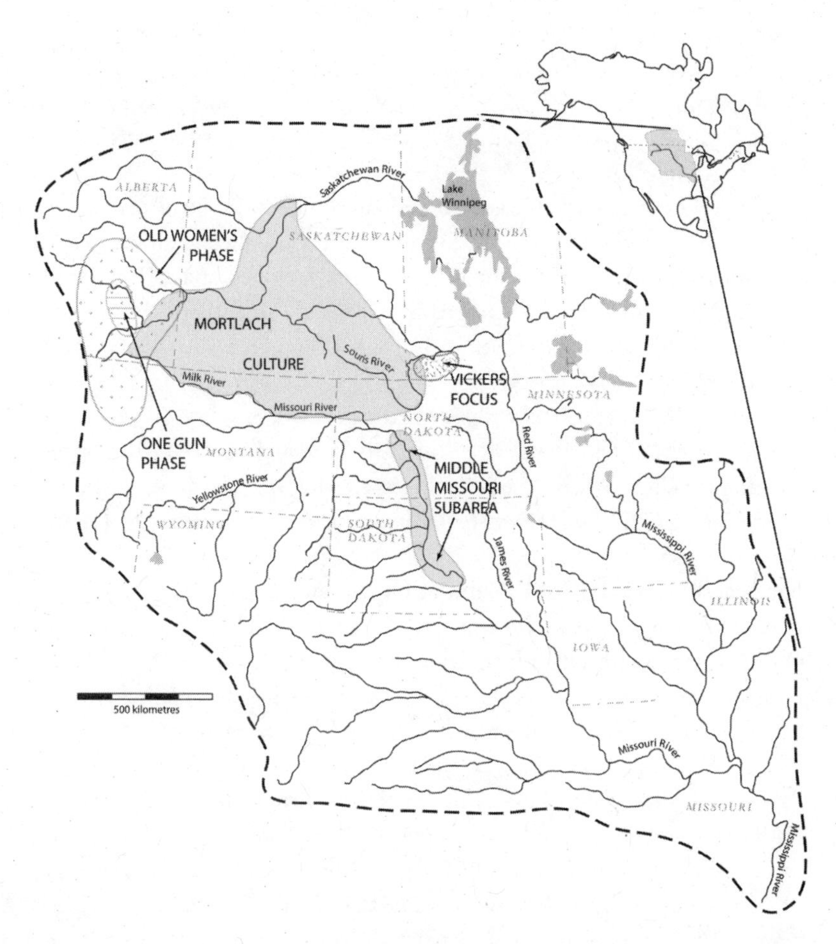

Source: Dale Walde, "Sedentism and Pre-Contact Tribal Organization on the Northern Plains: Colonial Imposition or Indigenous Development?" World Archaeology Volume 38 (2) (2006), Figure 5, p, 297.

Endnotes

1. Donald J. Lehmer, "Climate and Culture History in the Middle Missouri Valley," in *Reprints in Anthropology, Volume 8: Selected Writings of Donald Lehmer,* ed. W.R. Wood (Lincoln, NE.: J. & L. Reprint Company, 1977), 59–75.
2. J.R. Vickers, "Cultures of the Northwestern Plains: From the Boreal Forest Edge to Milk River," in *Plains Indians, a.d. 500–1500: The Archaeological Past of Historic Groups.* Ed. Karl Schlesier. (Norman: University of Oklahoma Press, 1994), 33.
3. Vickers, "Cultures of the Northwestern Plains," 6.
4. Ibid.
5. For a discussion of the benign climate during this time in Europe see, Brian Fagan, *The Great Warming: Climate Change and the Rise and Fall of Civilizations* (New York: Bloomsbury Press, 2008).

6. Reid A. Bryson and Wayne M. Wendland, "Tentative Climatic Patterns for Some Late Glacial and Post-Glacial Episodes in Central North America," in *Life, Land and Water: Proceedings of the 1966 Conference on Environmental Studies of the Glacial Lake Agassiz Region*, ed. William J. Meyer-Oakes (Winnipeg: University of Manitoba Press, 1967), 271–298.
7. A significant exception to the amelioration in climate with the Neo-Atlantic took place in the American southwest. During that period, the region was marked by drought and hardship for its inhabitants. Terry L. Jones, et al., "Environmental Imperatives Reconsidered: Demographic Crises in Western North America during the Medieval Climatic Anomaly [and Comments and Reply]," *Current Anthropology* 40 (1999), 137–170.
8. Bryson and Wendland, "Tentative Climatic Patterns," 294.
9. Renée Fossett, *In Order to Live Untroubled: Inuit of the Central Arctic, 1550–1940* (Winnipeg: University of Manitoba Press, 2001), 15–19.
10. Brian Fagan, *The Little Ice Age: How Climate Made History, 1300–1850* (New York: Basic Books, 2000).
11. James E. Fitting, "Regional Cultural Development, 300 B.C to a.d. 1000," in *Handbook of North American Indians: Volume 15. Northeast,* ed. Bruce Trigger (Washington: Smithsonian Institution Press, 1978), 44.
12. Dean R. Snow, *The Iroquois* (Oxford: Blackwell Publishers, 1994) 12–16.
13. Guy E. Gibbon, "Cultural Dynamics and the Development of the Oneota Life-Way in Wisconsin," *American Antiquity* 37 (1972), 168.
14. A major breakthrough in maize production occurred with the development of "Eastern 8-row or Northern Flint" around a.d. 800. Thomas J. Riley, Richard Edging, Jack Rossen, George F. Carte, Gregory Knapp, Michael J. O'Brien and Karl H. Scherwin, "Cultigens in Prehistoric Eastern North America: Changing Paradigms," *Current Anthropology* 31 (1990), 529.
15. It should be noted that a limited number of cultigens such as gourds and some seed producing plants were domesticated thousands of years ago. The adoption of maize during the Neo-Atlantic climate episode marked a profound shift in the importance of, and geographical extend of, plant domestication. Gayle J. Fritz, "Multiple Pathways to Farming in Precontact Eastern North America," *Journal of World Prehistory* 4 (1990): 387–435.
16. Dale Walde, "Sedentism and Precontact Tribal Social Organization on the Northern Plains: Colonial Imposition or Indigenous development?" *World Archaeology* 38 (2006), 296–297.
17. Biloine Whiting Young and Melvin L. Fowler, *Cahokia, the Great Native American Metropolis* (Urbana: University of Illinois Press, 2000), 316.
18. Thomas E. Emerson, *Cahokia and the Archaeology of Power* (Tuscaloosa: University of Alabama Press, 1997).
19. George R. Milner, "The Late Prehistoric Cahokia Cultural System of the

Mississippi River Valley: Foundations, Florescence and Fragmentation," *Journal of World Archaeology* 4 (1990), 2.

20. Archaeological Complexes "are groups of distinctive material remains that have been found at multiple sites in a given area . . . The material remains that typify a particular cultural complex include technologically and stylistically similar artefacts such as ceramic wares, points of particular types, and unique grave offerings. The distinctive material remains of a complex sometimes also include settlement traits such as certain kinds of residential lodges and mortuary features." Because archaeological nomenclature is not uniform, phases can also be identified as complexes. Michael Gregg, "Archaeological Complexes of the Northeastern Plains and Prairie-Woodland Border, a.d., 500–1500." In *Plains Indians, a.d. 500–1500: The Archaeological Past of Historic Groups,* ed. Karl Schlesier (Norman: University of Oklahoma Press, 1994), 72.
21. James B. Griffin, "Late Prehistory of the Ohio Valley," in *Handbook of North American Indians: Volume 15. Northeast,* ed. Bruce Trigger (Washington: Smithsonian Institution Press, 1978), 547–559.
22. James Warren Springer and Stanley R. Witkowski, "Siouan Historical Linguistics and Oneota Archaeology," in *Oneota Studies,* ed. Guy Gibbon (Minneapolis: University of Minnesota Publications in Anthropology Number 1), 69–83.
23. Duane Anderson, "Toward a Processual Understanding of the Initial Variant of the Middle Missouri Tradition: The Case of the Mill Creek Culture of Iowa," *American Antiquity* 52 (1987), 525–527.
24. Guy Gibbon, "Cultural Dynamics," 166. It should be noted that there is no consensus regarding the beginning of Mississippian influence on the Great Oasis Complex.
25. Brian J. Lenius and Dave M. Olinyk, "The Rainy River Composite. Revisions to Late Woodland Taxonomy," in *The Woodland Tradition in the Western Great Lakes: Papers Presented to Elden Johnson,* ed. Guy E. Gibbon (Minneapolis: University of Minnesota Publications in Anthropology Number 4, 1990), 80, fig. 8.3.
26. David Meyer and Scott Hamilton, "Neighbors to the North: Peoples of the Boreal Forest," in *Plains Indians, a.d. 500–1500: The Archaeological Past of Historic Groups,* ed. Karl Schlesier (Norman: University of Oklahoma Press, 1994), 112.
27. Meyer and Hamilton, "Neighbors to the North," 100.
28. Lenius and Olinyk, "The Rainy River Composite. Revisions to Late Woodland Taxonomy," 101. There is no consensus regarding the ethnic origins of the makers of Blackduck. Many American scholars assert that they were Siouan-speakers. In their discussion of the issue, Dyck and Morlan favour the Algonkian interpretation. Ian Dyck and Richard E. Morlan, "Hunting and Gathering Tradition: Canadian Plains," in *Handbook of North American Indians: Volume 13. Plains,* ed. Raymond J. DeMallie (Washington: Smithsonian Institution Press, 2001), 128–129.
29. Gregg, "Archaeological Complexes," 83.

30. R. Peter Winham and Edward J. Lueck, "Cultures of the Middle Missouri," in *Plains Indians, a.d. 500–1500: The Archaeological Past of Historic Groups,* ed. Karl Schlesier (Norman: University of Oklahoma Press, 1994), 159. The Siouan-speaking Mandan and Hidatsa may have originated from Great Oasis communities who developed a more complex social organization as a consequence of the wealth afforded by the combination of bison and floodplain agriculture in the Missouri basin.
31. W. Raymond Wood, "Plains Village Tradition: Middle Missouri," in *Handbook of North American Indians: Volume 13. Plains,* ed. Bruce Trigger (Washington: Smithsonian Institution Press, 2001), 190.
32. Lehmer, "Climate and Culture History in the Middle Missouri Valley," 59–73.
33. Walde, "Sedentism and Precontact Tribal Organization on the Northern Plains," 298.
34. Ibid., 305. It should be noted that bison jumps often led to the loss of large quantities of meat because flesh spoiled before it could be processed.
35. Ibid., 302–303.
36. Gregg, "Archaeological Complexes," 95.
37. Alwynne Beaudoin, "Climate and Landscape of the Last Two Thousand Years in Alberta," in *Archaeology in Alberta: A View from the New Millennium,* eds. Jack W. Brick and John F. Dormaar (Medicine Hat: The Archaeological Society of Alberta, 2003), 28.
38. Trevor Peck and Caroline Hudecek-Cuffe, "Archaeology on the Alberta Plains: The Last Two Thousand Years," in *Archaeology in Alberta: A View from the New Millennium,* eds. Jack W. Brick and John F. Dormaar (Medicine Hat: The Archaeological Society of Alberta, 2003), 90.
39. A "phase" in archaeology is defined as "an archaeological unit possessing traits sufficiently characteristic to distinguish it from all other units similarly conceived, whether of the same or other cultures or civilizations, spatially limited to the order of magnitude of a locality or region and chronologically limited to a relatively brief interval of time." Gordon R. Willey and Philip Phillips, *Method and Theory in American Archaeology* (Chicago: University of Chicago Press, 1963), 23.
40. Walde, "Sedentism and Precontact Tribal Organization on the Northern Plains,"-300.
41. Walde, "Avonlea and Athabaskan Migrations: A Reconsideration," 193.
42. Ibid., 193–194.
43. Walde, "Sedentism and Precontact Tribal Organization on the Northern Plains," 301
44. Margaret Hanna, *The Moose Bay Burial Mound* (Regina: Saskatchewan Department of Tourism and Renewable Resources, Saskatchewan Museum of Natural History Anthropological Series, No 3, 1976), 55.
45. Matthew Boyd, Clarence Surette, and Beverley (Bev) Alistair Nicholson, "Archaeobotanical evidence of prehistoric maize (*Zea mays*) consumption at the northern edge of the Great Plains," *Journal of Archaeological Science* 33 (2006), 1137.

46. Peck and Hudecek-Cuffe, "Archaeology on the Plains: The Last Two Thousand Years," 90; Walde, "Sedentism and Precontact Tribal Organization on the Northern Plains,"301.
47. Vickers, "Cultures of the Northwestern Plains," 28.
48. David Sauchyn and Alwynne Beaudoin, "Recent Environmental Change in the Southwestern Canadian Plains," *The Canadian Geographer* 42 (1998), 343.
49. Thomas J. Crowley, "Causes of Climatic Change over the Past 1000 Years," *Science* 289 (2000), 270–279.
50. Crowley, "Causes of Climatic Change over the Past 1000 Years," 270–279.
51. C. Richard Harington, ed. *The Year Without a Summer? World Climate in 1816* (Ottawa: Canadian Museum of Nature, 1992).
52. Ryan C. Bay, Nathan Bramall, and P. Buford Price, "Bipolar Correlation of Volcanism with Millennial Climate Change," *Proceedings of the National Academy of Sciences* 101 (2004): 6341–6345.
53. Brian Fagan, *The Little Ice Age,* 29.
54. Jared Diamond, *Collapse: How Societies Choose to Fail or Succeed* (New York: Penguin 2005), 267.
55. Fossett, *In Order to Live Untroubled,* 24.
56. Bryson and Wendland, "Tentative Climatic Patterns," 296.
57. Reid A. Bryson and Thomas J. Murray, *Climates of Hunger: Mankind and the World's Changing Climate* (Madison: University of Wisconsin Press, 1977), 47.
58. Rupert's Land Research Centre, *An Historical Overview of Aboriginal Lifestyles: The Churchill-Nelson River Drainage Basin* (Winnipeg: Rupert's Land Research Centre, University of Winnipeg, 1992), 31.
59. William D. Lipe, "The Depopulation of the Northern San Juan: Conditions in the Turbulent 1200s," *Journal of Anthropological Archaeology* 14 (1995), 143–169. Unlike eastern regions, the southwest underwent a lengthy period of desiccation during the Neo-Atlantic. Terry Jones, et. al., "Environmental Imperatives Reconsidered: Demographic Crises in Western North America during the Medieval Climatic Anomaly," *Current Anthropology* 40 (1990), 137–138.
60. George R. Milner, "The Late Prehistoric Cahokia Cultural System of the Mississippi River Valley: Foundations, Florescence, and Fragmentation," *Journal of World Prehistory* 4 (1990), 31–33.
61. Charles R. Cobb and Brian M. Butler, "The Vacant Quarter Revisited: Late Mississippian Abandonment of the Lower Ohio Valley," *American Antiquity* 67 (2002) 625–641.
62. Snow, *The Iroquois,* 31–76.
63. Bryson and Murray, *Climates of Hunger,* 33.
64. Gibbon, "Cultural Dynamics," 176.
65. Ibid., 178.
66. Anderson, "Toward a Processual Understanding," 528.

67. The term ecotone is a controversial one in the biological literature. It either describes the transitional zone between two distinct environments, or biomes, or it defines a distinct unit separate from the discrete biomes that it falls between. Robert E. Rhoades, "Archaeological Use and Abuse of Ecological Concepts and Studies: The Ecotone Example." *American Antiquity* 43 (1978): 608–614.
68. Bryson and Murray, *Climates of Hunger,* 29.
69. Ibid., 41.
70. Lauren W. Ritterbush, "Drawn By the Bison: Late Prehistoric Native Migration Into the Central Plains," *Great Plains Quarterly* 22 (2002), 259–270.
71. Ibid., 264.
72. Beth R. Ritter, "Piecing together the Ponca Past: Reconstructing Degiha Migrations to the Great Plains," *Great Plains Quarterly* 22 (2002), 271–284.
73. Lehmer, "Climate and Culture History in the Middle Missouri Valley," 63.
74. Connie A. Woodhouse and Jonathan T. Overpeck, "2000 Years of Drought variability in the Central United States," *Bulletin of the American Meteorological Society* 79 (1998), 2698–2699.
75. Gregg, "Archaeological Complexes," 80. Evidence for long-term drought in North Dakota is not unanimous. Schneider argued the analysis of Lake Moon sediment indicated the post a.d. 1200 period "was wetter and cooler than the preceding 1,500 years"; Fred Schneider, "Prehistoric Horticulture in the Northwestern Plains," *Plains Anthropologist* 47 (2002), 45.
76. The concentration of people peaked in the 15th century when as many as 1,000 people living in 90 dwellings gathered behind their protective walls. Douglas Bamforth, "Climate, Chronology, and the Course of War in the Middle Missouri Region of the North American Plains," in the *Archaeology of Warfare: Prehistories of Raiding and Conquest,* eds. Elizabeth N. Arkush and Mark W. Allen (Gainesville: University Press of Florida, 2006), 90–91.
77. Richard Krause, "Plains Village Tradition: Coalescent," in *Handbook of North American Indians: Volume 13. Plains,* ed. Raymond J. DeMallie (Washington: Smithsonian Institution Press, 2001), 196–197.
78. Douglas B. Bamforth and Curtis Nepstad-Thornberry, "Reconsidering the Occupational History of the Crow Creek Site (39BF11)," *Plains Anthropologist* 52 (2007), 157.
79. Bamforth, "Climate, Chronology, and the Course of War," 67.
80. Bamforth and Curtis Nepstad-Thornberry, "Reconsidering the Occupational History," 168.
81. John B. Gregg and Larry J. Zimmerman, "Malnutrition in Fourteenth-Century South Dakota: Osteopathological Manifestations," *North American Anthropologist* 7 (1987), 191–214.
82. Walde, "Sedentism and Precontact Tribal Organization on the Northern Plains," 301; Dale Walde, David Meyer, and Wendy Umfreed, "The Late Period on the Canadian and Adjacent Plains," *Revista de Arqueologica Americana* 9 (1995), 32.

83. Walde, "Sedentism and Precontact Tribal Organization on the Northern Plains," 301.
84. Douglas Bamforth, "An Empirical Perspective on the Little Ice Age Climatic Change on the Great Plains," *Plains Anthropologist* 35 (1990), 364.
85. Walde, "The Mortlach Phase," (Calgary: Ph.D. Diss., University of Calgary, 1994), 139.
86. Schneider, "Prehistoric Horticulture in the Northeastern Plains," 46.
87. Lehmer, "Climate and Culture History in the Middle Missouri Valley," 60.
88. Beverley (Bev) Alistair Nicholson, "Plains Woodland Influx and the Blackduck Exodus in Southwestern Manitoba during the Late Precontact Period," *Manitoba Archaeological Journal* 6 (1996), 69–86.
89. Beverley (Bev) Alistair Nicholson, Scott Hamilton, Garry Running, and Sylvia Nicholson, "Climatic Challenges and Changes: A Little Ice Age Response to Adversity—The Vickers Focus Forager/Horticulturalists Move On," *Plains Anthropologist* 51 (2006), 325.
90. Beverley (Bev) Alistair Nicholson and Scott Hamilton, "Cultural Continuity and Changing Subsistence Strategies During the late Precontact Period in Southwestern Manitoba," *Canadian Journal of Archaeology* 25 (2001), 69.
91. Nicholson et al., "Climatic Challenges and Changes," 325.
92. Catherine Flynn and E. Leigh Syms, "Manitoba's First Farmers," *Manitoba History* 31 (1996), 4–11.
93. Nicholson and Hamilton, "Cultural Continuity and Changing Subsistence Strategies," 69.
94. Nicholson et al., "Climatic Challenges and Changes," 325.
95. The effect of the Kuwae eruption (Volcanic Explosivity Index hereafter VEI 6) on the global climate may have been worsened by a major eruption (VEI 5) at Aniakchak, Alaska in approximately1450. Keith R. Briffa, Philip D. Jones, Fritz H. Schweingruber and Timothy J. Osborn, "Influence of Volcanic Eruptions on Northern Hemisphere Summer Temperature over the Past 600 Years," *Nature* 393 (1998), 453. The eruptions coincide with a sharp decline in stream flow in the North Saskatchewan River, a sign of greatly diminished glacial melt from cold temperatures. Roslyn A. Case and Glen M. MacDonald, "Tree Ring Reconstructions of Streamflow for Three Canadian Prairie Rivers," *Journal of the American Water Resources Association* 39 (2003), 710.
96. Nicholson et al., "Climatic Challenges and Changes," 329.
97. Ross Hassig, "The Famine of One Rabbit: Ecological Causes and Social Consequences of a Pre-Columbian Calamity," *Journal of Anthropological Research* 37 (1981), 172. I thank Berks Browne for providing me with this reference.
98. Sherburne F. Cook, "The Incidence and Significance of Disease among the Aztecs and Related Tribes." *The Hispanic American Historical Review* 26 (1946) 332.
99. Nicholson et al., "Climatic Challenges and Changes," 328

100. Ibid.
101. Nicholson and Hamilton, "Cultural Continuity and Changing Subsistence Strategies," 69. Another version posits that the Awaxawi remained near Devil's lake while the Hidatsa-proper, "travelled north, briefly living in a land with moose and a polar climate." Mary E. Malainey, "The Gros Ventre/Fall Indians in Historical and Archaeological Interpretation," *Canadian Journal of Native Studies* 25 (2005), 161.
102. Lehmer, "Climate and Culture History in the Middle Missouri Valley," 67.
103. Ibid., 329.
104. Walde, "Sedentism and Precontact Tribal Organization on the Northern Plains," 292.
105. Douglas Bamforth, "An Empirical Perspective," 364.
106. David Meyer, "People Before Kelsey: An Overview of Cultural Developments," in *Three Hundred Prairie Years: Henry Kelsey's "Inland Country of Good Report,"* ed. Henry Epp (Regina: Canadian Plains Research Center, 1993), 64.
107. R. Grace Morgan, *An Ecological Analysis of the Northern Plains as Seen through the Garratt Site* (Regina: University of Regina Occasional Papers in Anthropology No. 7, 1979), 111–113.
108. Ibid., 193.
109. Although there is no consensus in the literature about the true nature of bison migration, there is general agreement that the herds grew less predictable during the historic period. "Chase" hunting, the pursuit of the species from hunters mounted on horses is commonly acknowledged as the cause of the increasingly erratic behaviour of the herds. R. Grace Morgan, "Beaver Ecology/Beaver Mythology" (Edmonton: Ph.D. Diss., University of Alberta, 1991), 155–157.
110. Walde, "Sedentism and Precontact Tribal Organization on the Northern Plains," 302–303.
111. Morgan, *An Ecological Analysis,* 182.
112. Morgan, "Beaver Ecology/Beaver Mythology," 5.
113. Ibid., 27.
114. Ibid., 42.
115. David Smyth, "The Niitsitapi Trade: Euro-Americans and the Blackfoot Speaking peoples to the Mid 1830s" (Ottawa: Ph.D. Diss., Carlton University, 2001), iii.
116. Peter Fidler, *Journal of a Journey over Land from Buckingham House to the Rocky Mountains in 1792–93,* ed. Bruce Haig (Lethbridge: Lethbridge Historical Society, 1992), 10–12.
117. Smyth, "The Niitsitapi Trade," 306.
118. J. Rod Vickers and Trevor Peck, "Islands in a Sea of Grass: The Significance of Wood in Winter Campsite Selection on the Northwestern Plains," in *Archaeology on the Edge: New Perspectives from the Northern Plains,* eds. Brian Kooyman and Jane H. Kelley (Calgary: University of Calgary Press, 2004), 96.

119. Morgan, "Beaver Ecology/Beaver Mythology," 43–45.
120. Morgan, *An Ecological Analysis, 167–172.*
121. Walde, "The Mortlach Phase," 136.
122. Hubert G. Smith and Raymond W. Wood, eds. *The Explorations of the La Vérendryes in the Northern Plains, 1738–43* (Lincoln: University of Nebraska Press, 1980), 44.
123. Barbara Belyea, ed. *A Year Inland: The Journal of a Hudson's Bay Company Winterer* (Waterloo: Wilfred Laurier University Press, 2000), 112.
124. Smyth, "The Niitsitapi Trade," 109.
125. Morgan, "Beaver Ecology/Beaver Mythology," 165.
126. A.S Morton, *The Journal of Duncan M'Gillivray of the North West Company at Fort George on the Saskatchewan, 1774–5.* (Toronto: MacMillan of Canada, 1927), 166.

THE DRY BELT AND CHANGING ARIDITY IN THE PALLISER TRIANGLE, 1895–2000

GREGORY P. MARCHILDON, JEREMY PITTMAN, AND DAVID J. SAUCHYN

Abstract

Due to its average aridity as well as historical experience with prolonged droughts, the Dry Belt has long been identified as the most vulnerable sub-region within the larger Palliser Triangle in the Great Plains of southwestern Canada. Based upon 105 years of climate data, drought maps are redrawn based upon the historical record of drought within the Dry Belt. The result demonstrates that the size of the Dry Belt expands or contracts dramatically depending on the precise period being analyzed. Given current climate change scenarios, we should not be surprised to see a major expansion in the area of the Dry Belt in future decades similar to what was experienced between 1928 and 1938.

Sommaire

La zone sèche est depuis longtemps reconnue comme étant la sous-région la plus vulnérable du triangle de Palliser des Grandes Plaines du sud-ouest canadien à cause de son aridité moyenne et de son histoire de sécheresses prolongées. C'est en s'appuyant sur 105 ans de données climatiques que les cartes de sècheresse sont revues selon l'historique de sècheresse dans la zone sèche. Les résultats indiquent que l'étendue de la zone sèche augmente ou diminue de beaucoup selon la période spécifique analysée. S'appuyant sur les scénarios de changements climatiques actuels, il ne serait pas surprenant que l'étendue de la zone sèche s'accroisse de façon majeure dans les prochaines décennies par rapport aux cinquante dernières années, rappelant ce qui s'est produit entre 1928 et 1938.

Early History of the Dry Belt

The Dry Belt has long been identified as the most arid portion of the Palliser Triangle. Straddling the border of Alberta and Saskatchewan, this dry core is a prominent geographical and historical feature of the western prairies. Located in the rain shadow of the Rocky Mountains to the west and the Cypress and Sweet Grass Hills to the south, the Dry Belt receives, on average, less than 350 mm of precipitation per year, considerably below the average of the Palliser Triangle. In addition, it is subject to high moisture loss because of the warm and dry winter winds known as Chinooks, as well as summer heat waves of great intensity. Finally, the soils of the Dry Belt are light and have low water retention, thus making the area more sensitive to long periods of moisture deficiency.[1] As the epicentre of drought and depopulation before the Second World War, the Dry Belt is often portrayed as a cursed land, a destroyer of families, livelihoods and dreams, as illustrated in David Jones's classic history of the Dry Belt.[2]

The Dry Belt lies within the Palliser Triangle, itself long recognized for its aridity and limited biodiversity.[3] The Triangle is named after Captain John Palliser, the leader of the British North American Exploring Expedition of 1857–60 sponsored by the British government.[4] Palliser viewed the region as a northern extension of what he called the Great American Desert. His "triangle" is actually more of a parallelogram extending up from the current Canada-U.S. border (Figure 1). Covering more than 200,000 square km of southern Alberta and Saskatchewan (including a tiny bit of southwestern Manitoba), the Palliser Triangle encompasses what is now the single largest expanse of agricultural land in Canada.

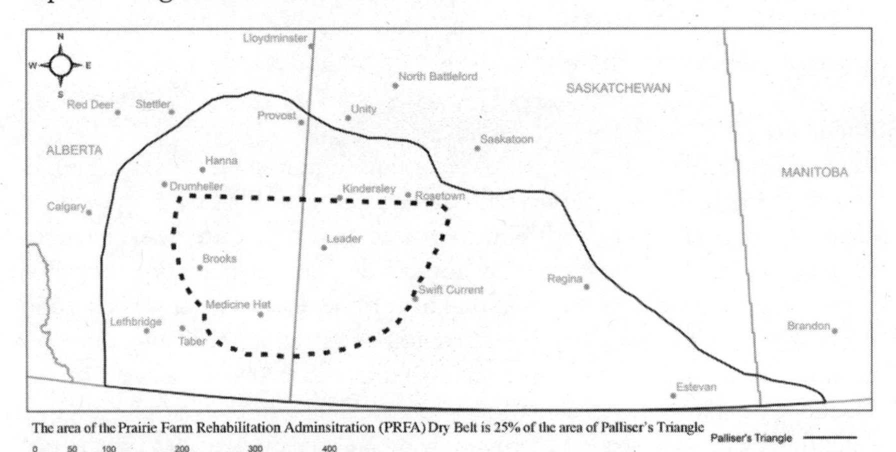

Source: J. Palliser, *Exploration-British North America. The Journals, Detailed Reports and Observations* (London: Eyre and Spottiswoode, 1863). W. Namanishen, Drought in the Palliser Triangle: A Provisional Primer (Regina: Prairie Farm Rehabilitation Administration, Agriculture and Agri-Food Canada, 1998), 2.

Figure 1. The PFRA Dry Belt and the Palliser Triangle.

Based on his observations of climate, vegetation and soil, Palliser concluded that the entire region was unsuitable for agriculture.[5] His findings were supported by Henry Youle Hind, a professor from the University of Toronto who had explored the western portion of the prairies for the United Province of Canada in 1858.[6] Hind referred to most of the Palliser Triangle as "Arid Plains," but he identified an arc of land directly to the north—"the fertile belt"—which he deemed as suitable for agriculture because it received more rainfall."[7]

What Palliser and Hind observed, however, was not merely the average aridity of the region, but its worst climatic feature—recurring drought.[8] From the mid-1850s until the mid-1860s, the southern Canadian plains were in the grip of one of the most prolonged droughts of the 19th century.[9] Based upon expedition recordings, the drought, at least during 1859 when Palliser traveled into the heart of the Dry Belt, was most severe in the western part of the prairies.[10]

Although most of the Palliser Triangle was opened for agricultural settlement in the late 19th century, the Dry Belt was settled by ranchers rather than farmers because of its high average aridity and continuing susceptibility to prolonged drought.[11] After the abnormally cold winter of 1905–06 which killed off approximately one-half of the cattle in the region, grain farmers began to move into the Dry Belt.[12] Initially, Dry Belt wheat farmers enjoyed bumper crops. Beginning in 1917, however, grain farmers suffered a series of drought years.

Since farmers on the Alberta side of the Dry Belt suffered the most, the Alberta government was eventually forced to come to the aid of their bankrupt municipalities and help move thousands of drought-stricken farmers out of the Dry Belt, as well as administer the social and physical infrastructure of the region as a Special Area.[13] During the Dirty Thirties, the droughts expanded in size and intensity (at least in Saskatchewan) precipitating a major population exodus.[14] Although there have been severe droughts since the 1930s, in particular in 1961, 1988 and 2001–02, farmers in the Dry Belt have not suffered as prolonged (multi-year) droughts as they experienced from 1917 until the late 1930s.

Villmow's Dry Belt

The location of the Dry Belt is not as precise as might be indicated by the Government of Canada's Prairie Farm Rehabilitation Administration (PFRA) in Figure 1.[15] The PFRA refers to the unique moisture characteristics of this region, and defines the boundaries of the Dry Belt based on a 350 mm precipitation isoline averaged over the period from 1961 to 1990. This raises the question of how, in more precise scientific terms, the Dry Belt should now be defined in light of this historical experience.

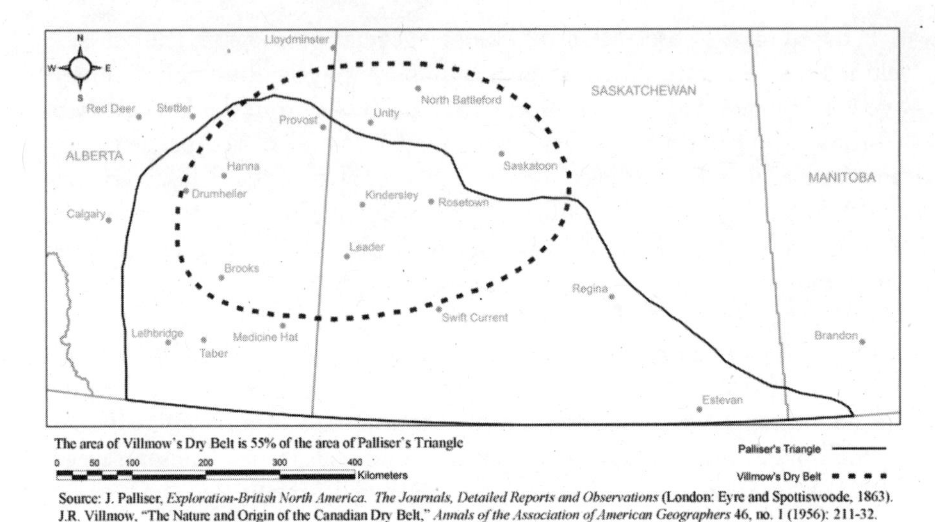

Source: J. Palliser, *Exploration-British North America. The Journals, Detailed Reports and Observations* (London: Eyre and Spottiswoode, 1863). J.R. Villmow, "The Nature and Origin of the Canadian Dry Belt," *Annals of the Association of American Geographers* 46, no. 1 (1956): 211-32.

Figure 2a. Villmow's Dry Belt and the Palliser Triangle.

In 1956, an American geographer, Jack Villmow, published an article in which he provided an apparently precise description of the Dry Belt (Figure 2a) based on the best climate data available at the time.[16] Villmow analyzed surface meteorological data in a number of ways. He collected data from different climate monitoring stations throughout the region, and mapped out climatic isolines based on 25 to 35 year averages in temperature (T) and precipitation (P) as well as seasonal variability in climate. He also classified the climate of the area following the method developed by the pioneering geographer and climatologist C. Warren Thornthwaite.[17] Of particular interest is Villmow's application of the Thornthwaite model of potential evapotranspiration (PET) to differentiate regions using isolines based on the ratio of PET to P. He found that much of the Dry Belt had a PET to P ratio less than or equal to 1.75, which corresponds to a P to PET ratio less than or equal to 0.57.[18] He was able to combine these results to describe the static boundaries of the dry belt shown in Figure 2a. The boundaries Villmow used for the Dry Belt, how-

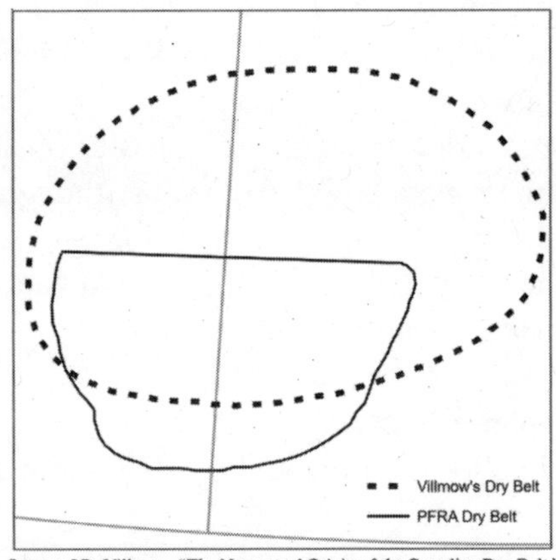

Source: J.R. Villmow, "The Nature and Origin of the Canadian Dry Belt," *Annals of the Association of American Geographers* 46, no. 1 (1956): 211-32. W. Nemanishen, *Drought in the Palliser Triangle: A Provisional Primer* (Regina: Prairie Farm Rehabilitation Administration, Agriculture and Agri-Food Canada, 1998), 2.

Figure2b. Villmow's Dry Belt and the PFRA Dry Belt.

ever, differ greatly from those used by the PFRA (Figure 2b). The Dry Belt, as defined by Villmow, has twice the areal extent of the PFRA's version.

Villmow's work contributed greatly to knowledge of the Canadian Dry Belt. Villmow also acknowledged that "moisture characteristics form the primary basis for the contention that the Dry Belt is a unique and distinctive climatic region."[19] However, moisture characteristics must include both the amount of precipitation distributed both spatially and temporally as well as a measure of temperature as it relates to evapotranspiration (moisture in soil and plant life lost to the atmosphere). As noted by Villmow, both precipitation and temperature vary greatly seasonally, inter-annually and spatially in the area.[20] Since the boundaries of the Dry Belt are defined by these variables, the question addressed below is whether the boundaries vary appreciably over time and space when averaged over different periods.

New Historical Maps of the Dry Belt

As illustrated in Figures 3 to 7, the boundaries of the Dry Belt vary significantly when based on mean conditions of different time periods. Drylands throughout the globe cannot be defined by static borders, and the Dry Belt is no exception.[21] To state the obvious, it is essential to specify the precise time period when mapping the Dry Belt because the boundaries will shift over time.

According to the United Nations Environmental Programme's (UNEP) *World Atlas of Desertification,* drylands are areas with an average annual P to PET ratio of less than 0.65.[22] This definition and the P/PET data for the Canadian prairies were used to construct the maps seen in Figures 3 to 6. This definition also corresponds well with Villmow's pioneering work, since he obtained P/PET values within this range for his Dry Belt. PET was calculated using the Thornthwaite method from a gridded climate database of monthly precipitation and temperature produced by Environment Canada.[23] Although this method is simple relative to more complex techniques developed during the past half century since Thornthwaite's work, the method has proven itself in terms of delivering accurate results. As a consequence, it has been employed in similar studies, namely the *World Atlas of Desertification.*[24] The technical aspects of this calculation are provided in Appendix 1.

The map produced by Villmow is difficult to reconstruct given the lack of precise time constraints. Villmow used 25 to 35 year averages, but this varied from station to station. Sometimes, even less than 25 to 35 year averages were used, but this is difficult to determine because Villmow did not provide the exact years that were incorporated into his averages. It almost appears as though Villmow may have assumed that time was unimportant in defining the Dry Belt.

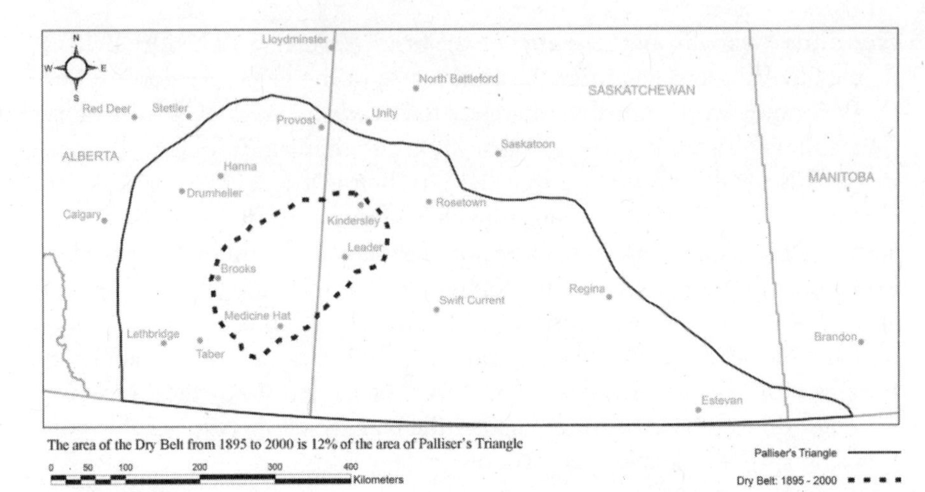

Source: J. Palliser, *Exploration-British North America. The Journals, Detailed Reports and Observations* (London: Eyre and Spottiswoode, 1863).

Figure 3a. The Dry Belt based on a P/PET <= 0.65 map for the period from 1895 to 2000 and the Palliser Triangle.

In his study for the PFRA, Walter Nemanishen acknowledges the significance of calculating the boundaries of the Dry Belt over specific time periods, but his exclusive use of precipitation to define the Dry Belt neglects the impact of evapotranspiration from the soil in determining the degree of aridity in the region. The use of the P/PET index accounts for this loss and provides insight into the amount of moisture available at the surface. Moreover, we were not limited to the 25 to 35 years worth of data that were available to Villmow. Our estimates are based upon historical data collected from 1895 until 2000. Different Dry Belt maps were produced corresponding to different periods of interest in the history of the area. Figure 3a illustrates the result for the full period from 1895 until 2000. Figure 3b shows this area in relation to the PFRA's Dry Belt and Villmow's Dry Belt respectively. The area with a P/PET < 0.65 is about half the size of the PFRA's Dry Belt and only about a quarter the size of Villmow's. What is noteworthy, however, is the extent to which the

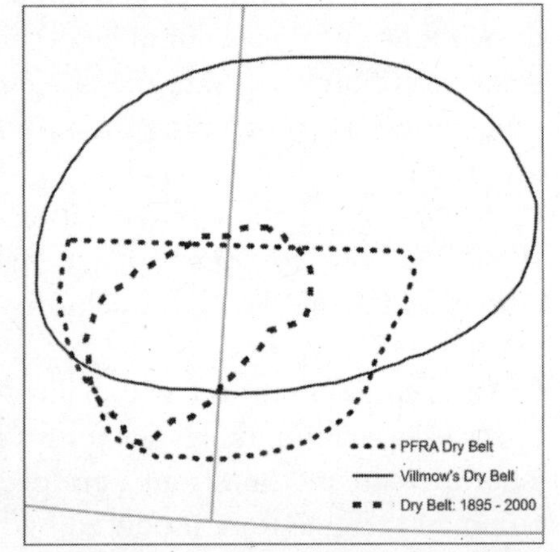

Source: J.R. Villmow, "The Nature and Origin of the Canadian Dry Belt," *Annals of the Association of American Geographers* 46, no. 1 (1956): 211-32. W. Nemanishen, *Drought in the Palliser Triangle: A Provisional Primer* (Regina: Prairie Farm Rehabilitation Administration, Agriculture and Agri-Food Canada, 1998), 2.

Figure 3b. Dry Belt based on a P/PET <= 0.65 map for the period from 1895 to 2000, Villmow's Dry Belt and the PFRA Dry Belt.

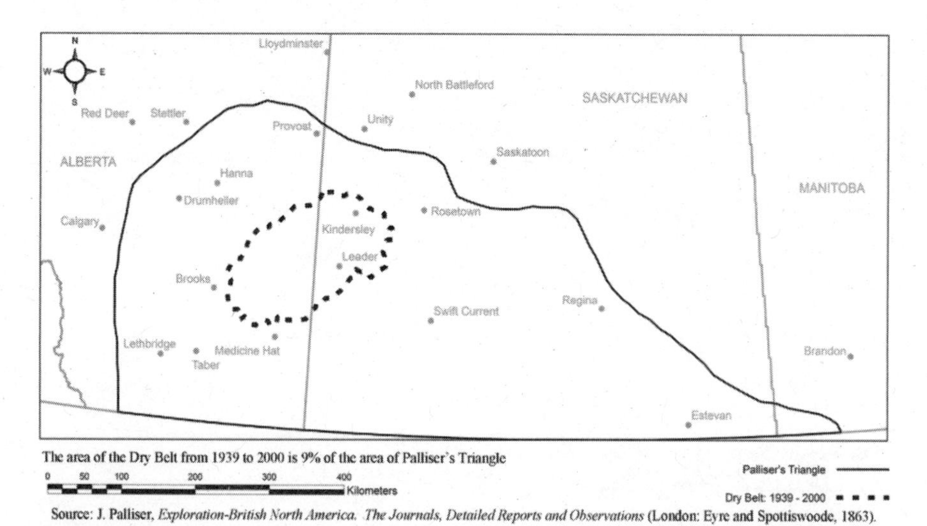

Source: J. Palliser, *Exploration-British North America. The Journals, Detailed Reports and Observations* (London: Eyre and Spottiswoode, 1863).

Figure 4a. The Dry Belt based on a P/PET <= 0.65 map for the period from 1939 to 2000 and the Palliser Triangle.

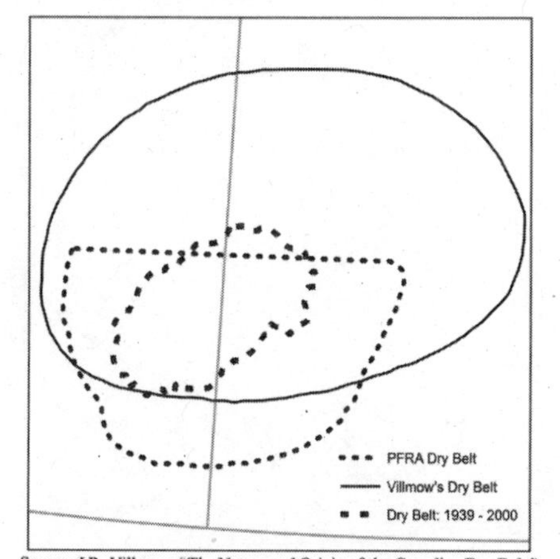

Source: J.R. Villmow, "The Nature and Origin of the Canadian Dry Belt," *Annals of the Association of American Geographers* 46, no. 1 (1956): 211-32. W. Nemanishen, *Drought in the Palliser Triangle: A Provisional Primer* (Regina: Prairie Farm Rehabilitation Administration, Agriculture and Agri-Food Canada, 1998), 2.

Figure 4b. Dry Belt based on a P/PET < = 0.65 map for the period from 1939 to 2000, Villmow's Dry Belt and the PFRA Dry Belt.

Dry Belt remains located almost entirely within the region originally identified by the PFRA and Villmow.

The Dry Belt for the post-Depression years from 1939 until 2000 produces a very similar result—a comparably shaped but slightly smaller region as seen in Figure 4a. Since 1939, this has remained the driest part of the Palliser Triangle, the sub-region least conducive to crop-based agriculture. Although crop farming continues in the region, it is far more restricted. Ranching has again supplanted farming as the dominant form of agriculture within the Dry Belt, particularly on the Alberta side of the border, partly due to the efforts of the Special Areas Board in facilitating the shift from grain farms to ranches or mixed farm-ranches in the 1930s.[25]

The areal extent of the Dry Belt during the post-Depression years is considerably smaller than those proposed by Villmow and the PFRA. It is less than 40% the size of the PFRA's Dry Belt and less than 20% the size of Villmow's. Figure 4b illustrates these differences.

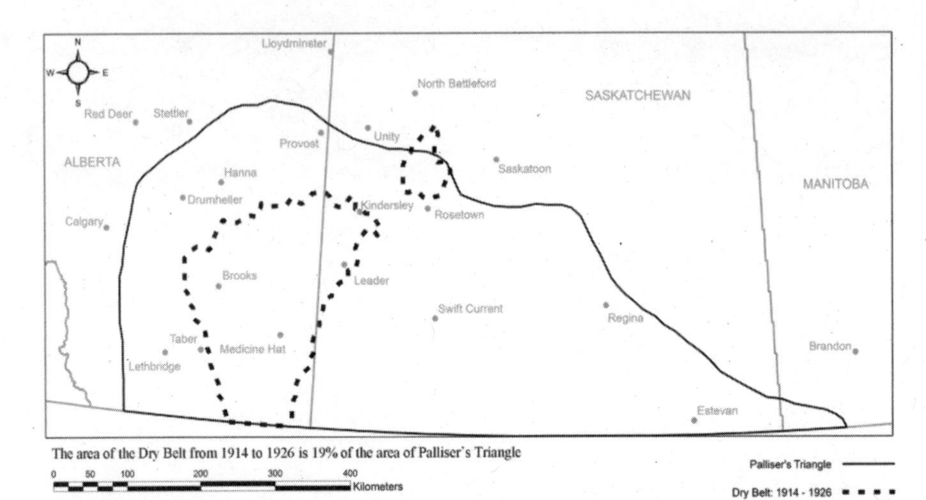

Source: J. Palliser, *Exploration-British North America. The Journals, Detailed Reports and Observations* (London: Eyre and Spottiswoode, 1863).

Figure 5a. The Dry Belt based on a P/PET <= 0.65 map for the period from 1914 to 1926 and the Palliser Triangle.

We should also expect the map of the Dry Belt to be quite different for the pre-Second World War era as a consequence of the prolonged droughts which devastated the farm families living in this part of the Palliser Triangle. In this case, it is worth examining the period beginning with the first recorded agricultural drought in 1914, and ending the year before the abnormally moist year of 1927. Figure 5a validates the considerable historical evidence of farmers in southwest Alberta suffering more from prolonged drought than their Saskatchewan neighbours over this time period. The Dry Belt was largely concentrated in southeastern Alberta, from south of Medicine Hat and Taber, Alberta, stopping short of Drumheller and Hanna, although it did swing northeast to include Leader and the region immediately west of Kindersley in Saskatchewan. In addition, the Dry Belt included a small patch in the central part of western Saskatchewan north of Rosetown and southeast of Unity.

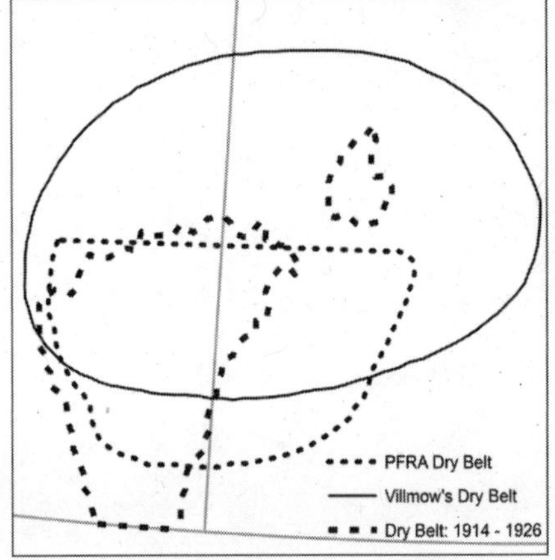

Source: J.R. Villmow, "The Nature and Origin of the Canadian Dry Belt," *Annals of the Association of American Geographers* 46, no. 1 (1956): 211-32. W. Nemanishen, *Drought in the Palliser Triangle: A Provisional Primer* (Regina: Prairie Farm Rehabilitation Administration, Agriculture and Agri-Food Canada, 1998), 2.

Figure 5b. Dry Belt based on a P/PET <= 0.65 map for the period from 1914 to 1926, Villmow's Dry Belt and the PFRA Dry Belt.

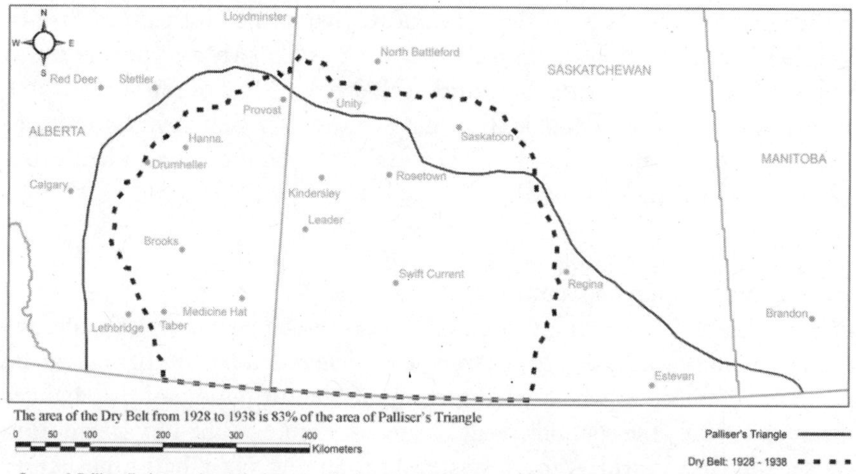

Source: J. Palliser, *Exploration-British North America. The Journals, Detailed Reports and Observations* (London: Eyre and Spottiswoode, 1863).

Figure 6a. The Dry Belt based on a P/PET <= 0.65 map for the period from 1928 to 1938 and the Palliser Triangle.

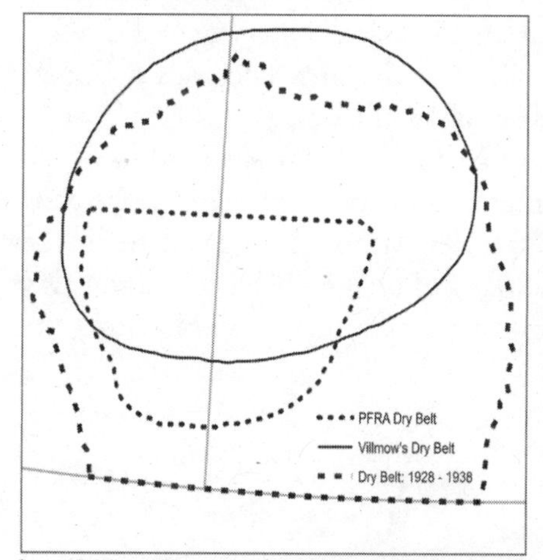

Source: J.R. Villmow, "The Nature and Origin of the Canadian Dry Belt," *Annals of the Association of American Geographers* 46, no. 1 (1956): 211-32. W. Nemanishen, *Drought in the Palliser Triangle: A Provisional Primer* (Regina: Prairie Farm Rehabilitation Administration, Agriculture and Agri-Food Canada, 1998), 2.

Figure 6b. Dry Belt based on a P/PET <= 0.65 map for the period from 1928 to 1938, Villmow's Dry Belt and the PFRA Dry Belt.

The areal extent of our Dry Belt during this period begins to match that of the PFRA's Dry Belt but still is considerably smaller than Villmow's Dry Belt. Figure 5b shows our Dry Belt to be about three quarters the size of the PFRA's Dry Belt and more than a quarter the size of Villmow's Dry Belt. In other words, there are major differences in the delineations of the boundaries between the Dry Belt for this period and these other two conceptions. The concentration of the drought during this period in Alberta, as discussed above, would not be easily inferred from the PFRA and Villmow maps but our map clearly shows the intensity of the drought in Alberta compared to Saskatchewan between 1914 and 1926.

The era of the Dirty Thirties is captured in Figure 6a. We began with 1928, because this was the year that drought returned to the Dry Belt after the unusually wet year of 1927. We ended in 1938 because this is the last recorded year of extreme drought until 1961. The impact of prolonged drought can be seen in the extent of the Dry Belt extending from Saskatoon in the northeast, the U.S. border south of Regina in the southeast and south of Taber in the southwest, almost as far west as Calgary and northwest as Stettler. This huge territory constitutes well over one-half of the area encompassed by the Palliser Triangle. Our map of the Dry Belt covers an area at least three times larger than PFRA's Dry Belt and about one and a half times larger than Villmow's Dry Belt.

Mapping Historical Change in the Dry Belt and Climate Change Implications

When the P/PET calculations are averaged over the entire period—1895 to 2000—the mean position of the boundaries delineated for this period can be used as a reference to which we can compare the magnitude of the changes observed for the other periods, as illustrated in Figure 7. The Dry Belt during the pre-Second World War era was more than one and a half times larger than the Dry Belt during the full reference period. Amazingly, the Dry Belt of the Dirty Thirties is nearly seven times larger than the Dry Belt in the reference period. For the post-Depression period, the Dry Belt shrinks to a point where it is only a little more than three-quarters the size of the reference. Once again, the climate variability of the area becomes apparent through the shifting temporal boundaries of the Dry Belt.

The historical redrawing of the Dry Belt illustrates the extent to which a geographical area determined by climate is not static even within the confines of little more than 100 years of time. Although referred to in the historical literature as a relatively fixed area of geographical space, the Dry Belt has expanded and contracted over the past century. Understanding the

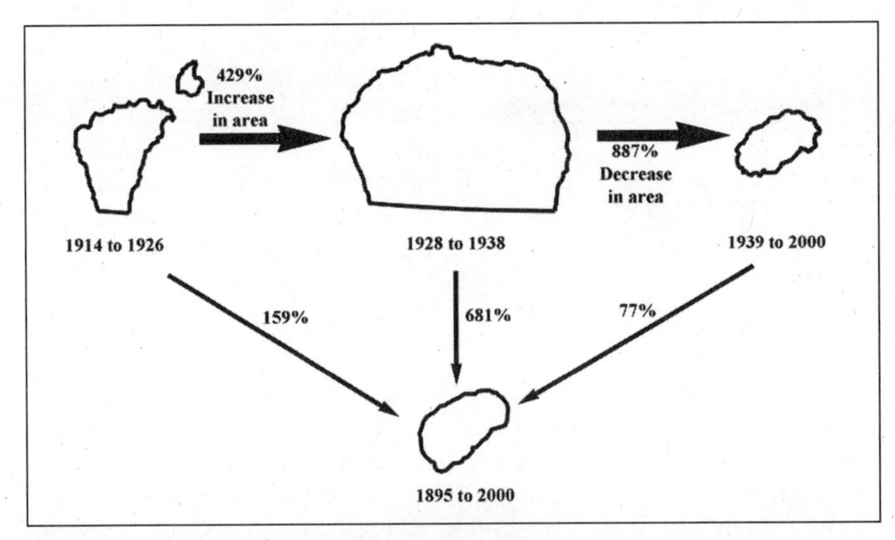

Figure 7. The areal change of the Dry Belt over three time periods relative to the reference period.

shifting nature of the Dry Belt is particularly important given the forecasts of current global climate change models in general, and climate change scenarios for the Great Plains of North America in particular.

Most global climate change scenarios predict increased drying in continental interiors and greater risk of droughts for the 21st century.[26] Based on the expert opinion of the United Nation's Intergovernmental Panel on Climate Change, there is a 66% probability of an increase in the area that could be affected by drought.[27] Using the history of the 20th century as a guide, this could mean that the size of the Dry Belt for a significant period of the 21st century could be as large as that experienced from 1928 until 1938. If this occurs, then almost all the Palliser Triangle would be unsuitable for most types of non-irrigated grain farming, and difficult for ranching without appropriate provision for stored water.

The climate history presented here reveals the Dry Belt to be the vulnerable core of the Palliser Triangle. It should not be surprising if the size of the Dry Belt increases dramatically in the 21st century as a result of climate change. Current climate change scenarios predict higher summer temperatures generating more evapotranspiration. As such, the Dry Belt must be targeted in a larger, risk-mitigation strategy to prepare for the impact of climate change in this century.

Appendix 1. Methodology for Creating the Dry Belt Maps

- The Dry Belt maps were derived from aridity index (P/PET) maps created for the region. A P/PET < 0.65 was used to delineate the boundaries of the Dry Belt. This threshold was chosen based on the UNEP's definition of drylands, which are those with an average annual P/PET < 0.65. (Source: United Nations Environment Programme (UNEP), *World Atlas of Desertification* (Edward Arnold, London, 1992))

- The Thornthwaite method of calculating PET was employed. Thornthwaite's equation is as follows:

$$PET = 1.6\,(10T/I)^{a}$$

Where:

- o PET = monthly potential evapotranspiration (cm)
- o T = mean monthly temperature (c)
- o $I = \sum_{m=1}^{n} i_m$, where n is equal to 12 (i.e. i is summed over the 12 months of the year)

- $i = (T/5)^{1.514}$
- $a = 6.75*10^{-7}I^3 - 7.71*10^{-5}I^2 + 1.79*10^{-2}I + 0.49$

(Source: C.W. Thornthwaite, "An Approach toward a Rational Classification of Climate," *Geographical Review* 38, no. 1 (1948): 55–94.)

- The PET values were then corrected for latitude by multiplying them by a factor corresponding to each month of the year and different intervals of latitude. (Source: V.M. Ponce, *Engineering Hydrology* (New Jersey: Prentice-Hall, Inc., 1989.)

- Potential evapotranspiration was calculated from mean monthly temperature and precipitation values in the Canadian Gridded Climate Database. This database was interpolated to a 50-km grid from climate station data found in the Canadian Climate Archive. An inverse square distance weighting scheme was applied to obtain the values at a 50-km spatial resolution true at 60°N on a polar stereographic secant projection aligned with 111°W. (Source: R. Hopkins, Canadian Gridded Climate Data, Environment Canada (2001.))

- Aridity indices (P/PET) were calculated at a monthly time scale for each grid point and then averaged over the year. Values for each period of interest were then created by averaging the aridity indices corresponding to the particular years of interest. The gridded points were then loaded into a GIS, each with their corresponding values of P/PET. Rasters were interpolated using the P/PET calculations for each original grid point and specific time period as the Z-value. This raster was further interpolated to a spatial resolution of 5 km by Kriging to smooth the boundaries between raster classes.

- The rasters were then classified with a class break at P/PET = 0.65. The resulting maps showed all values below or equal to 0.65 in one color and all values above 0.65 in a different color.

- New polyline shape files were created and loaded into the GIS. These shape files were edited to contain the spatial boundaries of the class break at P/PET = 0.65 for the different periods of interest. The boundaries of the Dry Belt at different temporal resolutions were then completed.

Endnotes

1. G.P. Marchildon, S. Kulshreshtha, E. Wheaton, and D. Sauchyn, "Drought and Institutional Adaptation in the Great Plains of Alberta and Saskatchewan, 1914–1939," *Natural Hazards* 45, no. 3 (2008): 3991-411.
2. D.C. Jones, *Empire of Dust: Settling and Abandoning the Prairie Dry Belt* (Calgary: University of Calgary Press, 2002, orig. 1984).
3. D. Lemmen and L. Dale-Burnett, "The Palliser Triangle," in K. Fung (ed.), *Atlas of Saskatchewan: Second Edition* (Saskatoon: University of Saskatchewan, 1999), 41.
4. I.M. Spry (ed.), *The Papers of the Palliser Expedition, 1857–1860* (Toronto: Champlain Society, 1968). Also see I.M. Spry, *The Palliser Expedition: An Account of the John Palliser's British North American Expedition, 1857–1860* (Toronto: Macmillan, 1963).
5. Lemmen and Dale-Burnett, "The Palliser Triangle," 41.
6. H.Y. Hind, *Narrative of the Canadian Red River Exploring Expedition of 1857 and of the Assiniboine and Saskatchewan Expedition of 1858* (Edmonton: Hurtig, 1971, orig. 1860). Also see W.L. Morton, *Henry Youle Hind, 1823–1908* (Toronto: University of Toronto Press, 1980).
7. D. Hayes, *Historical Atlas of Canada: Canada's History Illustrated with Original Maps* (Vancouver: Douglas and McIntyre, 2002), 207.
8. According to Gerald Davidson, the recurring cycle of drought as much as high average aridity defines the prairie region: G.T. Davidson, "An Interdisciplinary Approach to the Role of Climate in the History of the Prairies," in R. Wardaugh (ed.), *Toward Defining the Prairies: Region, Culture, and History* (Winnipeg: University of Manitoba Press, 2001), 201.
9. A.B. Beaudoin, "What They Saw: The Climatic and Environmental Context for Euro-Canadian Settlement in Alberta," *Prairie Forum* 24, no. 1 (1999): 1–40. D.J. Sauchyn and W.R. Skinner, "A Proxy Record of Drought Severity for the Southwestern Canadian Plains," *Canadian Water Resources Journal* 26, no. 1 (2001): 253–72.
10. W.F. Rannie, "Summer Rainfall on the Prairies during the Palliser and Hind Expeditions, 1857–59," *Prairie Forum* 31, no 1 (2006): 17–38.
11. *Historical Atlas of Canada: The Land Transformed, 1800–1891* (Toronto: University of Toronto Press, 1993), plate 42. *Historical Atlas of Canada: Addressing the Twentieth Century* (Toronto: University of Toronto Press, 1990), plate 17.
12. O.E. Baker, "Agricultural Regions of America: Part VI—the Spring Wheat Region," *Economic Geography* 4, no. 4 (1928): 399–433. D. Breen, *The Canadian Prairie West and the Ranching Frontier, 1874–1924* (Toronto: University of Toronto Press, 1983), 141–2.
13. G.P. Marchildon, "Institutional Adaptation to Drought and the Special Areas of Alberta, 1909–1939," *Prairie Forum* 32, no. 2 (2007): 251–72. D.C. Jones and R.C. Macleod (eds.), *We'll all be Buried down Here: the Prairie Dryland Disaster, 1917–1926* (Calgary: Historical Society of Alberta, 1986). J. Gorman, *A Land Reclaimed:*

A Story of the Special Areas in Alberta (Hanna, AB: Special Areas Board, 1988).

14. Marchildon et al., "Drought and Institutional Adaptation in the Great Plains of Alberta and Saskatchewan, 1914–1939."
15. W. Nemanishen, *Drought in the Palliser Triangle: A Provisional Primer* (Regina: Prairie Farm Rehabilitation Administration, Agriculture and Agri-Food Canada, 1998), 2.
16. J.R. Villmow, "The Nature and Origin of the Canadian Dry Belt," *Annals of the Association of American Geographers* 46, no. 1 (1956): 211–32.
17. C.W. Thornthwaite, "An Approach toward a Rational Classification of Climate," *Geographical Review* 38, no. 1 (1948): 55–94.
18. Villmow, "The Nature and Origin of the Canadian Dry Belt."
19. Ibid.
20. Ibid.
21. United Nations Environment Programme (UNEP), *World Atlas of Desertification* (Edward Arnold,London, 1992.)
22. Ibid.
23. R. Hopkins, *Canadian Gridded Climate Data* (Ottawa: Environment Canada, 2001). Thornthwaite, "An Approach toward a Rational Classification of Climate."
24. UNEP, *World Atlas of Desertification.* W.C. Palmer and A.V. Havens, "A Graphical Technique for Determining Evapotranspiration by the Thornthwaite Method," *Monthly Weather Review* 86 (1958): 123–8.
25. Marchildon, "Institutional Adaptations to Drought and the Special Areas of Alberta, 1909–1939," *Prairie Forum* 32, no. 2 (2007): 251–71.
26. R. Watson and the Core Writing Team, eds., *Climate Change 2001: Synthesis Report, Intergovernmental Panel on Climate Change* (Cambridge: Cambridge University Press, 2001).
27. Intergovernmental Panel on Climate Change, *Climate Change 2007: The Physical Science Basis—Summary for Policy Makers* (Geneva: IPCC Secretariat, 2007).

COMPARATIVE WATER GOVERNANCE IN THE FOUR WESTERN PROVINCES

MARGOT HURLBERT

Abstract

Water governance in the Western Canadian provinces is experiencing a transformation. A new model of water governance, embracing principles of community participation and shared responsibility is evolving. This model replicates the user-based model of water governance endorsed by many international water experts. From this review of Western Canadian water governance, it is concluded the greatest challenge is the integrated management of both the quantity and quality of water.

This paper will explain, outline and critique some of the models adopted by the Western Canadian provincial governments for managing water resources by analyzing their legislative framework. In assessing the 'governance of water resources' focus will be on quantity allocation decisions, but implications of these decisions on water quality will be considered. There are primarily three models: user-based management, government agency management, or a market-based management system. The Western Canadian provinces' water models, as outlined in legislation, reflect aspects of these three models. After outlining these aspects, the provincial models will be assessed based on principles of water management developed by the World Water Council. These principles are accountability, participation in decisions by all stakeholders, predictability, transparency, and decentralization. Overall, the water models reflected in the Western Canadian provinces' legal frameworks evidence these principles, and show advancement because of initiatives to adopt more user-based water governance practices (the water model most conducive to these principles of best practice). This paper will conclude, however, that improvements can still be made especially in the area of integrated water management.

Sommaire

La gouvernance de l'eau dans les provinces de l'Ouest canadien est en pleine transformation. Un nouveau modèle de gouvernance de l'eau qui repose sur la participation communautaire et de partage des responsabilités est en évolution. Ce modèle répète le modèle de gouvernance de l'eau qui s'appuie sur les utilisateurs pour la gestion et qui est endossé par plusieurs experts internationaux de l'eau. Cet examen de la gouvernance de l'eau dans l'Ouest canadien conclut que le plus grand défi concerne la gestion de la quantité et de la qualité de l'eau.

Cet article donne un bref compte rendu, ainsi qu'une explication et une évaluation des divers modèles de gouvernance de l'eau adoptés par les gouvernements des provinces de l'Ouest canadien en analysant leur cadre législatif. L'évaluation de la gestion des ressources en eau se fera principalement par rapport aux décisions prises quant à l'allocation de l'eau, mais les implications de ces décisions sur la qualité de l'eau seront aussi prises en considération. Il existe trois principaux modèles de gouvernance : le premier favorise la gestion par les utilisateurs, le deuxième favorise une gestion effectuée par un organisme gouvernemental, et le troisième favorise une gestion axée sur le marché. Les modèles de gouvernance de l'eau, comme le prévoient les lois, reprennent des éléments de ces trois modèles. Un bref compte rendu de ces éléments sera suivi par une évaluation des modèles de gouvernance provinciaux en s'appuyant sur les principes de gestion de l'eau établis par le Conseil mondial de l'eau. Ces principes sont : la responsabilité, la participation de tous les intervenants quant à la prise de décisions, la décentralisation, la transparence et la prévisibilité. En général, les modèles de gouvernance de l'eau que l'on retrouve dans les cadres législatifs des provinces de l'Ouest canadien reprennent ces principes et montrent des progrès grâce à des initiatives favorisant l'adoption de pratiques de gouvernance de l'eau axée sur la gestion par les utilisateurs, ce modèle étant le plus propice aux principes de pratiques exemplaires. L'article conclut néanmoins qu'il y a encore place à l'amélioration quant aux approches intégrées à l'égard de la gestion de l'eau.

1. Introduction

It is increasingly being recognized that Canada's available freshwater is limited, and policy makers are facing increasing pressures to keep clean fresh water available to all those who need it.[1] Issues related to water, its scarcity and governance, are ranked most important with respect to climate change and its potential impact on the Prairies.[2] Given these factors, the review of water governance becomes increasingly important in a proactive strategy responding to Canada's future water requirements. The assessment of the legal framework establishing governance of Canada's water resource is necessary to ensure this framework adequately reflects Canadian goals and priorities, and can respond to increasing demands in an efficient, effective manner. In Canada the legal framework setting the rules of water governance is predominately, although not exclusively, determined by provincial governments and, as a result, different water management models exist in each province.

This paper will examine the legal framework established by statute in the Western Canadian provinces of Manitoba, Saskatchewan, Alberta, and British Columbia. First, three predominant water governance models illustrated in the provincial frameworks will be discussed, and then these frameworks will be assessed using the best practices of water governance developed by the World Water Council.

Western Canadian water governance, established through provincial legislation, heralds from many different sources and influences. Water governance initially was determined by court decision, or common law, which is referred to as the riparian water laws of Britain, where laws developed on a case-by-case basis over several hundred years in a land of relative water abundance. Because the common law riparian doctrine could not meet the development needs of Canada and British Columbia, and meet the challenge of a semi-arid climate, Canada, British Columbia, and later the Prairie Provinces, enacted statutes replicating portions of the United States' judicially established prior appropriation system of "first come, first right."[3] Canada codified these priorities in statute under a government allocation by way of license.[4] It was not until 1930 that Manitoba, Alberta and Saskatchewan had natural resources transferred to them by the federal government in the Natural Resources Transfer Agreements. Water was confirmed as part of the transfer in these Agreements by *The Natural Resources Transfer (Amendment) Act*.[5] In 1931, the Prairie provinces enacted their own provincial water laws. While many of the main features of the original federal water law still apply, including Crown ownership and allocation of interest by license,[6] the Prairie provinces have diverged in some aspects of their respective water management regimes. These will be illustrated in the following water models.

2. Water Models

There are three major alternatives to the governance of water rights and interests. Generally these models relate to the bundle of property rights associated with water, i.e. whether it is owned privately, as public property, or common property. In Canada, because the Crown owns all water, and water rights are allocated by license, this property ownership distinction is not applicable; however, the property distinction is illustrative as parallels can be seen in the characteristics of bundles of water rights received by way of water license. Based on the three models of property rights, the three institutional models are:

- Government agency management, generally associated with water regarded as public property—Government defers its authority for the management of water to an agency which assumes authority for directing who does, and does not, receive water rights in accordance with bureaucratic policies and procedures.
- User-based management, generally associated with water regarded as common property—Water users, or those with license or rights to water join together and coordinate their actions in managing water resources. Decision making is collective among users.
- Market, generally associated with water owned as private property—Water is allocated and reallocated through private transactions. Users can trade water rights through short term or long term agreements or temporary or permanent transfers, reallocating rights in response to prices.[7]

All of these models are used in Western Canada in a variety of combinations. While no one model is used exclusively, this practice is consistent with water management in other countries. Within the same river basin, or even within an irrigation system, there may be user-based management within and between some groups of users, transfers between individual farmers occurring through market-type mechanisms, and government agencies administering allocation of water resources. Examples illustrating these models in Western Canada will be provided.

2a. Government Agency Management

All four western provinces have significant aspects of government agency management of water, partly due to the Crown ownership of water and consequent necessity to license rights of access to and use of water. The three most important aspects of government agency water governance are management of license rights, setting water right priorities, and conflict resolution. All aspects entail significant government agency involvement; each will be discussed in turn.

Saskatchewan and Manitoba provide the best, most comprehensive and consistent examples of a government agency management of water. Saskatchewan moved to the Crown corporation model in 1984, with what is now the Saskatchewan Water Corporation tasked with the management of water rights.[8] The Saskatchewan Watershed Authority has the ability to issue a water license for a period of time and on terms and conditions it considers appropriate.[9] A license may be refused, but the reasons for such refusal do not appear in the legislation, and it would appear to be within the Saskatchewan Watershed Authority's discretion.[10] In Manitoba, a government department, the Manitoba Water Stewardship, has been created for this task[11] and is legislated mandate for the granting, revocation and amendment of licenses are set out in *The Water Rights Act*.[12]

Alberta and British Columbia have substantive aspects of government agency based management of water, and both provide for Crown ownership of water and the granting and administration of license interests.[13] The Alberta government has similar discretion to fail to renew a license, or add conditions,[14] and water management orders could be made restricting a licensee's practices.[15] Similar restricting provisions do not exist in British Columbia.

All four western provinces have a legislated scheme of priority for water uses and diversions administered by a government agency. These schemes generally involve some combination of grandfathering of rights issued pursuant to predecessor legislation, priorities based on a combination of principles of first in time or purpose of use (except Saskatchewan), and an ability of the provincial government to amend or cancel water rights (except British Columbia).

In Manitoba, priorities are established between licenses based on date of submission of the application, and may be denied if the licence would negatively affect an aquatic ecosystem.[16] Priorities of licensed rights established by legislation are domestic, municipal, agricultural, and industrial, followed by irrigation and other purposes, in that order.[17] If a licensee applies, and has a priority of use greater than licenses already allocated (and there is no further water to allocate), the minister may issue the licence to the new applicant and cancel or restrict the rights of previous licences; [18] further, any unutilized allocated water rights may, after a period of one year, be reacquired by the minister.[19] British Columbia has a very similar provision for priority.[20] The remaining two western provinces have since diverged.

As there isn't a statutory scheme of water rights in Saskatchewan, it is unclear what the priority of water rights will be in the event of a conflict. Policy provides for priority of municipal water use, a policy that is set by the Saskatchewan Watershed Authority and, as a result, can be amended by the Saskatchewan Watershed Authority; therefore, the administration of priorities would appear to be left to the discretion of officials at the Saskatchewan Watershed Authority.[21] The current Act, unlike that of Alberta, Manitoba and British Columbia, makes no provision for priority of water license or types of use. Because of this, provisions described in the next section relating to water conflict may be particularly germane in respect of Saskatchewan. It may be that today it is extremely rare and unusual to find cases in which water licenses are not renewed, or in which the powers of cancellation are exercised. There is, however, a growing sentiment that water shortages, and thus conflicts, will be increasingly prevalent.[22] It has yet to be seen how this power will be exercised, and if it will be exercised in a manner acceptable to parties involved during times of water shortage. Significant unrest may occur if licensees view their license as entitlements. Any attempt to remove or limit these water rights may be met with a degree of outrage making it difficult, or impossible, to exercise the statutory powers.[23] If this is the case, this power will provide little flexibility in managing water in times of shortage.

In Saskatchewan, priorities for water rights and water diversions granted pursuant to previous Saskatchewan water acts are preserved; however, they can be amended, cancelled or suspended by the Saskatchewan Watershed Authority.[24] It is argued by some that the Saskatchewan Watershed Authority's ability to cancel any rights granted prior to its creation in 1984 does not exist.[25] This view is based on the failure of *The Saskatchewan Watershed Authority Act, 2005* language to expressly specify that rights granted prior to 1984 may be amended or canceled.[26] The language, however, is sufficiently broad on a modern, liberal interpretation of the statue to allow for the amendment or cancellation of license interests prior to 1984.[27]

Alberta has a scheme of water priorities particularly applicable to its industrial and agricultural base. There are four types of water rights: existing licenses, household users, exempted agricultural users, traditional agricultural users.[28] Licensees and traditional agricultural users have priority amongst themselves according to the priority number assigned to their license or registration (generally the date of filing). Household users have priority over other diversions.[29] Old water licenses are explicitly grandparented in the legislation, and are explicitly not subject to any provisions in the *Water Act* which are inconsistent with the term or condition of the grandparented license.[30] These licenses can be subject to a holdback of license allocation in order to return water to in-stream needs; however, this is a voluntary mechanism.

The Western provinces all have legislated mechanisms for resolution of water conflict in the first instance by government. The Saskatchewan Act contains a detailed mechanism to appeal decisions of the Saskatchewan Watershed Authority for cancellations and refusals of licenses, including requirements of notice of decisions by the corporation and opportunities for submissions.[31] The Act contains a provision for informal and formal complaints in respect of other person's actions taken in relation to water pursuant to the Act. These complaints can be ultimately appealed by aggrieved parties to the Water Appeal Board, and then the Court of Queen's Bench.[32] The Act does not contain principles to be followed in the resolution of such disputes and complaints. These disputes presumably are decided bureaucratically without statutory guidelines.

In Alberta, if two or more persons are unable to resolve a complaint or dispute, they may request the Director (as designated pursuant to legislation) to review the matter.[33] The matter may, by agreement of the parties, be sent for mediation. The Director may make a water management order in the event the Director is of the opinion it is required for various reasons, such as remedying an adverse effect on the aquatic environment, human health, property or public safety, where drilling is causing an adverse effect on groundwater, or if the diversion of water should be suspended or to resolve priorities disputes.[34] Compensation is payable in the amount the Director considers appropriate.[35] Appeals of decisions of the Director can be made to the Environmental Appeals Board.[36]

Manitoba provides that disputes over compensation for government expropriation are pursuant to *The Arbitration Act,* and for suspension or cancellation of a license or other Ministerial decisions to the Municipal Board.[37] Similar provisions exist in British Columbia's legislation for appeals of decisions of the comptroller, the regional water manager or an engineer to the Environmental Appeals Board.[38]

2b. User-Based Management

Although promising aspects of user-based management are being incorporated into water governance frameworks (which will be discussed subsequently) the best example of user-based management of water licenses, or collective water user decision making, is irrigation districts. Alberta, Saskatchewan and British Columbia allow for the formation of irrigation districts, or water user groups, which allow collective water user decisions in respect of their irrigation area. Alberta irrigation legislation is consolidated in *The Irrigation Districts Act*,[39] through which each irrigation district is run by a Board of Governors and subject to the supervision of a provincially appointed Irrigation Council. While the Board of Governors regulates by bylaw the supply and distribution of water to water users and conditions of supply, the irrigation district applies for and holds a license on behalf of all users in its district. In British Columbia, water users' communities can be formed upon application to the government licensor.[40] The water users' community, once formed, has the exclusive control and operation of the works constructed or used under the licenses mentioned in its certificate of incorporation. The water users' community appoints a manager to carry out directions set out in resolutions passed at general meetings of the water users' community.[41]

Whereas both Alberta and British Columbia allow irrigation communities or districts to manage the water allocated to the irrigation communities or districts, Saskatchewan retains its overarching government agency management of water allocations. As such, irrigation groups have a lesser mandate as users of water in the irrigation district, and must receive a certificate of water allocation directly from the government. Instead of an irrigation district holding one water license and allocated rights amongst its members, each person must apply to the minister responsible for the Act for a water certificate after obtaining a water services agreement from the Irrigation District. *The Irrigation Act* then provides for the establishment of irrigation districts within Saskatchewan and their governance and dissolution.[42] Irrigation Districts are corporations, and every district consumer, or person, with a water service agreement within an irrigation district is a member and entitled to receive notices of meetings and vote. The Irrigation District establishes a board which makes decisions, such as those setting the tariff of charges to its members. Manitoba does not have a Manitoba act specifying rules respecting irrigation; however, it does provide for a license for irrigation purposes.[43]

Principles of user-based management are also being incorporated into the Government Agency management of water. In British Columbia, the comptroller of water rights or regional water manager issues licenses, but provisions exist for publication of license applications and objections by any affected licensees or riparian owner.[44] Although discretion is granted in respect of the issuance of the license, conflicts between users and problems of water quality aren't determined by these officials; instead, the minister may designate the area for the purpose of developing a water management plan to address these issues.[45] This is a user-based management response to conflict management.

In Saskatchewan, one or more advisory committees may be appointed by the Saskatchewan Watershed Authority for a specific period and specific purpose.[46] These committees can be for the purpose of advising on any activities of the Saskatchewan Watershed Authority. There is no legal requirement that the Saskatchewan Watershed Authority follow the committees' advice, so this may be only passing lip service to user-based management; however, it could also be an important first step in greater user-based water management. Manitoba also provides for the designation of a water planning authority for a watershed (many of which are Conservation Districts as well) whose task is to set a watershed management plan,[47] and a water council to oversee the development and implementation of watershed management plans and advise on various water related issues.[48] Again, there is no legal requirement to follow the approved watershed management plan, but a regulation may be passed requiring it to be considered prior to any prescribed decision or approval under any specified act or regulation.[49]

The only avenue for public consultation and participation in water management in Alberta is if the minister directs the development of a water management plan.[50] Then, an integrated approach may be adopted and other persons, local authorities, or agencies, may cooperate on any activities. The plan must include a summary of issues and matters or factors to be considered in deciding whether to issue water licenses, or to approve a transfer of a water license. This policy reinforces Alberta's adoption of a market-based solution to water management.

2c. Market-based Management

Alberta has led the provinces in the development of a water market where transfer of water rights is allowed in accordance with an approved water management plan, or by Cabinet order in the absence of such a plan.[51] Applications for transfer must be made to the Director, and will only be allowed if one of these two conditions is met and the license is in good standing. Further restrictions on transfer are that there is no significant adverse effect on the aquatic environment or the rights of others (agreements in writing from the other users are required if their rights are affected). The proposed transfer will be reviewed and considerations such as existing potential or cumulative effects on the aquatic environment; any applicable water conservation objective, hydraulic, hydrological and hydro geological effects; effects on other users and licensees; public safety; and, any other matters may be taken into account. The application may be referred for comment to other agencies, and will require public notice and public review as appropriate.[52] If it is in the public interest to protect the aquatic environment or implement a water conservation objective, 10% of the allocation of water under a license being transferred can be withheld.[53] In the South Saskatchewan River Basin in Alberta, for example, there is a South Saskatchewan Basin Water Management plan which allows the Director to consider applications to transfer water allocations within the basin.

3. Assessment of Legislative Frameworks of Water Management

It is apparent from reviewing the three main water governance models, and how these models are replicated in the legislative framework of the Western Canadian provinces, that, although the government agency model is predominant, significant aspects of user-based water governance are being incorporated into the government agency model and, in Alberta, into a market-based management model. In order to assess the legislative framework of Western Canadian water governance, the criteria of the international water industry will be employed. These criteria are the principles developed by the World Water Council, Water Action Unit, as principles to help guide the assessment and reformation of water institutions internationally.[54]

For the purpose of this paper an analysis will be carried out of the respective provincial legislative frameworks and their provisions. This review will be effected without data on actual practices and decisions made implementing and effecting these legislative frameworks. It is acknowledged this analysis is limited by the fact it is based solely on the provisions and language of the legislation governing water management.[55] The principles for analyzing the legislative framework are as follows:

- Decentralization and subsidiarity—delegation of responsibility and authority of water management to the lowest feasible level. This process involves managing surface waters at the catchment's level with involvement of all stakeholders whereby land and water resources should be managed at the basin level by integrating groundwater in order to maximize and share water benefits.
- Accountability—in water management, conservation and service delivery.
- Participation by all stakeholders—whether public or private, communities or non-governmental organizations (NGOs).
- Predictability—all laws and regulations should be applied fairly and consistently.
- Financial sustainability—through recovery of both operational and capital costs.
- Transparency—clear policies, rules, regulations, and decisions, with information available to the public.[56]

International water researchers have found these principles of good governance to assist in the fair, effective and environmentally sensitive management of water.[57] The principles will be modified somewhat in this analysis for applicability to the situation of Western Canadian provinces. An overview of the assessment of the three models of water management is illustrated in Table 1.

The user-based management model scored highest of all models; however, there is still a need to safeguard for predictability of water decisions which can be done with an explicit decision making process, the provision of required information to the public, and clear guidelines and policies. Although government agency management initially scores low, this is partly due to the generic, flexible nature of the wording of the governing legislation; for example, decision making responsibility resides with the government

PRINCIPLE	USER-BASED MANAGEMENT	GOVERNMENT AGENCY MANAGEMENT	MARKET-BASED MANAGEMENT MODEL
Decentralization and subsidiarity	Evidence of decentralization, not subsidiarity	Potential challenges identified	Potential challenges identified
Accountability	Local accountability	Challenges in respect of quality	Dependent on market rules
Participation	Local participation	Limited participation	Market participants only
Predictability	Predictable, given limited scope	Potential challenges identified	Predictable, to an extent
Financial Sustainability	Beyond scope of paper	Beyond scope of paper	Beyond scope of paper
Transparency	Yes, internally	Yes	Potential challenges identified

Table 1. Water Management Assessment—Overview.

which results in a low score in the category of decentralization. It may be, however, that actual water governance in the government agency model is effected differently through consultation, participation and accommodation. Similarly, market-based management ranks very low, but this may be indicative of its inapplicability as a comprehensive water management solution (which these principles were developed to assess). Relying on a market for a commodity satisfying a basic human need which has traditionally been a public property resource in Canada, may also be ideologically difficult for some. Amongst certain classes of users, such as irrigators or industrial users, however, commodification of a water right may be possible and add value, if transactions are enforceable and supported by sufficient numbers of users and market transactions. The creation and support of a market may require significant government involvement to overcome possible challenges in meeting the criteria, and to ensure comprehensive water management in the interests of all stakeholders and all types of uses.

Based on a legislative review, and organized by criteria, more detail of this analysis follows. This detail and discussion of the strengths and weaknesses of the three models of water governance reflected in Western Canadian legislation is important in ensuring any potential challenges are appropriately remedied either through legislation, policy or practice, and understanding the policy implications inherent in the choice of water governance model.

3a. Decentralization and Subsidiarity

International research has shown decentralized natural resource management, or delegation of decisions to the lowest feasible level yields economically efficient and socially equitable and environmentally sustainable results.[58] Decentralization does not entail a lack of involvement in water management by government; in fact, clear government legislation and policy and a close collaboration with government by users are required to set guidelines and parameters of decentralization. Local management can implement a restorative response to conflict and possibly prevent conflict over water; however, national programs for sustainability and services to assist local empowerment and enforce local decisions will be required.[59] An overarching factor is that the related encompassing component of decentralized governance is subsidiarity, or holistic water planning. Community planning in relation to water governance involves all aspects of water governance from allocation decisions, groundwater licensing decisions, to decisions regarding source pollution and activities affecting water quality. Integrated water resources management promotes the coordinated, sustainable development and management of water, land, and related resources to maximize equitable economic and social development.

Decentralization and subsidiarity are important for two main reasons:

i. Decentralized management decisions and planning allows for local community practices and values which are then adopted and embraced in practice. Community participation ensures community commitment;
ii. Decentralization also allows a three-part economic analysis which incorporates externalities which might otherwise be lost in the cost benefit analysis: A conventional top-down economic perspective reflects prices paid and relative values of inputs and outputs; a bottom-up perspective that reflects the true value to the community and its residents of what might be otherwise marginal resources to outsiders; and lastly, a sideways interaction of economic interventions with non-economic values such as health benefits from improved water quality.[60]

The decentralization and subsidiarity of Western Canadian water governance models will be reviewed in respect of water institutions and water quality.

The user-based management model best reflects the principle of decentralization as actual users make decisions (the lowest feasible stakeholder group), whereas a market-model of water governance does not provide for decentralization, except in the initial phase of establishing the water management plan which establishes the parameters and conditions applicable to water trades. After these parameters and conditions are set, water trades can occur with very little community involvement. The government-agency model predominantly employed by the Western provinces does not achieve a goal of decentralization as the power for water allocation decisions resides in the government department or agency managing water; however, with a population of one million people, there is a certain characteristic of decentralization in the single Saskatchewan government agency of the Saskatchewan Watershed Authority. This population may not support any more decentralization; therefore, together with local advisory committees, perhaps this is the best water model in the circumstances.

An important first step in decentralization has also occurred in British Columbia and Alberta with water management plans mandated to resolve conflicts in British Columbia, for example, with enforcement through regulation. This move reflects a shift from a government agency management structure to a more decentralized model of shared responsibility between government and the community, and empowerment of community decisions making. It is unclear, however, if plans developed will be appropriate in relation to water basin or catchments, or will apply to only one portion thereof. In the latter case, it is unclear what mechanism or institution will coordinate these plans on an appropriate basin level.

The particular constitutional setting in relation to water creates challenges. As an example, provincial responsibility for water management results in a somewhat fragmented approach to water management, which essentially results in certain decentralization. Although some benefit of local decision making results from this provincial jurisdiction, multiple provincial jurisdictions being responsible for a water basin creates significant challenges which will be discussed further in this paper.

Decentralized water quality planning is also in the very formative stages and not yet reflected consistently in legislation. At the user-based level, the integration of water quality is just beginning where, although irrigation districts and water user groups are not mandated to govern water quality, they do handle issues such as salinity with water quality ramifications. These issues, however, would be confined to their irrigation community. Local advisory committees in Saskatchewan do not have legislated authority to make binding declarations respecting water allocations, let alone water discharges and land use. In Alberta, water management plans have not yet provided guidance in respect of water quality decisions,[61] while in British Columbia provisions are made to enforce water management plans by legislation, and it appears there is a possibility water quality may be included in the plan; however, whether this power is used in respect of land use, farm practices, non-point source pollution and source pollution has yet to be seen. Communities are in the formative stages of developing plans with the Town of Langley developing a pilot plan.[62]

Interestingly, the inter-jurisdictional Prairie Provinces Water Board (formed by Saskatchewan, Alberta, and Manitoba) has in its mandate the integrated management of water and related resources. The Master Agreement provides:

> The parties agree for the future application of the provisions of the Master Agreement . . . to work together and to co-operate . . . for the integrated development and use of water and related resources to support economic growth according to selected social goals and priorities and to participate in the formulation and implementation of comprehensive planning and development programs . . . [63]

A duty to promote, through consultation and the exchange of information, the recognition of the interdependence of quality and quantity of water in the management of watercourses is a duty of the Board.[64] This is an important acknowledgement in support of this criteria of subsidiarity made by the three Prairie provinces and the federal government, albeit the mandate of the Board is currently only consultative.

3b. Accountability

This best practice relates to the water governance model employed being answerable to all water users in respect of (i) service delivery, (ii) conservation, and (iii) water management. As actual service delivery evaluation would require an audit, close examination of this issue is beyond the scope of this paper. In respect of access to service delivery, all provinces have human rights legislation which protects against discriminatory access to services and is subject to the equality provisions of s. 15 of the *Charter of Rights*[65] thereby disallowing inequality in any legislation pertaining to water and rights to access.

Although full-scale water conservation through such specific implementation plans as demand-side management has not yet occurred, all Western Canadian provinces have developed water conservation plans. This is promising evidence the issue is being reviewed, monitored and planned for, albeit not necessarily reflected in legislation. Both Manitoba and Alberta have developed comprehensive water strategies which include conservation as a key strategy.[66] In Saskatchewan, the Saskatchewan Watershed Authority consulted, and in November, 2006, released the Saskatchewan Water Conservation Plan.[67] British Columbia consulted and released a Water Conservation Strategy, and is currently developing an implementation plan.[68] Though few provinces have embarked on watershed protection through legislation which links land use, water allocations and water quality together,[69] these water strategies, albeit not yet reflected through legislation, show promise.

Generally, the accountability of Western Canadian provincial water institutions is good, but accountability is challenging because of fragmentation in respect of water quality and management of inter-jurisdictional issues. Significant water institutions are managed by government departments or crown corporations under the oversight of democratically elected officials. Legislation clearly outlines complaint procedures in respect of allocation decisions and resulting conflicts, and provides mechanisms for their resolution;[70] however, in respect to day-to-day management decisions of a Government Agency model, some difficulty to customize decisions to particular conditions may occur because of the standardizing of policy and procedures. As decisions are generally at the discretion of agency officials applying their interpretation of the standard policies, there is a danger of relatively little communication and accountability to users.[71] In respect of a market-based model, the accountability of water transactions will be as good as the rules set up to govern the market transactions. In Alberta, the water management plans allowing market transactions provide that the Director consider transactions as provided for in the plan. Specifying conditions of transfer, and ensuring transferees have binding obligations to account for their actions, will be paramount.In respect of user-based models, decisions of the user-based group will be subject to administrative law procedures for review of its decisions. Any additional legislative or policy requirements to ensure procedural and substantive fairness, and prevent the possibility of perceived unfairness, are always important in any local management group.

Accountability is not so clear when examining allocation decisions and their affect on quality, the fragmented nature of water quality governance, and, finally, on the legislated obligation in respect of water quality. Although allocation decisions reside in one accountable water institution, it is not clear the effects on water quality are always considered by this one institution. Other entities often make decisions which have effects on water quality without the involvement of the institution responsible for water quality.

All provinces, except British Columbia, allow for suspension and revocation of water allocation licenses for such reasons as the 'public interest' in Saskatchewan, the protection of aquatic ecosystems in Manitoba, and both of these reasons in Alberta.[72] Although what qualifies as 'public interest' or 'protection of the aquatic ecosystem' is based on the opinion of the government officials, these decisions are subject to review but are not actionable for negligence.[73] Provision does exist in the legislation for notification of, and possibly non-issuance of, a license in Alberta if *The Environmental Protection and Enhancement Act* has not been complied with.[74]

Generally responsibility for water quality rests with the minister responsible for the environment. This is the same ministry responsible for water allocations in the case of Alberta, Saskatchewan, and British Columbia where the Minister of Environment is responsible for both water allocation decisions around environmental discharges and environmental permitting of new development. This approach would arguably further accountability as decision making in regard to water quality and quantity would be cohesive. Manitoba has separate ministries responsible for water allocation decisions and the environment, the licensing and permitting of industrial development, and discharges affecting water quality.[75] Arguably, the cohesive management of water allocation decisions and water quality would be challenging.

Further fragmentation in respect of water quality governance occurs in two ways. First, few provinces have embarked on watershed protection which links land use, water allocations, and water quality. These initiatives, however, are starting to occur.[76] Second, the minister responsible for the environment of a province often is not responsible for significant development which has ramifications for water quality management. This is reflected in several major areas:

- Right to farm legislation, and the permitting of waste from intensive livestock operations by the Minister of Agriculture (not the Minister of Environment), and the protection of intensive livestock operations from suits for nuisance by neighbours;[77]
- Oil and gas well activity in Alberta;[78]
- Forestry in British Columbia;[79]
- Regulation of non-point source pollution such as pesticides.[80]

As a result, it is hard to think a Minister of the Environment with responsibility for water quality would have the ability to effectively manage the issue. Further exasperating this situation is the fact that, although provinces have many environmental laws in place, considerable concern exists in the lack of enforcement of environmental laws and prosecution for environmental offences.[81] Instead of command and control legislation and regulation with difficult and costly enforcement, smart regulation and policies are needed with financial incentives to reduce harm to the environment and improve water quality.

Some cause for concern still exists as Alberta and British Columbia's legislation does not impose any obligation on the government officials to ensure a healthy environment.[82] Saskatchewan's statutory language assigns responsibility for enhancing and protecting the quality of the environment on the minister responsible for the environment. Manitoba's legislation specifically establishes the aims and object of the department to be to protect the quality of the environment, protect environmental health of present and future generations of Manitobans, and provides the opportunity for all citizens to exercise influence over the quality of their living environment.[83] These two provinces' obligations, however, may just be semantics as the government is exempt from any civil liability in respect of these statutory duties.[84]

Obligations exist for municipalities providing potable water to attain quality standards.[85] This 'end of the pipe' responsibility ensures accountability, but is a costly method of water management when the provincial and federal governments have the jurisdiction to manage both source and non-source activities affecting water quality.

The inter-jurisdictional nature of Western Canadian water governance affects water accountability in two major respects. The first is determining responsibility between the provinces and federal government in respect of water issues, and the second resolving issues of quality and quantity between provincial and federal governments. The constitutional arrangement of the provinces in relation to water has accountability implications; therefore, at times the constitutional arrangement is used by each level of government as a reason not to take accountability for an environmental issue. Often governments blame one another for environmental problems and the other's lack of action in regard to a specific problem;[86] however, this position is misleading as a constitutional analysis of water issues will provide an answer to jurisdiction which may point to one or the other level of government, and perhaps in certain situations, both levels of government.[87]

The topic of water spans several heads of legislative power assigned to the federal and provincial governments in the Canadian Constitution. The provincial government has powers which relate to water, including property (generally including water in its definition).[88] The federal government has certain powers in relation to water, albeit historically somewhat more limited than the provinces.[89] Limits would include powers in relation to water allocation in order to facilitate navigation, and the water quality and quantity to maintain and preserve fish and their habitat. The federal government takes control of water once it crosses an inter-provincial or international boundary in accordance with the federal head of power relating to inter-provincial works and undertakings.[90]

It is apparent different aspects of water governance are dealt with by either the provincial or federal governments and sometimes both (as with water quality). Very complicated legal rules exist for determining if a matter is of federal or provincial jurisdiction, and if it can be affected by both pieces of legislation. On some of the rules, legal scholars disagree[91] as often it is somewhat unclear which head of power a matter falls in, and a detailed constitutional analysis ensues. If not identified within a specific head of power, the federal government will have jurisdiction either through the opening words of s. 91 of the Constitution Act, 1867, which assign the federal government power over all matters not expressly assigned to the provinces, or through the broad head of power of peace, order and good government ("POGG"), or matters of a national concern.[92] If provinces are unable to deal with an issue because it crosses boarders, such as marine or water pollution, the federal government will be found to have jurisdiction as it is a matter of national concern.[93] Based on this shared jurisdictional approach, the Canadian constitutional framework does provide accountability by establishing the responsible government in relation to a water issue.

The dynamics of water and its travel across borders creates accountability issues. An upstream provincial government which imposes significant externalities of pollution, scarcity and water fluctuations on downstream users currently has a significant advantage. Downstream users and their governments don't have accessible legal mechanisms to remedy this activity, and little political leverage exists to negotiate a solution. In the case of *Interprovincial Cooperative*,[94] Manitoba environmental legislation, which created a statutory right of action against pollution originating upstream in the neighbouring provinces of Saskatchewan and Ontario, was found unconstitutional. The extraterritorial origin of the pollution removed it from the legislative reach of Manitoba, despite the injury to Manitoba's fishery.[95] The time delay of legal and negotiated resolution of transboundary water issues significantly affect responsiveness to water issues and conflicts.[96] A solution will require a model of inter-jurisdictional accountability and coordinated water management.

Cooperative solutions are always easier when parties work from common needs and values. Often, there are perceived or actual distinctions in values reflected in provincial and municipal institutions, laws, rules, and infrastructure; consequently, the value of certain uses of water to one community may be very different to another community. This difference may be accentuated when communities are represented by separate provincial governments, and even national governments, with very different political mandates and agendas.

The current *Master Agreement on Apportionment* between Canada, Alberta, Saskatchewan, and Manitoba contains a strict formula of sharing water.[97] In the event of a severe water shortage, the inability of Saskatchewan residents to have drinking water will be inconsequential as the formula is the only mechanism of allocation. This strict formula was developed partly as a response to inability for Saskatchewan and Alberta to agree on what developments should occur, and a mandate change several decades ago. This historical impasse for agreeing on developments affecting water should not be forgotten as water shortages loom on the horizon.[98] Research confirms having mechanisms in place to respond to issues is important in responding to issues and potential conflicts.[99]

3c. Participation

The advantages of user-based management are legitimacy and community acceptance of decisions because of community participation. The principles behind decisions are based on custom, local knowledge and experience which makes them reflective of the community and the community values they affect. As well the embracement of the community in decisions reduces enforcement costs. Because local community values are incorporated into decisions, a comprehensive set of factors are considered when making a decision. In such a manner, there are less 'externalities' or factors not accounted for in decision making. Because actual users make decisions, these decisions can be adaptable and flexible, reacting to changes quickly and generally in a cost-effective manner.

User-based management is difficult if users do not know one another and lack existing relationships. If the social capital or relationships of the community with outside agencies and institutions, and between members of the community, do not exist, decisions of a user-based management group may be difficult to achieve and implement, and not reflective of a community consensus. It is important to assess whether sufficient community or social capital exists to support user participation and effect user-based management. If not, development of this social capital and community may be required.

Government agency management scores low in relation to civic participation. This is partly due to the nature of governing legislation which vests all decision-making authority in the government department or crown corporation tasked with water management in Saskatchewan and Manitoba. Generally the government department or crown will set standard procedures and policy for the entire province. Because of the standardizing process, some difficulty in customizing these procedures and policies to particular conditions may occur. Decisions are generally at the discretion of agency officials applying their interpretation of the standard policies, and participation of users is limited.[100]

Government agency management is often criticized for being unclear, uncertain and unaccountable to the public because of the reliance on technical water experts. The area of water management is a complicated, complex information intensive area dependent on technical reports of hydrologists and engineers. Allocation decisions are often framed primarily in technical terms of engineering procedures such as irrigation schedules and reservoir operation rules. Because of this, water decisions are often entrusted to government officials who are relied on for their expertise. This is the antithesis of public participation.

Meaningful public involvement requires the development of a public understanding of these technical issues. Armed with this capacity proper questions can be asked, development priorities set, and community impacts assessed in respect of water decisions. Developing this capacity is time consuming and expensive. However, not developing this capacity makes any public consultations or input of the new watershed advisory committees in Saskatchewan and conservation districts in Manitoba of limited value.

Saskatchewan Watershed Authority has incorporated some participation of the community in water management through the local watershed advisory committees. These committees have no legislated mandate or jurisdiction to ensure their decisions are implemented. Consulting with the public in setting policies and practices, and providing local advisory committees with enforceable mandates and the capacity to meaningfully participate in decisions would contribute to public participation and counter the government agency management model's weaknesses.

British Columbia and Alberta allow for water management plans to be developed at the community level. These provisions are laudable, and the only causes for concern are that it is unclear if the advice of local advisory committees must be given weight in decisions, how often the plan will be revisited after written, and who, exactly, will determine water management plans. Without proper funding and capacity building of these groups and plans by government, and on going re-evaluation of the water management plans, their utility may be compromised.

The market-based management model approach allows for open participation. Any barriers would be structural and include such factors as inadequate capital and unavailability of water licenses which favour large, well established water users, and exclude small, first time water users from participation.[101]

Currently, institutions such as the Prairie Provinces Water Board and the International Joint Commission exist to manage inter-jurisdictional water issues among Canadian provinces, and between Canada and the United States. Although provincial stakeholder participation is ensured, the extent of community participation in these bodies is uncertain, a factor which may have implications for the community acceptance of these bodies' resolution of issues.[102]

3d. Predictability

Most aspects of water allocation management in the Western Canadian provinces appear very clear and predictable. Manitoba's and British Columbia's legislation sets out procedures for obtaining water licenses, priorities, and disputes resolution procedures,[103] while Alberta legislation clearly specifies priority of licenses and how decisions will occur in respect of transferring water licenses. Although Saskatchewan legislation is very clear on who is making decisions, the Saskatchewan Watershed Authority, factors taken into account in decision making are not as clear in the legislation. The lack of direction respecting priority of licenses in the Saskatchewan legislation may result in a lack of certainty, potential for arbitrariness, and unpredictability.[104]

Generally, in a government agency model, standard procedures and policy will be set, and because of the standardizing process, some difficulty to customize these procedures and policies to particular conditions may occur. The government agency generally has extensive technical expertise and decisions are framed in technical terms. Decisions are generally at the discretion of agency officials applying their interpretation of the standard policies. Often there is relatively little communication to users.[105] As such predictability may be compromised. Because of the reduction of conflicts or issues to technical solutions, agency administration is often criticized for being unclear, uncertain, and unpredictable.

The market-based management model employed by Alberta is not a true laissez-faire market with vendors and purchasers conducting transactions purely based on market rules; a certain amount of oversight is retained in the review of these transactions and, as such, the predictability of a market model reduced somewhat.[106] The factors taken into account when reviewing the transaction, however, are quite specific which increases the predictability of this discretionary review.

Because of the limited mandate of managing their users' licenses, user-based management as practiced in the Western Canadian provinces is quite predictable. A risk does exist that the users group may make decisions without clear communicated policy and a version of 'sand box' politics ensues, reducing predictability. More research would need to occur to determine if this is, in fact, a concern.

3e. Inter-jurisdictional Water

When water management issues span provincial or international boundaries their predictability is lessened, partly due to the constitutional analysis and lack of inter-provincial legal remedies discussed earlier in this paper. Layers of government, institutions (both government and non-government) over and above water users of the river basin add complexity and challenge, reducing predictability. It is important to recognize this challenge and the current institutions (such as the Prairie Provinces Water Board) which were formed for the specific purpose of meeting this type of challenge. Creating new institutions and new water rules may not be the answer, but empowering, strengthening, and supporting existing networks of government and non-government institutions may be the right approach.

An accurate determination of financial sustainability, or whether there is recovery of both operational and capital costs in water management and governance, is beyond the scope of this paper. It is important to point out, however, that it is generally thought Canada under prices its water resource and fails to recognize the full external costs of water.

The market-based management model is thought to be a solution to the more efficient allocation and pricing of water. The statutory provisions allowing transfer are touted by some researchers, and the Alberta government, as advancing the goals of efficient allocation of water interests and conservation in encouraging the transfer of surplus interests. This process is also described as creating a non-regulatory method of reducing wasteful use by creating an incentive to save water and transfer its marginal value for compensation.[107] Many would argue the market tool does not capture the community value of water, nor does it facilitate political and ethical considerations in allocation decisions. The risk of the market is that impacts on third parties not party to a market transaction are neglected, and third parties have difficulty enforcing their interests in a court of law.

Amongst certain classes of water users, such as irrigators or industrial users, the commodification of water may add value if supported by sufficient numbers of users, market transactions, and enforceability. This creation and support of a market may require significant government involvement to ensure comprehensive water management in the interests of all stakeholders and all types of uses. It may be that the cost of creating, regulating and administering such a water market offsets any efficiency created by a market. In a government agency or user-based model, agreeing on what externalities exist respecting provincial and community management of water and its uses, and then how to ensure adequate internalization of these, by those receiving the benefit and not downstream users, in either monetary payments or in-kind services is a monumental task with little precedent of success.[108]

3f. Transparency

All Western Canadian provinces employ the government agency model and have legislation containing clear rules for water license priorities and the procedure for applications for water licenses.[109] There are also obligations on government officials managing water to communicate decisions to applicants in respect of their decisions in relation to determining conflicts.[110] Overall, Western Canadian provinces employ a high degree of transparency through their government agency model.

In respect of the market model employed in Alberta it is unclear if obligations exist to publish decisions in respect of allowing a transfer. There is a clear provision to have a public review of a proposed transfer of allocation of water under license;[111] therefore, it would be presumed that notification of the Director's decision would naturally follow, even without a legislated obligation.

User-based decisions respecting water management will be transparent to all users as obligations of management are imparted on the user community with some powers of delegation to a manager. Decisions of the irrigation community are based on majority decisions and obligations of notice of meetings to users exist.[112] What is not apparent in all Western Canadian provinces is whether any water stakeholder who is not an irrigation user has an ability to receive notice of decisions made by the irrigation district or communities.[113]

Transparency and the disclosure of water quality tests is starting to become a statutory obligation. In Saskatchewan, if the results of drinking water tests are outside specified parameters, labs conducting water quality tests must notify the Minister of the Environment and the water consumers.[114] Similar public notification obligations exist in other provinces.[115] As well, once a year the supplier of drinking water must provide its consumers with notification of the quality of water in comparison with the levels set by regulation and their compliance with the sample submission requirement in their permit, or present an annual report of the quality of drinking water.[116]

The Western Canadian provinces have employed a combination of the three water models and in their selection have achieved a degree of success in meeting the criteria of the World Water Council. This analysis has illustrated that user-based governance is most consistent with these World Water Council criteria, but is not without considerable challenge in achieving integrated water management. A careful well-thought path for future changes to Western Canadian water governance is warranted, and great care should be taken in the determination of appropriate water management models. In order to ensure proper decisions, consultation with all stakeholders and participation of the community is essential. Participatory forums in which representatives of water users are brought together to provide knowledge and understanding in the evaluation of water governance, and its applicability to their particular circumstance, is optimal. These forums would legitimize existing water rights and management practices, and provide insightful consultation and guidance of any needed changes or outstanding issues. Research shows that, allowing non-governmental organizations, research organizations, and a variety of other groups to collaborate in evaluating changes in water management tend to perform better than those controlled exclusively by government at any level.[117]

4. Conclusion

Examples of the three international models of water governance can be found in the Western Canadian provinces' legislative frameworks, and this pattern of employing a combination of models is consistent with good water governance. Irrigation councils in Alberta and Saskatchewan, and water users' associations in British Columbia, reflect a user-based management model with local community participation. Saskatchewan has a predominantly government agency-based management system, but all Western provinces employ this model to an extent. Alberta has all three types of water management models, including a market-based management model with transferable water licenses.

Support for user-based management is increasing internationally as the most appropriate water governance model because it allows participation through decentralization, and achieves socially equitable and environmentally sustainable results. The Western Canadian provinces are incorporating more user-based governance into their water frameworks. This move is promising and consistent with this study's positive evaluation of the user-based model based on the World Water Council criteria of good water governance; however, the market-based management model and government agency management model should not be automatically dismissed based on this evaluation. Each governance model has advantages and disadvantages: the market model can ensure the efficient use of the water resource but may exclude some from participating; the government agency model has the benefits of transparency, accessibility to water as a "right," and accountability, but may not allow community and user participation in decisions leaving these to technical experts. The user-based management model has the advantage of reflecting community values and norms in decisions but lacks centralized planning.

The applicability of the proper combination of models for a specific province or water basin should be determined based on the appropriateness of the model for the water resource in the area and characteristics of the community and water stakeholders. In Saskatchewan, given its small population, government agency-based management may be most appropriate with the appropriate measures to protect against any potential disadvantages of not allowing for participation, decentralization and subsidiarity. In Alberta, irrigation farmers may be best served by the ability to participate in a water market in respect of their water allocations, while municipal drinking water is excluded and protected from this market.

It is important to note that, in assessing the Western Canadian provinces' water governance and the three models of water management (albeit modified in regards to the Crown ownership of water) which are government agency management, user-based management and market-based management, no one model is problematic. Each model has the potential for certain challenges; the recognition of these challenges is the important first step for their mitigation. In the assessment of Western Canadian water governance it is evident the biggest challenges are meeting the best practice of decentralization and subsidiarity, the delegation of responsibility and authority of water management to the lowest feasible level, and managing water on a catchments level with the involvement of all stakeholders.

In Western Canada the allocation of water rights has evolved through practice and statutory provision for several decades; water quality has only more recently entered the political and legislative sphere, typically in a separate forum of legislation dealing with environment. This history creates challenges in integrating institutional structure for dealing with both water quality and quantity. The current trend to decentralization and user-based governance requires attention to both the integration of water quality management in order to maximize decentralization benefits, and the minimization of decentralization costs.

The importance of careful thought and community participation in decisions with implications on both water quality and quantity needs to be reinforced to communities; thereafter, grassroots decisions of these communities must be allowed in respect of both water quality and quantity. This requirement may include environmental permitting of industrial development, to be determined through a careful consideration of the communities' views of effects of an industrial development on water quality and quantity. The legal framework to allow for this is in place, what is needed is fine tuning to ensure all development decisions which may impact water quality and quantity are reviewed by the affected communities and, specifically, by those individuals involved in community water governance. If significant investment by government, institutions, and community is made in local advisory committees, increased participation in environmental assessment procedures, coupled with the development of water conservation plans in Saskatchewan and water management plans in Alberta and British Columbia, this transformed user-based, integrated water management framework could be a reality.

Endnotes

1. The Conference Board of Canada, *InsideEdge* (Winter 2005): 16.
2. Climate Change Impacts and Adaptation Directorate (CCIAD), *Climate Change Impacts and Adaptation: A Canadian Perspective—Water Resources* (Ottawa: Natural Resources Canada, 2002). The Honourable Tommy Banks, Chair, *Water in the West: Under Pressure, Fourth Interim Report of the Standing Senate Committee on Energy, the Environment and Natural Resources* (Ottawa: Senate Committee on Energy, the Environment and Natural Resources, November, 2005) available at: www.senate-senat.ca/EENR-EERN.asp
3. David R. Percy, "Water Rights in Alberta," *Alberta Law Review* 15 (1977): 142.
4. *North-West Irrigation Act,* S.C. 1894, c. 30. This is consistent as well with the *Territories Real Property Act,* S.C. 1886, c. 51. In b.c. similar provisions were made in the Land Ordinance, 1865, and the *Water Privileges Act,* 1892, S.C. 1894, c. 30.
5. S.C. 1938, c. 36.
6. s. 2 of *The Water Rights Act,* C.C.S.M. c. W80, s. 3 of *The Water Act,* R.S.A. 2000, c. W-3, s. 2 of *The Water Act,* R.S.b.c. 1996, c. 483., s. 38 of *The Saskatchewan Watershed Authority Act, 2005,* S.S. 2005, c. S-35.03. These acts will be cited by name hereafter except to differentiate between b.c. and Alta. the title *The Water Act (b.c.), supra,* and *The Water Act (Alta.), supra* will be used respectively.
7. Bryan Randolph Bruns and Ruth Meinzen-Dick, "Frameworks for Water Rights: An Overview of Institutional Options," in Bryan Randolph Bruns, Clausia Ringler and Ruth Meinzen-Dick (eds.), *Water Rights Reform: Lessons for Institutional Design* (Washington, D.C.: International Food Policy Research Unit, 1995), available at: www.loc.gov/catdir/toc/ecip062/2005030412.html
8. The Water Corporation Act, S.S. 1983–84, c.W-4.1, s. 42. Now *The Saskatchewan Watershed Authority Act, 2005,* S.S. 2005, c. S-35.03.
9. s. 50 of *The Saskatchewan Watershed Authority Act, 2005, supra.*
10. s. 52, ibid. An example of factors for the cancellation, amendment and suspension of a license by the corporation are the failure to comply with terms or conditions of a license, usage of the water for a purpose other than that specified on the license, contravention of any provision of the Act, or if Cabinet considers it to be in the public interest. (s. 54, ibid.).
11. s. 54, ibid.
12. *Supra.* Rights of suspension and revocation of licenses are available if in the minister's opinion aquatic ecosystems need to be protected and maintained (s. 9.2, ibid.).
13. s. 32 (Administration), s. 3 (Crown Ownership) and s. 49 (licensing) *The Water Act (Alta.), supra* and s. 2 (Crown ownership), 6 (licensing), *The Water* Act, *(b.c.), supra.*
14. s. 60 not in public interest or adverse affect on aquatic environment and s. 60 and 51(3) *The Water Act (Alta.), supra.* Rights pursuant to a license may be suspended or restricted if needed for water level/in-stream flow or protection and maintenance of an aquatic ecosystem. s. 9.1 and 9.2 of *The Water Rights Act, supra.*

15. s. 97 of *The Water Act (Alta.), supra.*
16. s. 8(1) and s. 9.1(2) of *The Water Rights Act, supra.*
17. s. 9, ibid.
18. s. 14(1) ibid.
19. s. 15, ibid. This legislative scheme is very similar to the federal water scheme in existence for all Prairie provinces in 1920. The *Irrigation Act,* S.C. 1920. c. 55, s. 5.
20. s. 15 *The Water Act (b.c.), supra.*
21. Historically, Saskatchewan had a priority scheme similar to that of Manitoba. s. 14 of R.S.S. 1953, c. 48 and Gisvold, Per, *A Survey of the Law of Water in Alberta, Saskatchewan and Manitoba* (Ottawa: Prairie Farm Rehabilitation Administration and Economics Division, Canada Department of Agriculture, 1956), 29.
22. CCIAD, *Climate Change Impacts and Adaptation: A Canadian Perspective—Water Resources, supra* and The Honourable Tommy Banks, Chair, *Water in the West: Under Pressure, Fourth Interim Report of the Standing Senate Committee on Energy, the Environment and Natural Resources, supra.*
23. David R. Percy, "The Limits of Western Canadian Water Allocation Law," *Journal of Environmental Law and Practice* 14 (2004): 315.
24. s. 41(1) and (2) of *The Saskatchewan Watershed Authority Act, 2005, supra.* Many large prairie water licenses were originally granted before 1930 by the federal government so it will be with some difficulty that provincial legislation will be able to affect these licenses; Percy, "The Limits of Western Canadian Water Allocation Law," *supra.*
25. Percy, "The Limits of Western Canadian Water Allocation Law," *supra.*
26. s. 41(1) *supra,* and also its predecessor legislation, s. 79(1) of *The Saskatchewan Watershed Authority Act,* S.S. 2002, c. S 35.02.
27. s. 41(1) provides " . . . any right, privilege or authority granted to any person, or any rights, privilege or authority of any person preserved or recognized, pursuant to any Act or former Act or Act or former Act of the Parliament of Canada and existing on the day on which this Act comes into force that entitles the person to divert, use or store water . . . remains in full force and effect unless and until it is amended, cancelled or suspended by the corporation pursuant to this Act." Such an interpretation is consistent with s. 10 of *The Interpretation Act, 1995,* S.S. 1995, c. I-11.2, which provides that every enactment be interpreted as being remedial and given a fair, large, and liberal construction and interpretation to best ensure the attainment of its object.
28. Part 3 of *The Water Act (Alta.), supra.*
29. However, this priority is subject to the Director making a water management order if the household diversion has a significant adverse effect on the aquatic environment or on a licensee or traditional agriculture user. Ss. 23 and 27 of *The Water Act (Alta.), supra.*
30. s. 18(2) of *The Water Act (Alta.) supra,* and see Nigel Banks, "Water Law Reform in Alberta: Paying Obeisance to the 'Lords of Yesterday,' or Creating a Water

Charter for the Future?" (1995) 49 *Resources: The Newsletter of the Canadian Institution of Resources Law* 1, which argues these license are not subject to environmental restrictions.

31. s. 53, 54, 69 of *The Saskatchewan Watershed Authority Act, 2005, supra.*
32. s. 83, ibid.
33. s. 94 of *The Water Act (Alta.), supra.*
34. s. 97 of *ibid.* An inspector may issue a water management order in respect of priorities. s.32, *ibid.*
35. s. 158, ibid.
36. Part 9 of ibid.
37. *The Arbitration Act,* C.C.S.M., c. A.120. See s. 19 and s. 24 of *The Water Rights Act, supra.*
38. s. 92 of *The Water Act (b.c.), supra.* Note there are some exceptions where decisions such as canceling in whole or in part a licence for failure to pay rentals or comply with the *Act* are not appealable.
39. *The Irrigation Districts Act,* R.S.A. 2000, c. I-11.
40. Part 3, s. 51 of *The Water Act (b.c.), supra.*
41. s. 53, ibid.
42. S.S. 1996, c. I-14.1
43. s. 1 and 9(5) of *The Water Rights Act, supra.*
44. ss. 10 and 12 of *The Water Act (b.c.), supra.*
45. s. 62 of *The Water Act (b.c.), supra.*
46. s. 19 of *The Saskatchewan Watershed Authority Act, 2005, supra.*
47. s. 14–23 of *The Water Protection Act,* C.C.S.M. c. W65. This Act allows a Conservation District to be designated as a watershed planning authority pursuant to s. 14(i). This arguable allows integration of issues.
48. s. 24–28 of *The Water Protection Act, supra.*
49. s. 23 *The Water Protection Act, supra.*
50. s. 9(1) of *The Water Act (Alta.), supra.*
51. s. 81 of *The Water Act (Alta.), supra.*
52. These requirements are not part of the legislated rules, but posted as policy by Alberta Environment on their web site at: www.gov.ab.ca/env/water
53. s. 83 of *The Water Act (Alta.), supra.*
54. A report prepared as a follow up to a World Water Forum organized by a World Water Council outlined these principles for assessing challenges and initiating change of water institutions. World Water Council Water Action Unit, *World Water Actions, Making Water Flow for All* (London, England: Earthscan Publications Ltd, 2003): 22; David B. Brooks, *Water, Local-Level Management* (Ottawa: International Development Research Center, 2002).
55. The assessment of the water models and the legislative framework of the Western Provinces utilizing the models could be approached in many ways. Ideally one

would assess whether the needs of water users were achieved or perhaps if the residents and stakeholders relying on the water resource had their water goals or strategy met based on their choice of water model and its implementation. In the past few years the Western Provinces have developed water strategies which would allow this type of review. However, research on user and stakeholders' perceptions and experience in relation to water management would have to be compiled, studied, and assessed. This data would allow an accurate conclusion to be drawn on the effectiveness of both the water model(s) chosen and its implementation. Alberta's "Water for Life" strategy was developed and can be found at: http://www.waterforlife.gov.ab.ca/html/background2.html . See further strategies in section 3(b)(ii)

56. World Water Council Water Action Unit, *World Water Actions, Making Water Flow for All, supra.*
57. Brooks, *Water, Local-Level Management, supra:* 4.
58. Ibid.: 5. See also Mizanur Muhammed Rahaman and Olli Varis, "Integrated Water Resources Management: evolution, prospects and future challenges," (2005) Sustainability: Science Practice, & Policy 1, available at: http://ejournal.nbii.org
59. Ibid., 40.
60. Ibid., 88–89.
61. The South Saskatchewan River Basin Water Management Plan provides for conservation, transfers, and the creation of a Interbasin Water Coordinating Committee, but does not contain mention of water quality issues, other than to state it should be studied further. The plan is available at: http://www3.gov.ab.ca/env/water/regions/ssrb/plan.html
62. Warren Dixon, A. B. Badelt, V. Cameron and M. Robbins, "Pilot Water Management Plan for the Township of Langley," Canadian Water Resources Association, British Columbia Branch, 2006 Conference Proceedings, Water Under Pressure, Balancing Values, Demands and Extremes: 239.
63. Prairie Provinces Water Board, *The 1969 Master Agreement on Apportionment and Bylaws, Rules and Procedures,* (Ottawa: Prairie Provinces Water Board, October, 2003), paragraph 13.
64. Clause 8(f) of Schedule E, ibid.
65. Part I of *The Constitution Act, 1982,* being Schedule B to the *Canada Act 1982* (U.K.), 1982, c. 11.
66. Manitoba's can be found at: http://www.gov.mb.ca/waterstewardship/waterstrategy/pdf/index.html and Alberta's at: http://www.waterforlife.gov.ab.ca/. It is noteworthy that Manitoba's strategy embraces planning the water resource for the next generation; Alberta's focuses on sustainable industry.
67. http://www.swa.ca/WaterConservation/default.asp; see the Saskatchewan Water Conservation Plan available at: www.swa.ca
68. www.env.gov.bc.ca/wsd/plan_protect_sustain/water_conservation/wtr_cons_strategy/implement.html

69. s. 25 of British Columbia's *Drinking Water Protection Act,* S.b.c. 2001; c. 9 provides drinking water officers broad powers to prohibit any activity that may pose a significant risk to drinking water. Also Manitoba's Water Strategy has as a goal the review and consolidation of all water legislation.
70. See section 2(b) of this paper.
71. Bruns and Meinzen-Dick, "Frameworks for Water Rights," *supra.*
72. See section 2(a) of this paper.
73. An Act must be shown to be in "bad faith," s. 24.1 *The Water Rights Act, supra,* and not actionable if done in good faith, s. 157 of *The Water Act, supra* and s. 95 of *The Saskatchewan Watershed Authority Act, 2005, supra.*
74. Notification in respect of allocation applications in Saskatchewan, see s. 61 of *The Saskatchewan Watershed Authority Act, 2005, supra. The Environmental Protection and Enhancement Act,* S.A. 1992, c. I-E3.3. s. 16(1) of *The Water Act (Alta.), supra.*
75. Manitoba has a minister responsible for Conservation and a Minister of Water Stewardship.
76. s. 25 of British Columbia's *The Drinking Water Protection Act,* S.b.c. 2001, c.9 provides drinking water officers broad powers to prohibit any activity that may pose a significant risk to drinking water.
77. s. 3(1) of *The Agricultural Operations Act,* R.S.S. 1978, c. A-12.1. Although s. 21 provides a minister shall not approve a plan if pollution of ground or surface water shall occur. See Manitoba's *Livestock Manure and Mortalities Management Regulation,* Reg. 42/98 which deals with water quality issues. No discharge of livestock manure is allowed into surface or groundwater (s. 11).
78. *Environmental Assessment (Mandatory and Exempted Activities) Regulation,* Alta., Reg. 111/93.
79. *The Forest Practices Code of British Columbia,* S.b.c. 1994, c. 41 is exempted in its water legislation from water management plans. S. 65(2)(a) of *The Water Act (b.c.), supra.*
80. David R. Boyd, *Unnatural Law, Rethinking Canadian Environmental Law and Policy,* (Vancouver: UBC Press, 2003): 36–41. Most scientists believe current water quality problems are largely caused by non-point source pollution such as agricultural runoff, urban runoff, airborne water pollution and the failure to adequately regulate the use of toxic chemicals by the *Canadian Environment Protection Act, 1999,* S.C. 1999, c. 33.
81. Ibid., 34 and 238.
82. See s. 5 of *The Environmental Management Act,* S.b.c. 2003. c. 53 which states the duty, powers, and function of the minister extend to any matter regarding the management, protection and enhancement of the environment among other things. However, no set standard of quality exists. It is interesting to note the inconsistent language of "wise use" in Alberta's legislation. The purpose of *The Environmental Protection and Enhancement Act,* R.S.A. 2000, c. E-12 is the protection, enhancement and wise use of the environment. The purposes of economic

growth and ecosystem integrity are both stated. This language also appears in the clause 12(i) outlining the duties of the Minister of the Environment:
shall generally do any acts the Minister considers necessary to promote the protection and wise use of the environment for the benefit of the people of Alberta and future generations.

83. s. 2(1) of *The Environment Act,* C.C.S.M., C. E-125.
84. s. 46 of *The Environment Act, supra* of Manitoba. Even British Columbia legislation establishes immunity: s. 61 and 129 of *The Environmental Management Act, supra.*
85. *Potable Water Regulation,* A.R. 122/93; now A.R. 277/2003, S.29 (4) of Saskatchewan's *The Water Regulations,* 2002, E-10.21, Reg 1.
86. Steven A. Kennett, *Managing Inter-jurisdictional Waters in Canada : A Constitutional Analysis* (Calgary: Canadian Institute of Resource Law, 1991): 89.
87. Water management was not treated as a single topic in the Canadian Constitution. In 1867 when the *British North America Act* (Now *The Constitution Act, 1867* (U.K.), 30 & 31 Victoria, c. 3.) was negotiated and agreed to, water and water scarcity in the Prairie provinces, pollution and environmental degradation were not planned for in the negotiations.
88. These headings include publicly owned lands, mines, minerals and royalties, property and civil rights, local works and undertakings, and natural resources which included the right to make laws in relation to the development, conservation and management of non-renewable natural resources and forestry resources in the province. It is through the first heading "lands" that the provincial jurisdiction to water primarily resides. In traditional Canadian common law, water rights transferred with the land with which it was associated. "Land" is defined as "every species of ground, soil or earth whatsoever, as meadows, pastures, woods, moors, waters, marshes, furs and heath." Earl Jowitt, *The Dictionary of English Law* (London: Sweet & Maxwell Limited, 1959): 1053.
89. These include federal lands (national parks, Indian reserves), trade and commerce, navigation and shipping, seacoast and inland fisheries, works for the general advantage of Canada, entering into treaties, and matters not specifically assigned to the provinces. s. 91 of *The Constitution Act,* 1982, *supra.* The federal government is responsible for ensuring the safety of drinking water within areas of federal jurisdiction, such as national parks and Indian reserves and water quality in respect of inter-jurisdictional waters. The *Canada Water Act,* R.S.C. 1985, C. C-11.
90. s. 91(29) and 91(10) the declaratory power of *The Constitution Act, 1982, supra.* Further the federal government could utilize its "spending power" and lastly its "peace, order and good government" power in the introductory words of s.91 to assert jurisdiction to water. Kennett, *Managing Interjurisdictional Waters in Canada,* 23–28.
91. The inter-jurisdictional immunity test is very troubling and contentious for both academic commentators and courts alike. Some such as Professor Peter P. Hogg

argue that a lesser test of "impairment" is possible in the even the legislation only indirectly affects the federal head of power. See Hogg, *Constitutional Law of Canada,* Student Edition (Toronto: Carswell, 2002). This seems improbable in light of decisions such as *Ordon Estate* v. *Grail* [1998] 3 S.C.R. 437.

92. s. 91 provides, "It shall be lawful for the Queen, by and with the Advice and Consent of the Senate and House of Commons, to make Laws for the Peace, Order, and good Government of Canada, in relation to all Matters not coming within the Classes of Subjects by this Act assigned exclusively to the Legislatures for the Provinces; and for greater Certainty, but not so as to restrict the Generality of the foregoing Terms of this Section, it is hereby declared that (notwithstanding anything in this Act) the exclusive Legislative Authority of the Parliament of Canada extends to all Matters coming within the Classes of Subjects next hereinafter enumerated . . ."
93. *R.* v. *Crown Zellerbach* [1988] 1 S.C.R. 401 and *Canada (A.G.)* v. *Hydro-Quebec,* [1997] 3 S.C.R. 213.
94. *Interprovincial Co-operative* v. *R.* [1976] 1 S.C.R. 477.
95. Similarly, the Mathias Colomb Band in Manitoba continues to have an unresolved outstanding claim for similar water issues, now decades old, unresolved by both the courts and negotiation. *Mathias Colomb Band of Indians.* v. *Saskatchewan Power Corp.* (1994) 13 C.E.L.R. (N.S.) 245 (Man. C.A.) For some of the legal hurdles of the provinces and the federal government suing on another see Kennett, *Managing Interjurisdictional Waters in Canada,* 174; and see Hogg, *Constitutional Law of Canada, Student Edition,* 289. Unless an agreement is constitutionalized, this is an ultimate constraint on its enforceability.
96. This issue currently exists in relation to the high level of nutrients in Lake Winnipeg as a result of non-point source pollution from run off into waterways throughout Alberta, Saskatchewan, Ontario and Manitoba, which drain into Lake Winnipeg, and the also Alberta tar sands pollution and use of water in relation to affects on Saskatchewan. See http://www.cbc.ca/manitoba/features/lakewinnipeg/creek.html, and Lake Winnipeg Stewardship Board, *Recommending Solutions to reduce Nutrient Loading to Lake Winnipeg,* found at: www.redriverbasincommission.org/Conference/22nd; and "Oilsands tapping Sask water supply," November 14, 2006, the Leader-Post: A1.
97. Water is shared such that 50% of flows must be passed to Saskatchewan, who in turn must pass the same proportion to Manitoba. Prairie Provinces Water Board, *The 1969 Master Agreement on Apportionment and Bylaws, Rules and Procedures, supra.*
98. In respect of the South Saskatchewan River Basin, the 50% flow requirement occurs after certain needs are met in Alberta jurisdictions. Prairie Provinces Water Board, *The 1969 Master Agreement on Apportionment and Bylaws, Rules and Procedures, supra,* the Preface.

99. W. Adger, "Social Aspects of Adaptive Capacity," in J. Smith, R. Klein and S. Huq (eds.), *Climate Change, Adaptive Capacity and Development* (London: Imperial College Press, 2003), 29–49.
100. Bruns and Meinzen-Dick, "Frameworks for Water Rights," *supra.*
101. This criticism could also be leveled against the irrigation users example cited as illustrating the user-based management model. Brooks, *Water, Local-Level Management, supra,* 26 and 47.
102. Ultimately there will not be any legal enforceability of these decisions. See Kennett, *Managing Interjurisdictional Waters in Canada: A Constitutional Analysis, supra,* 174.
103. See section 2(a) of this paper.
104. Percy suggests this state is "unprecedented" in modern water law in North America. See "Water Rights Law and Water Shortages in Western Canada," *Canadian Water Resources Journal* 11, no. 2 (1986): 19–20.
105. Bruns and Meinzen-Dick, "Frameworks for Water Rights," *supra.*
106. Part 5 Division 2 of *The Water Act (Alta.), supra.*
107. Percy, "The Limits of Western Canadian Water Allocation Law," *supra.*
108. There have been successful negotiations in other industries such as the electricity industry which has border supply agreements along the Saskatchewan/Manitoba border and the Alberta/Saskatchewan border. However, with deregulation of the electrical industry and a competitive market place, these cooperative agreements have become problematic. See also Steven A. Kennett, *Managing Interjurisdictional Waters in Canada, supra,* 95.
109. See part 2(a) of this paper.
110. Ibid.
111. s. 81(6) of *The Water Act (Alta.), supra.*
112. See Part B of *The Water Act (b.c.), supra.*
113. No provisions in British Columbia legislation provide for this; however, Saskatchewan's *The Irrigation Act, 1996, supra* (s. 20) and Alberta's *The Irrigation Act, supra* (s. 41(3)) provides for annual business reports and financial statements to be filed with the Minister.
114. s. 39(8)(9) of *The Water Regulations, 2002* R.S., c. E-10.21 Reg 1.
115. s. 14 *The Drinking Water Act,* S.b.c. 2001, c. 9.
116. Ibid., s. 44(1): An annual report is also prepared on the state of drinking water quality in Saskatchewan and tabled in the Legislative Assembly. S. 19 of *The Environmental Management and Protection Act, 2002,* S.S. 2002, c. E-10.21. British Columbia has the same provision in s. 4.1 of *The Drinking Water Act,* R.S.b.c. and Manitoba in s. 5 of *The Drinking Water Safety Act,* S.M. 2002, c. 3.
117. Brooks, *Water, Local-Level Management, supra.*

A PARTICIPATORY MAPPING APPROACH TO CLIMATE CHANGE IN THE CANADIAN PRAIRIES

LORENA PATIÑO AND DAVID A. GAUTHIER

Abstract

Human-induced global warming and climate change are predicted to have serious impacts on water availability and quality, particularly in the dryland areas of the Canadian prairies. The sustainability of rural communities depends, in part, on the capacity of government institutions to address the current and future vulnerabilities to climate change. The integrated water resource management (IWRM) methodology is grounded in the principles of good governance, and employs participatory approaches. Numerous agencies and sectors are involved at diverse jurisdictional levels, resulting in separate federal, provincial, and municipal water systems. This convergence in the management of water complicates the operationalization and implementation of a sustainable IWRM approach. This paper proposes a public participation geographic information systems (PPGIS) approach to water resource management in order to foster dialogue and cooperation, thus promoting participation and good water governance practices.

Sommaire

On prévoit que le réchauffement de la planète et les changements climatiques d'origine anthropique vont avoir des impacts sérieux sur la disponibilité et la qualité de l'eau, particulièrement dans les zones arides Prairies canadiennes. La viabilité future des collectivités rurales dépend en partie de la capacité des institutions gouvernementales d'adopter des mesures pour faire face à la vulnérabilité actuelle et future aux changements climatiques de ces collectivités. Dans le contexte de gestion intégrée des ressources en eau (GIRE) et en s'appuyant sur les principes de bonne gouvernance et d'approche participative, la complexité et les difficultés associées à l'opérationnalisation et l'implémentation d'une approche durable de GIRE résident dans la coexistence de nombreux organismes, secteurs et échelons administratifs impliqués dans la gestion des ressources en eau. De nombreux établissements interagissent à divers niveaux d'administration, avec comme résultats des réseaux d'eau de compétence fédérale, provinciale et municipale qui diffèrent. Cet article suggère une approche s'appuyant sur des systèmes d'information géographique (SIG) participatifs pour la gestion des ressources en eau afin de promouvoir le dialogue, la communication et la coopération entre les partis impliqués, favorisant ainsi la participation et de bonnes pratiques de gestion des ressources en eau.

Introduction

Human-induced climate change is expected to reduce the quality and quantity of water resources, particularly in areas already facing water stress.[1] The Canadian prairies have been identified as vulnerable to the impacts of climate change, primarily because this area is dominated by an agricultural sector that relies on already scarce water resources.[2] In addition, prairie rural communities have become increasingly prone to the negative impacts of climate change as a result of social and economic factors in the agricultural sector.[3]

The sustainability of rural communities in the Canadian prairie drylands, under the forecasted impacts of climate change on water resources, depends, in part, on the capacity of government to address the current and future vulnerabilities of those communities. Public policy is generally developed at regional, provincial, or federal levels, but climate change weaknesses are site-specific.[4] Consequently, future policies addressing climate change and water issues may be limited when attending to concerns for rural communities.

The water crisis in many parts of the world may more accurately be identified as a crisis of governance.[5] Integrated water resource management (IWRM) emphasizes a basin-wide approach "under the principles of good governance and public participation."[6] In addition to considering multiple dimensions of sustainability (e.g., social, economic, and biophysical), there is a need to integrate the multiple institutional levels (e.g., global/national to local, government to civil, and formal to informal) involved in water governance.[7] This unified approach would aim at the development of strategies and policies that allow for the incorporation of local knowledge, values, conditions, and information derived from both qualitative and quantitative modes of inquiry.[8]

Institutional Water Context of the South Saskatchewan River Basin

Encompassing 168,000 square kilometres, the South Saskatchewan River Basin (SSRB) contains five major watersheds. Its natural delineation extends across Alberta and Saskatchewan, providing a diverse institutional and jurisdictional setting. Water management is a provincial government mandate, and the provinces are increasingly adopting watershed-based approaches, involving multiple stakeholders in the aspects of provincial water resource management. Hurlbert depicts the complex legal institutional water structures involved in the management, decision-making, monitoring, and enforcement of water resources in Saskatchewan and Alberta.

The 1987 *Federal Water Policy* calls for joint and co-operative management approaches with the provinces, fostering integrated water management based on a watershed approach.[9] The 1985 *Canada Water Act* promotes the idea of agreements with the provinces in water quality matters of urgent national concern.[10] However, the decentralized administrative system and the numerous agencies and sectors involved in the management of water pose serious challenges to the water governance process.

At the federal level alone, there are at least 19 departments that have some involvement in water management.[11] Since the creation of the 1987 federal water strategy, many authors have depicted Canadian water management as fragmented, with water institutions lacking clear roles and mandates.[12] These problems underscore the difficulties of operationalizing and implementing a sustainable IWRM approach within the context of the legal water institutional system in the SSRB.

At the provincial level, there are five Saskatchewan and three Alberta legal-governmental provincial institutions involved in water management. Responsibility for waterworks and potable water rests with local governance, and active day-to-day management of water is increasingly assumed by smaller local institutions. In addition to these legal institutions, there is a large number of non-governmental and civic organizations, with a variety of scopes and interests, that are involved in water governance at all jurisdictional levels.[13] Figure 1 shows some of the institutions involved in the governance of water at different jurisdictional levels in the SSRB.

An Example of the Local Institutional Level in the SSRB: The Blood Tribe Community

Magzul's study of Alberta's Blood Tribe reveals the expansive challenge of adapting water management to climate change.[14] Figure 2 summarizes the Blood Tribe community's vulnerability to climate change, with respect to water resources. This case illuminates the diversity of jurisdictions, institutions, and instruments involved in water governance in the SSRB. Magzul also exposes the abundant perspectives and interpretations of climate change and water issues in the SSRB. The IWRM commitment to good governance and participatory processes requires approaches that recognize the multi-dimensional value of water[15] by fostering the communication of value-laden

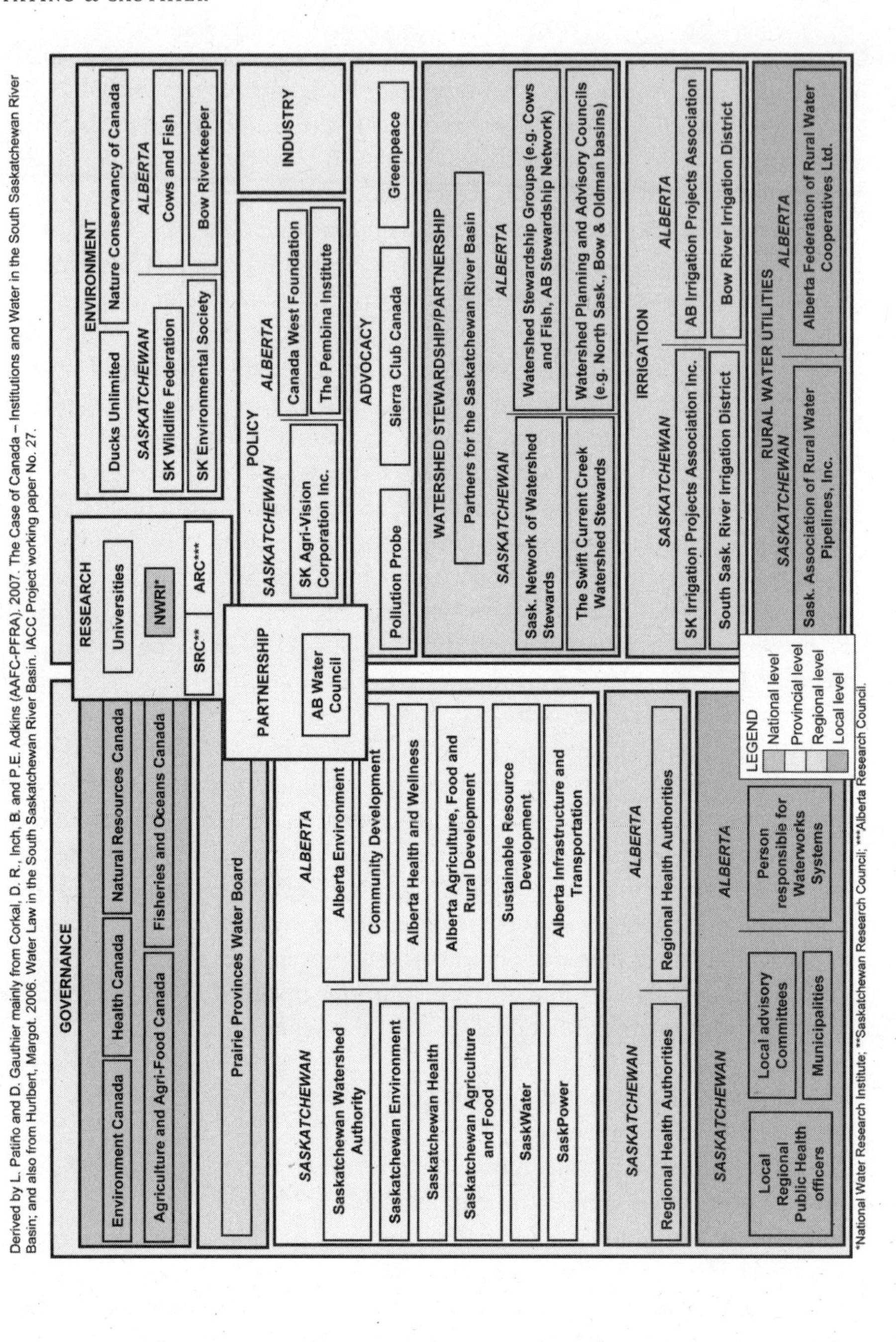

Figure 1. Some Important Institutions in the Saskatchewan Institutional Water Map, South Saskatchewan River Basin.

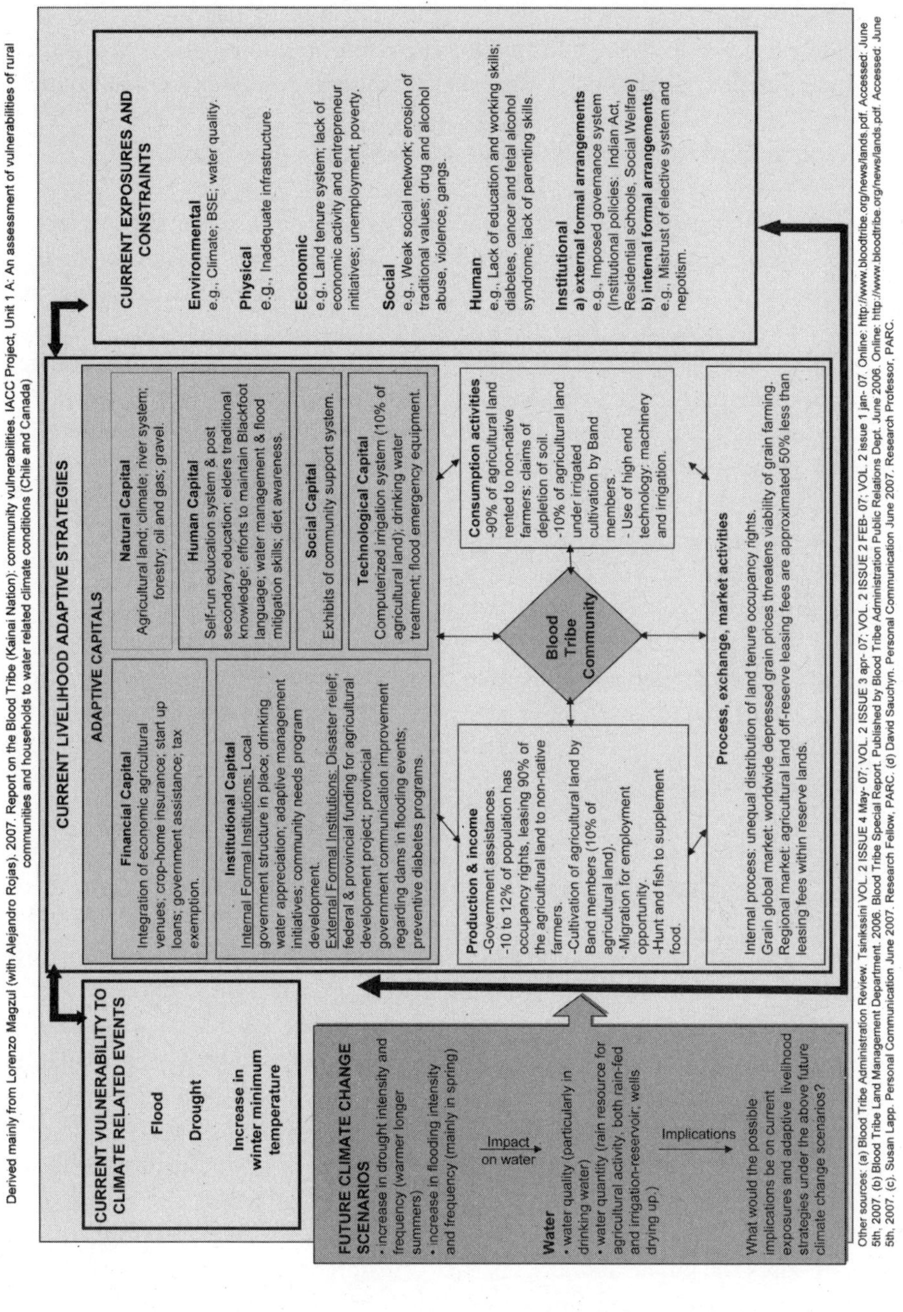

Figure 2. Blood Tribe Community Vulnerability to Climate Change Focused on Water.

perspectives. Such an approach might be useful in regions like the SSRB, where a de-centralized administrative system, multi-sectoral environment, and numerous viewpoints of water make responding to climate change difficult.

Geographic Information Systems (GIS) and Public Participation (PP)

The GIS in PPGIS

Generally, a geographic information system (GIS) refers to a computer system (i.e., a combination of software and hardware) that can manage and analyze spatial information. The Environmental Systems Research Institute (ESRI) defines a GIS as "a system for management, analysis, and display of geographic knowledge, which is represented using a series of information sets such as maps and globes, geographic data sets, processing and work flow models, data models, and metadata."[16] The typical outputs of GIS are maps that are sometimes accompanied by tabular information.

GIS use has been characterized as an expert-driven approach, and heavily criticized as a controlling tool for government surveillance.[17] Referring to the case of India, Puri and Sahay argue that governmental GIS applications tend to be categorized as techno-centric approaches, requiring a technical expert to upload scientific data to the GIS about an application domain (e.g., soil degradation).[18] Such an application encompasses a spatial analysis of various digital area maps, according to a particular application domain. Generally, the output of the analysis is used in the decision-making process regarding the corresponding application domain in a top-down approach.

The PP in PPGIS

Some authors refer to public participation as a horizontal relationship where different realities and different modes of enquiry are acknowledged. This horizontal connection facilitates dialogue among multiple perspectives to promote knowledge exchange, establishing a co-learning process.[19] Public participation also emphasizes collective and voluntary effort in the sharing of risks, responsibilities, resources, and benefits. Under participatory approaches, stakeholders aim at auto-development and acquiring a place in the decision-making process.[20]

In public participation, it is important to clarify 'who' is the public and 'what' constitutes participation (i.e., purpose of participation). There are a number of dimensions and typologies to consider when determining who composes the public. Potential participants might be culled from a particular group (e.g., only visible minority seniors) or a complex public with individuals of all ages and ethnic backgrounds. The literature regarding the purpose of participation identifies a spectrum of participatory scales oriented towards power, conflict resolution, administration, and planning processes. These participatory scales range from participation that aims at educating and informing, requires active and constant involvement, fosters social interaction and empowerment, or results in manipulation. 'Who' is the public, and 'what' type of participation are significant in that the aims and outcomes of the participatory process depend on the particular combination of public and purpose of participation.[21]

Participatory approaches enable the process of knowledge and information exchange.[22] In addition, because they can improve communication between governmental institutions and the rural communities they serve, participation provides significant potential to develop adaptation policies and strategies tailored to local vulnerabilities.[23]

Public Participation Geographic Information Systems (PPGIS)

Public participation geographic information systems (PPGIS) combine participation and technologically-based spatial analysis.[24] A GIS contributes a visual language and spatial analysis capabilities to participation,[25] and participation provides a social venue and multi-perspective approach for using the GIS in the decision-making process.[26] Increasingly, GIS applications incorporate participatory and bottom-up approaches, and also integrate top-down and bottom-up venues, exemplified by the research of Puri and Sahay in India, and Sedogo and Groten in Burkina Faso, Africa.[27]

In 1996, Schroeder introduced PPGIS as "a variety of approaches to make GIS and other spatial decision-making tools available and accessible to all those with a stake in official decisions."[28] Since then, PPGIS practitioners have expanded the scope of their applications, but have not yet reached a consistent definition of PPGIS.[29] PPGIS applications range from community development and empowerment,[30] urban planning[31] and regional planning,[32] to land use management[33] and environmental planning.[34]

Despite a lack of definitional consensus, PPGIS practitioners generally agree on the main characteristics and potentialities of PPGIS.[35] PPGIS are grounded in the use of geo-spatial information management tools (e.g., digital maps, aerial photography, satellite imagery, and GIS). These tools compose peoples' spatial knowledge in the forms of maps and visual images, providing the interactive venue for discussion, information exchange, analysis, advocacy, and support in decision-making.

PPGIS is multidisciplinary in that it integrates outside expert knowledge with local knowledge. PPGIS also integrates multiple realities and diverse forms of information by fostering social learning and supporting two-way communication. PPGIS assists broad participation across diverse socio-economic contexts, locations, and sectors,[36] and builds on stakeholders' participation to foster spatial learning, analysis, decision-making, and

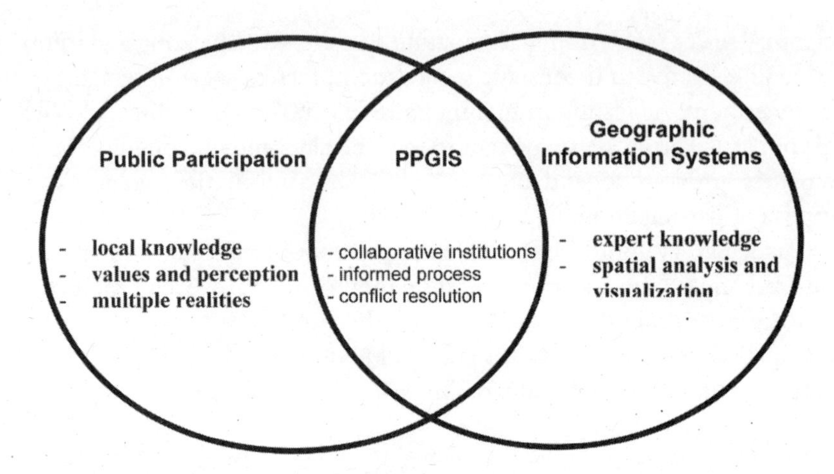

Figure 3. Public participation geographic information systems: an integration approach to the multiple dimensions of knowledge.

action.[37] PPGIS promote a flexible governance framework, unifying local participation and regional planning in a bottom-up and top-down scheme. In the Canadian context, data has the potential to be transformed by scaling up local participatory information for regional purposes, and scaling down regional information for local use.[38]

Figure 3 shows the integrative potential of PPGIS. While participatory approaches have been developed within the social sciences in order to capture values and perceptions, GIS technology has mainly emerged within the sphere of computer-based and biophysical sciences to assist in the search for generalities and objectivity. PPGIS has been helpful in conflict resolution, reaching consensus, and producing an informed process.[39] As well, using PPGIS builds collaborative institutions that benefit from social interaction and technological advances to resolve problems.[40]

Incorporating PPGIS to Climate Change Impacts on Water Under an IWRM Perspective

This paper proposes a PPGIS approach to water resource management in order to foster dialogue and cooperation, thus promoting participation and good water governance practices. Figure 4 shows a PPGIS model adapted from Patiño and Gauthier.[41] The model depicts the integration of the multiple realities of water resource management. The model is "based on information sharing in which the meaning of the information is attributed through an open dialogue that allows an informed process and acknowledges multiple realities, hence facilitating a reflective experience that aims towards the construction of a more integrated knowledge."[42]

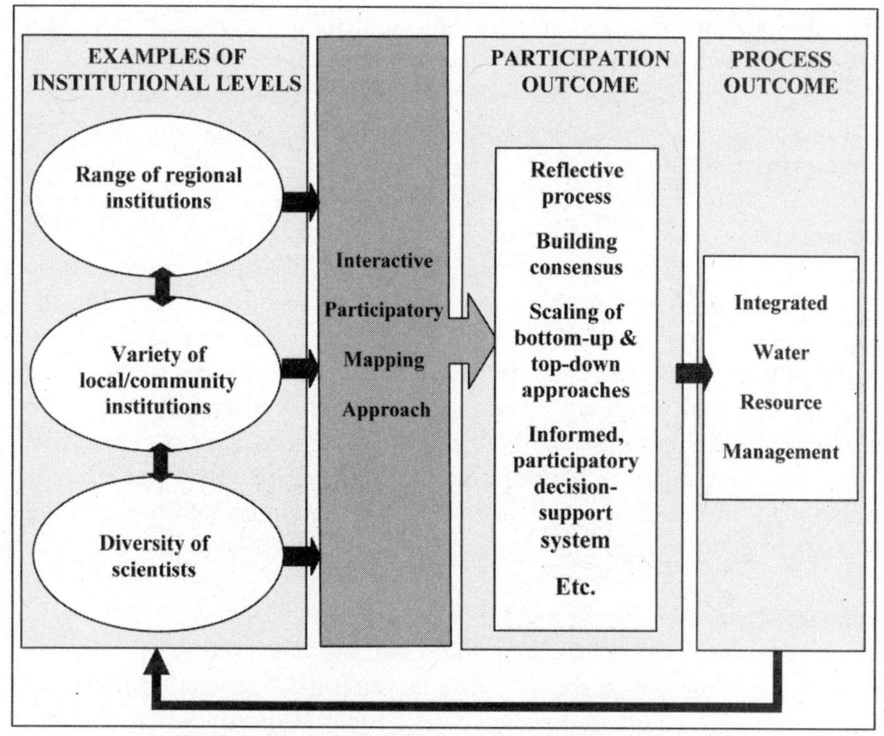

Adapted from: Patiño and Gauthier. 2006. *Participatory mapping to assess institutional vulnerability and adaptation to climate change* [online]. IACC Project working paper [cited 13 January 2007]. Available from World Wide Web: (http://www.parc.ca/mcri/papersHID.php).

Figure 4. PPGIS model for integrating multiple realities of water resource management using participatory mapping techniques.

In this PPGIS model, each institutional level enjoys its own view of how water resources ought to be managed. In addition, each institutional level is also comprised of multiple publics. Regional institutions, include federal and provincial governments, that are each responsible for the formulation of water policies, strategies, standards, and regulations. In addition, federal and provincial governments encompass a number of departments with mandates addressing social, economic, and environmental concerns. Similar complexity occurs at the community and scientific levels.

Federal and provincial institutions, generally responsible for setting guidelines, standards, and regulations for water management, tend to construct water resource knowledge using sectoral and top-down approaches.[43] Adopting PPGIS in an inter-sectoral and inter-institutional environment streamlines the process of identifying and interpreting connections within the multiple dimensions of water.[44] This model also facilitates meaningful dialogue, allowing the integration of knowledge and different realities. This knowledge integration furthers the process of understanding water resources and management at institutional levels, accounting for the dynamic character of knowledge construction.

Conclusion

This paper has drawn upon GIS capabilities and participatory approaches to develop a model for integrating the different perspectives involved in the management of water within the SSRB. By relying on the use of a visual language in order to present different realities, the model aspires to the IWRM principles of good governance and participation that assists the integration of meaningful information and multiple interpretations, and uses visual language to present different realities. Through fostering an IWRM approach, the water sector increases its possibilities for being better prepared to adapt and cope with the forecasted impacts of climate change.

Acknowledgements

The authors wish to express their gratitude for the invaluable contribution to this paper provided by the following Institutional Adaptation to Climate Change project members in alphabetical order: Darrell Corkal, Margot Hurlbert, Suzan Lapp, Lorenzo Magzul, Gregory Marchildon, and David Sauchyn. The authors would also like to thank Angela Scott of Regina, Saskatchewan, for her skillful copy-editing of the paper. Finally, the authors also wish to acknowledge the financial support provided by the Institutional Adaptation to Climate Change project funded through the Social Science and Humanities Research Council of Canada (SSHRC) Multi-collaborative Research Initiative (MCRI) program, as well as funding provided by SSHRC Standard Operating Grant program to D. Gauthier.

Endnotes

1. Tim Barnett, Robert Malone, William Pennell, Detlet Stammer, Bert Semtner, and Warren Washington, "The Effects of Climate Change on Water Resources in the West: Introduction and Overview," *Climate Change* 62 (2004): 1–11; Stewart Cohen and Tanuja Kulkarni (eds.), "Water Management & Climate Change in the Okanagan Basin," Environment Canada & University of British Columbia, Project A206, submitted to the Adaptation Liaison Office, Climate Change Action Plan, Natural Resources Canada, Ottawa, 2001.

2. Environment Canada, "Threats to Water Availability in Canada," 2004, National Water Research Institute, Burlington, Ontario. NWRI Scientific Assessment Report series No. 3 and ACSD Science Assessment Series No. 1.
3. David Gauthier and Allison McPhee, *Prairie Rural Communities and Issues of Climate Change,* Prairie Adaptation Research Centre Final Report, Regina, Saskatchewan, 2002. Available at: http://www.parc.ca/research_pub_communities.htm (cited 15 March 2007).
4. Richard Klein and Joel Smith, "Enhancing the Capacity of Developing Countries to Adapt to Climate Change: A Policy Relevant Research Agenda," in J.B. Smith, R.J.T. Klein, and S. Huq (eds.), *Climate Change, Adaptive Capacity and Development* (London: Imperial College Press, 2003), 317–335; Barry Smit and Olga Pilifosova, "From Adaptation to Adaptive Capacity and Vulnerability Reduction," in J.B. Smith, R.J.T. Klein, and S. Huq (eds.), *Climate Change, Adaptive Capacity and Development,* 9–25.
5. The Global Water Partnership, *Towards Water Security: A Framework for Action.* Prepared for presentation at the Second World Water Forum convened by the World Water Council, and the Ministerial Conference convened by the Dutch Government at The Hague, the Netherlands, March 17–22, 2000. Available at: http://www.gwpforum.org/gwp/library/sec2.pdf (cited 9 April 2007); Willem Alexander, *No Water No Future: A Water Focus for Johannesburg* [online]. Contribution of HRH the Prince of Orange to the Panel of the UN Secretary General in Preparation for the Johannesburg Summit, 2002, World Summit on Sustainable Development. Available at: http://www.nowaternofuture.org/pdf/NoWaterNoFuturre%20Final%20Version.pdf (cited 15 April 2007); Tarisai Garande and Suzan Dagg, "Public Participation and Effective Water Governance at the Local Level: A Case Study from a Small Under-Developed Area in Chile," *Environment, Development and Sustainability* 7 (2005): 417–431.
6. Muhammad Mizanur Rahaman and Olli Varis, "Integrated Water Resources Management: Evolution, Prospects and Future Challenges," *Sustainability: Science, Practice, & Policy* 1(1) (2005): 15.
7. Satish Puri and Sundeep Sahay, "Participation Through Communicative Action: A Case Study of GIS for Addressing Land/Water Development in India," *Information Technology for Development* 10 (2003): 179–199; Lorena Patiño and David Gauthier. *Participatory Mapping to Assess Institutional Vulnerability and Adaptation to Climate Change,* IACC Project working paper, 2006. Available at: http://www.parc.ca/mcri/papersHID.php (cited 13 January 2007).
8. Patiño and Gauthier, *Participatory Mapping.*
9. Environment Canada, Federal Water Policy, 1987. Available at: http://www.ec.gc.ca/water/en/info/pubs/fedpol/e_fedpol.pdf (cited 22 March 2007).
10. Department of Justice Canada, Canada Water Act (R.S., 1985, c. C-11), 2007. Available at: http://laws.justice.gc.ca/en/ShowFullDoc/cs/C-11///en (cited 13 April 2007).

11. Darrell Corkal, Bruce Inch, and Phillip Atkins (Agriculture and Agri-Food Canada-Prairie Farm Rehabilitation Administration—hereafter AAFC-PFRA), *The Case of Canada—Institutions and Water in the South Saskatchewan River Basin.* SSHRC-MCRI Institutional Adaptation to Climate Change Project, 2007. Available at: http://www.parc.ca/mcri/ass_gov_initw.php (cited 12 April 2007).
12. Corkal, Inch, and Atkins, *The Case of Canada.*
13. Hurlbert, *Water Law in the South Saskatchewan River Basin*; Corkal, Inch, and Atkins, *The Case of Canada.*
14. Lorenzo Magzul, *Report on the Blood Tribe (Kainai Nation): Community Vulnerabilities.* IACC Project, Unit 1 A: An Assessment of Vulnerabilities of Rural Communities and Households to Water Related Climate Conditions (Chile and Canada), IACC Project working paper, 2007. Available at: http://www.parc.ca/mcri/comm_vul_assw.php (cited 21 March 2007).
15. Helen Ingram, "Water as a Multi-Dimensional Value: Implications for Participation and Transparency." In response to the paper by Macro Shouten and Klaas Schartz, "Water as a Political Good: Implications for Investments," *International Environmental Agreements* 6 (2006): 429–433.
16. Environmental Systems Research Institute, *GIS: Getting Started,* 2006. Available at: http://www.esri.com/getting_started/index.html (cited 5 April 2007).
17. Renee Sieber, "Public Participatory Geographic Information Systems: A Literature Review and Framework," *Annals of the Association of American Geographers* 96:3 (2006): 491–507.
18. Puri and Sahay, "Participation Through Communicative Action."
19. Renée I. Boothroyd, Stephen B. Fawcett, and Pennie G. Foster-Fishman, "Community Development: Enhancing the Knowledge Base Through Participatory Action Research," in Participatory *C*ommunity *R*esearch: *T*heories and *M*ethods in *A*ction (Washington D.C.: American Psychological Association, 2004), 37–52; M. Montero, "Appendix B. New Horizons for Knowledge: The Influence of Citizen Participation," in *Participatory Community Research: Theories and Methods in Action* (Washington D.C: American Psychological Association, 2004) 251–262.
20. Michael Pimbert, *Institutionalising Participation and People-Centered Processes in Natural Resource Management. Research and Publications Highlights.* (London: International Institute for Environment and Development, 2004), 36.
21. Marc Schlossberg and Elliot Shufford, "Delineating 'Public' and 'Participatory' in GIS," *URISA Journal* 16(2) (2005): 15–26.
22. Boothroyd, Fawcett, and Foster-Fishman, "Community Development"; Montero, Appendix B. *New Horizons for Knowledge.*
23. Patiño and Gauthier, *Participatory Mapping.*
24. Daniel Weiner, Trevor M. Harris, and William J. Craig, *Community Participation and Geographic Information Systems* (Paper presented at the ESF-NSF Workshop on access to geographic information and participatory approaches using geographic

information, Spoleto), 2001. Available at: http://www.iapad.org/publications/ppgis/Community%20Participation%20and%20Geographic%20Information%20Systems.pdf; Schlossberg and Shufford, "Delineating."

25. Schlossberg and Shufford, "Delineating."
26. Sieber, "Public Participatory Information Systems"; Giacomo Rambaldi, Mike McCall, Daniel Weiner, Peter Mbile, and Peter Kyem, *Participatory GIS,* 2004. Available at: http://www.iapad.org/participatory_gis.htm (cited 26 March 2006); Stacy Warren, "The Utopian Potential of GIS," *Cartographica* 39–1 (2004): 5–17; Sieber, "Public Participatory Geographic Information Systems."
27. Puri and Sahay, "Participation Through Communicative Action"; Laurent G. Sedogo and Sussane M. E. Groten, "Integration of Local Participatory and Regional Planning: A GIS data Aggregation Procedure," *GeoJournal* 56–2 (2002): 69–81.
28. Sieber, "Public Participatory Information Systems," 492.
29. Harlan Onsrud and Max Craglia, "Introduction to the Special Issues on Access and Participatory Approaches in using Geographic Information," *URISA Journal* 15 (2003): 5–7.
30. Giacomo Rambaldi, Mike McCall, Peter Kyem, and Daniel Weiner, "Participatory Spatial Information Management and Communication in Developing Countries," *The Electronic Journal on Information Systems in Developing Countries* 25–1 (2006): 1–9.
31. Samuel Dennis, "Prospects for Qualitative GIS at the Intersection of Youth Development and Participatory Urban Planning," *Environment and Planning* A 38–11 (2006): 2039–2054.
32. Sedogo and Groten, "Integration of Local Participatory and Regional Planning."
33. Peter Kyem, "Of Intractable Conflicts and Participatory GIS Applications: The Search for Consensus Amidst Competing Claims and Institutional Demands," *Annals of the Association of the American Geographer* 94–1 (2004): 37–57; Puri and Sahay, "Participation Through Communicative Action."
34. Jonathan Ball, "Towards a Methodology for Mapping 'Regions for Sustainability' Using PPGIS," *Progress in Planning* 58 (2002): 81–140.
35. Daniel Weiner, Trevor M. Harris, and William J. Craig, *Community Participation and Geographic Information Systems.* (Paper presented at the ESF-NSF Workshop on access to geographic information and participatory approaches using geographic information, Spoleto, 2001); Onsrud and Craglia, "Introduction to the Special Issues."
36. Weiner, Harris and Craig, *Community Participation*; Sieber, "Public Participatory Information Systems Across Borders"; Rambaldi, McCall, Weiner, Mbile, and Kyem, *Participatory GIS*; Warren, "The Utopian Potential of GIS."
37. Rambaldi, McCall, Weiner, Mbile, and Kyem, *Participatory GIS.*
38. Laurent G. Sedogo and Sussane M. E. Groten, "Integration of Local Participatory and Regional Planning: A GIS Data Aggregation Procedure," *GeoJournal* 56–2 (2002): 69–81.

39. Robert Feick and Brent Hall, "Balancing Consensus and Conflict with a GIS-Based Multi-Participant, Multi-Criteria Decision Support Toll," *GeoJournal* 53 (2001): 391–406; Kyem, "Of Intractable Conflicts"; Warren, "The Utopian Potential of GIS."
40. Kyem, "Of Intractable Conflicts."
41. Patiño and Gauthier, *Participatory Mapping*.
42. Patiño and Gauthier, *Participatory Mapping*, 21.
43. Corkal, Inch, and Atkins (AAFC-PFRA), *The Case of Canada*.
44. Patiño and Gauthier, *Participatory Mapping*, 21.

PART 2:
THE SOUTH SASKATCHEWAN RIVER BASIN

THE NATURAL CHARACTERISTICS OF THE SOUTH SASKATCHEWAN RIVER BASIN: CLIMATE, GEOGRAPHY AND HYDROLOGY

BRENDA TOTH, DARRELL R. CORKAL, DAVID SAUCHYN,
GARTH VAN DER KAMP, AND ELISE PIETRONIRO

Abstract

The South Saskatchewan River Basin (SSRB) is an extensive watershed characterized by significant variability in its natural environment. The basin's headwaters originate in the eastern slopes of the Canadian Rockies, and experience considerably more precipitation than does the more arid downstream area in the Prairies. Temperature and precipitation regimes have a wide seasonal variation throughout the year. Such climate variability contributes to specific localized soil and surface water regimes. This paper analyzes the hydrology of the SSRB, focusing on the basin's main rivers. When human intervention is removed from the recorded flows on the river system it is evident that mountain runoff from the eastern slope of the Rockies into the river system is the main source of water, supplying more than 75% of the natural runoff within the basin while less than 25% of the basin's runoff is generated from the Prairies. Between 40% and 50% of the SSRB is classified as non-contributing as it does not generate runoff into rivers or streams in an average year. The semi-arid Prairie Ecozone portion of the SSRB is heavily reliant on local snow, rain, and groundwater for its water supply, and is naturally exposed to a high risk and frequency of drought. Water quality in the SSRB is also highly variable. Water quality in the main rivers is generally good, low in minerals, but is exposed to contamination by land-use activities and the natural environment (e.g., sediment, bacteria). Thus portions of the main rivers are under stress, as classified by Canadian water indices. Ground water aquifers have much less risk of contamination from land-use, and are less prone to bacterial contamination, unless the aquifer is shallow or under the influence of surface water infiltration (e.g., flooded wells). However, ground water is generally more mineralized from natural geologic conditions, and the mineral content places limitations on its suitability for use (e.g., most groundwater cannot be used for irrigation).

The SSRB's natural climate variability has had significant historical impacts on the hydrology of the basin, thus hydrological variability is a natural feature of the SSRB. Since the last of the continental glaciers melted away from the prairies about 10,000 years ago, droughts and floods have been a recurrent natural phenomenon. Recorded and inferred droughts are documented for the past 1,500 years. Historical records show a decreasing trend in stream flows and lake levels. Global warming and future climate change scenarios suggest the region may be exposed to an even greater range of variability for the climate, hydrology and water quality within the basin. Climate-induced water stress is expected to place increasing pressures on ecological and human adaptations in the SSRB.

Sommaire

Le basin de la rivière Saskatchewan-Sud (BRSS) est un grand bassin hydrographique caractérisé par une grande variabilité dans son environnement naturel. Les cours supérieurs du bassin ont leurs origines dans le versant est des Rocheuses canadiennes et reçoivent beaucoup plus de précipitations comparativement à la zone plus sèche en aval située dans les Prairies. Les régimes de température et de précipitation ont des variations saisonnières importantes tout au long de l'année. La variabilité du climat contribue à des régimes spécifiques d'eau de surface et de sol localisés. Cet article analyse l'hydrologie du BRSS, en mettant l'accent sur les principales rivières du bassin. Lorsque l'intervention de l'homme est soustraite des débits enregistrés du réseau hydrographique, il est évident que la source principale d'eau provient du ruissellement du versant est des Rocheuses qui atteint le réseau hydrographique. Celui-ci fournit 75 % du ruissellement naturel à l'intérieur du bassin tandis que moins de 25 % du ruissellement du bassin provient des Prairies. De plus, de 40 à 50 % du BRSS est classifié comme ne produisant pas de ruissellement atteignant les cours d'eau lors d'une année moyenne. La partie du BRSS appartenant à l'écozone des prairies, une région semi-aride, dépend beaucoup de la neige et de la pluie qui tombent sur la région, ainsi que des eaux souterraines pour son approvisionnement en eau puisqu'elle est naturellement prédisposée à des fréquences et des risques élevés de sècheresse. La qualité de l'eau dans le BRSS est aussi très variable. La qualité de l'eau des rivières principales est bonne, ayant un contenu faible en minéraux. Cependant, elle est sujette à être contaminée par l'utilisation des terres et par l'environnement naturel (p. ex., sédiments, bactéries). Des parties des rivières principales sont sous stress, tel que défini par l'Indice canadien de la durabilité des ressources hydriques. Les nappes phréatiques sont moins à risque d'une contamination causée par l'utilisation des terres ou d'une contamination bactérienne, à moins que la nappe ne soit que peu profonde ou qu'elle soit infiltrée par de l'eau de surface (p. ex., puits inondés). Cependant, l'eau souterraine a un contenu plus élevé en minéraux à cause des conditions géologiques naturelles, ce qui s'avère être une contrainte quant à son utilisation (p. ex., la majeure partie des eaux souterraines ne peut être utilisée pour l'irrigation).

La variabilité naturelle du climat du BRSS a eu un impact historique sur l'hydrologie du bassin, et donc la variabilité hydrologique est une caractéristique du BRSS. Depuis la fonte du dernier glacier continental, il y a environ 10 000 ans, les sècheresses et les inondations sont devenues des phénomènes naturels récurrents. Toutes les sècheresses des 1500 dernières années qui ont été enregistrées ou celles auxquelles l'histoire fait référence sont documentées. Les dossiers historiques indiquent une tendance à la baisse des débits des cours d'eau et du niveau des lacs. Le réchauffement de la planète et les scénarios de futurs changements climatiques indiquent que la région pourrait être exposée à une encore plus grande variabilité climatique hydrologique ainsi qu'à une plus grande variabilité de la qualité de l'eau dans le bassin. Il est prévu que le stress hydrique d'origine climatique exercera des pressions croissantes sur la capacité à s'adapter de l'environnement et de l'homme dans le bassin de la rivière Saskatchewan-Sud.

Introduction

The North and South Saskatchewan Rivers carry runoff from the eastern slopes of the Rocky Mountains into the Saskatchewan-Nelson River systems that drain much of the Prairie Provinces into Hudson Bay (Figure 1). The South Saskatchewan River Basin (SSRB) is one of Canada's larger watersheds (between 165,210 and 167,765 km^2, depending on the data source)[1] experiencing the widest variability of the natural characteristics of climate, geography and hydrology. The South Saskatchewan River joins with the North Saskatchewan River just east of Prince Albert, Saskatchewan, making the SSRB part of the larger Saskatchewan River Basin (406,000 km^2).[2] The geography varies from rugged Canadian Rocky Mountains to expansive prairie grasslands and boreal forest. The climate differs across the basin and exhibits wide seasonal fluctuations. The hydrology of the basin is distinguished by relatively few major rivers that flow eastward to extensive areas of flat topography dry-land interspersed with ponds, wetlands and lakes, in what is sometimes referred to as *prairie pothole* topography. The major rivers within the SSRB are the Bow, Oldman, Red Deer and South Saskatchewan Rivers. The South Saskatchewan River (SSR) forms at the confluence of the Bow and Oldman Rivers approximately 70 km east of Medicine Hat, Alberta (Figures 2 and 3), and the Red Deer River joins the SSR just east of the Alberta-Saskatchewan provincial border.

Due to its very flat, poorly drained topography and the variable climate, the Prairies are naturally prone to floods and droughts. The exposure to highly variable runoff conditions in the Prairie portion of the SSRB affects the basin's available water supply and increases dependence on the more reliable Rocky Mountain precipitation as a source of water. Because surface water in the SSRB is derived largely from snowmelt, it is highly sensitive to climate variability and the SSRB is therefore one of Canada's watersheds that is most vulnerable to drought. The relationships among the natural characteristics of the climate, geography, and hydrology of the South Saskatchewan River Basin are described in this paper.

The paper begins with an overview of the basin's climate and geography, presents an analysis of the average annual natural and naturalized (i.e., without human intervention) flow of the rivers within the basin, describes the unique hydrologic characteristics of the Prairie Ecozone Region, and provides an overview of surface and ground water quality and contamination threats. We conclude with a section that identifies some of the region's historic exposures to drought, caused by the highly variable characteristics of the South Saskatchewan River Basin.

Methodology for Naturalized Flow Determination and Data Sources

As a basis of the description and analysis of climate, geography, and hydrology several sources and analysis techniques were utilized. Climate data were obtained from Environment Canada's climate archive. Geographic data were obtained from the EcoAtlas of Canada, the Water Survey of Canada (WSC) within Environment Canada, as well as Agriculture and Agri-Food Canada's Prairie Farm Rehabilitation Administration (PFRA).

Surface water hydrology data were obtained from the WSC database, HYDAT. WSC measures water levels and computes flow at numerous hydrometric gauging sites. The flow at any WSC gauge can be *natural flow* or *non-regulated flow* where there are no diversions or human use upstream of the gauge. As the flow progresses downstream, there can be more influence on the flow by human use, and the resulting flow is termed *regulated flow*. To understand water flows within the SSRB watershed without human intervention, hydrologic data was analyzed for the South Saskatchewan River, and its contributing streams and rivers. Alberta Environment and Saskatchewan Watershed Authority (SWA) have undertaken the analysis to provide the *naturalized flows* by removing the effect of humans on the natural water supplies. The determination of naturalized flows provides a better understanding of the natural hydrological characteristics both temporally and spatially within the basin. This information guides governments, water managers and water users in making decisions on water allocations and distribution for human needs and maintaining flows to preserve aquatic ecosystems. Naturalized flows were provided in a weekly database and determined through calculations that consider the known water diversions for human uses, the influence of evaporation from any constructed reservoirs and the flow that is returned to the rivers after human uses. Average daily, monthly and annual naturalized flows were computed from the Alberta Environment and SWA naturalized flow databases for the period from 1961 to 1990, a commonly-used baseline hydrologic record.

To more fully understand the relative contribution of different portions of the basin to total flow, an estimate was made of the flow contributed by the Montane-Boreal and the Prairie Ecozone portions of the SSRB. The EcoAtlas of Canada provides a map of the ecozones and this information was used to calculate ecozone areas within the basin. The flow contributions from these ecozone areas were determined by selecting stream gauges near the borders of the ecozones and using the flow and contributing areas for each of these gauges to calculate total flows and normalized flows (average annual flow volume expressed as mm over the area of the basin). The total contributing area above the Lake Diefenbaker Reservoir, as given in the HYDAT database was used to quantify the runoff from the contributing areas in the Prairie Ecozone.

Climate

Precipitation and temperature are the key characteristics of climate in any given region. The potential for evaporative demand, or evapotranspirative demand when plant growth is present, is termed *potential evapotranspiration* (PET), and is a function of temperature and humidity of the air. PET is the amount of water that would either evaporate from the soil surface, or be used by plant growth in an environment with no water limitations. Those areas where the PET exceeds the precipitation (P) suffer from a deficit in precipitation, of which there are various classes. Within the SSRB, the environment is classified as semi-arid, dry-subhumid, sub-humid, or humid with the following definitions[3]:

- Semi-arid when P is less than ½ PET ($P<0.5PET$);
- Dry-subhumid when P ranges from 0.5 to 0.6PET;
- Subhumid when P ranges from 0.6 to 1.0PET;
- Humid when P is more than PET ($P>PET$).

The largely semi-arid climate of the SSRB reflects the location of the basin in the rain shadow of the Rocky Mountains in the continental interior.

Of Canada's regions with extensive weather monitoring, the central prairies, which includes the SSRB, has the most variability of natural precipitation from year to year and from place to place. An analysis of precipitation by Longley shows the coefficient of variation (standard deviation/mean) was greater than 25% during the period from 1900–1950.[4] Total annual precipitation ranges from less than 300 mm on the plains to 1,200 mm in the mountains (Figure 3). Climate data from Environment Canada's 1961–1990 climate normals shows the percentage of precipitation that falls as snow ranges from 24% at Red Deer to 40% at Banff (Table 1).

The mean annual temperature ranges from approximately 2° C to 6° C with extreme differences between maximum summer temperatures (+42° C) to minimum winter temperatures (-51° C) (Table 1).

Potential evapotranspiration (PET), estimated using the Penman method, also varies accordingly, ranging from approximately 450 mm to over 1,000 mm.[5] Figure 4 is a map of the *aridity index*, or P/PET. The largest area of the basin is classified as semi-arid, with potential evapotranspiration more than double the available precipitation. These natural conditions are key factors affecting water scarcity in the South Saskatchewan River Basin, impacting both ecological and human water needs.

Geography, as Described by Ecozones

Associated with the diverse physiography and climate of the SSRB, there are three distinct Ecozones (Figure 5), each with a different hydroclimate and role in the regional water balance.[6] Most of the flow in the rivers of the SSRB is generated in the **Montane Cordillera Ecozone** from the melt of snow and ice. This ecozone represents approximately 10% of the entire area of the basin.[7] In this rugged mountain landscape of folded and faulted sedimentary rock, temperature and precipitation vary with elevation and exposure to sun and wind. Annual precipitation ranges from a high of 1,200 mm in the upper elevations, to less than 500 mm in the lower elevations in the rain shadow of the mountains. Vegetation is stratified according to elevation. The Montane low- to mid-elevation (400–1,500 m) forests are dominated by lodgepole pine, white spruce, aspen, poplar and Douglas fir. The dominant trees in the subalpine forest are Engelmann spruce and subalpine fir. Above the limit of tree growth, herbaceous alpine occurs in places underlain by permafrost. At high elevations, Engelmann spruce and alpine fir make up the old growth forest.[8]

The **Boreal Plains Ecozone**, occupies a small proportion (approximately 8%) of the SSRB and is located mainly west and north of Calgary, with a small portion at the eastern end of the SSRB in central Saskatchewan where the

North and South Saskatchewan Rivers join to form one river.[9] In the Boreal Plains the mean annual precipitation ranges from 450 to 650 mm. The growing season precipitation (June, July and August) is approximately 200 mm, and the mean summer temperature is approximately 15° C. The land cover is a mix of agricultural land, aspen forest, mixed wood forest (white spruce, aspen and balsam poplar), jack pine forest, lodgepole pine, grasslands, peat lands and boreal wetlands. Soils include Black and Dark Gray Chernozems, Gray and Gleyed Luvisols, Brunisolic and Gleysolic soils, ideally suited for fertile farmland that produces cereal crops, oilseeds, and hay.

The **Prairie Ecozone** occupies 80% of the SSRB,[10] covering an area of about 135,000 km^2. The mean annual precipitation in the Prairie Ecozone ranges from 300 to 450 mm and the growing season precipitation ranges from 160 to 240 mm. Mean annual temperature ranges from 1.8 to 4° C with summer growing temperatures averaging 17° C. The soils are primarily Brown, Dark Brown, and Black Chernozemic with a thick zone of A-horizon soil (topsoil) from the accumulation of organic matter.[11] While the natural vegetation in the Prairie Ecozone is a mainly grasslands with some aspen forest and shrub cover, suppression of natural fire hazards has caused the encroachment of aspen and shrubs into the grasslands.

The Prairie Ecozone supports some of the most productive agricultural land in Canada, growing spring wheat, cereal grains and oilseed food crops, and hay as a feed crop for livestock. Livestock production also relies on grazing of natural pasture lands. John Palliser, one of the first European explorers to the Canadian Prairies around the southern Alberta-Saskatchewan border, defined what is now referred to as the Palliser Triangle (Figure 5). Palliser described the area as "desert or semi-desert in character, which can never be expected to become occupied by settlers."[12] However, his exploration took place during an extensive multi-year period of drought in the late 1850s. Palliser did not understand the range of climate variability of the Prairies, nor the unique characteristics of the Chernozemic soils in the Prairies which would actually prove to be assets that would support the development of agriculture in the basin. The Chernozemic soils of the SSRB are a product of the clay-rich glacial deposits that cover most of the region. Together with the climate and the native land cover of the Prairies these soils play a vital role in the natural environment. Chernozemic soils develop primarily in areas such as grasslands where approximately 50% of the biomass is below the surface, dying off each winter. This decaying biomass contributes to a deep soil profile, often exceeding one meter, with high organic content and water holding capacity. The top one meter of the soil profile can store between 100 and 300 mm of water for plant growth. Thus, Chernozemic soils have a large water-holding capacity and if sufficiently recharged, can retain moisture available for plants, even during hot dry spells in the summer, and sometimes through to the next growing season. These soils are particularly suitable for selected agricultural crops.

Agriculture in the Prairies has successfully been adapted to utilize the natural available precipitation from the moisture retained in the soil. Dryland (non-irrigated) agricultural practices rely on understanding and using practices that take advantage of the *vertical water balance* (infiltration of precipitation that results in retained soil moisture against evaporative demand). Such practices include the planting of drought tolerant crops on suitable soils, and utilizing conservation operations that retain soil organic matter and preserve soil moisture.[13] And yet while dryland agriculture has proven to be successful within the SSRB, it will always remain at risk from its exposure to natural climate variability in the form of droughts and floods.

Surface Water Hydrology of the South Saskatchewan River Basin

Natural Flow in the Major River System

The surface water flows of the SSRB are defined by topography and climate, and are influenced by landscape, vegetation and human activities (e.g., water diversions). There are four major sub-basins that effectively contribute water flow into the South Saskatchewan River; the Bow River, the Oldman River, the Red Deer River, and the South Saskatchewan River sub-basins together form the South Saskatchewan River Basin. Some secondary sub-basins (i.e., Bigstick Lake and Seven Persons Creek in Figure 2) contribute flow into the South Saskatchewan River system primarily during wet years while Swift Current Creek generates flow into the South Saskatchewan River in most years. Presently, there is little water development demand on the Red Deer River and demand is about 20–30% of the runoff of the sub-basin. However, water demand within the Red Deer sub-basin is likely to increase in the future a result of increased population, industrial development, and irrigation. There is substantial demand due to significant water development or water diversions (primarily for irrigation) on both the Oldman and the Bow River systems in Alberta, and some significant diversions from Lake Diefenbaker in Saskatchewan.

Natural flows were analyzed by removing human intervention for the major river system of the SSRB. The naturalized flow hydrographs for each of the sub-basins is presented in Figure 6. The annual hydrograph pattern for the river in each sub-basin has a similar pattern. Precipitation falling as snow in the colder winter months is stored, and does not normally produce stream flow until after the spring melt season (exceptions to this pattern result when melting occurs during warmer than normal winters or with rain on snow events). Spring snowmelt produces what is referred to as a spring *freshet;* water runoff from the melting snow is the most significant event of the hydrological year and is recorded as a large peak flow for a relatively short period of time during late spring and early summer in the hydrographs for each sub-basin.

The average peak flow rates during the spring freshet range from 120 cubic meters per second (cms) in the Red Deer sub-basin, to approximately 850 cms on the South Saskatchewan River flowing into Lake Diefenbaker. The average peak flow rate in the Oldman and Bow River sub-basins are approximately equal, ranging from 300 to 400 cms. The peak flow rate occurs slightly earlier in the Bow River (May to June) than it does at the mouth of the Oldman and at the flow into Lake Diefenbaker (June). The dual peak hydrograph is a unique feature exhibited only in the Red Deer sub-basin. The Red Deer basin has a small percentage of area at high elevation and the snowmelt from this area contributes to the June peak in the flow hydrograph while the earlier peak is runoff is from the significantly larger prairie region which peaks in April. Similarly, the higher elevation snow melt runoff is responsible for the June peaks in the hydrographs for the Oldman, Bow River and the South Saskatchewan (see the maximum and minimum hydrograph peaks).

Figure 6 also shows the range of flow rates between the first and third quartile flows, with the difference between both representing the range of variability of half of the naturalized flows. The variability between maximum and minimum flows indicates that the greatest range of contributions occurs principally during the period from March to July (there is a time lag for the influence of mountain snowmelt). The variability in the periods not influenced by snowmelt is likely accounted for by rain and soil moisture conditions (e.g., precipitation infiltration into dry soils means less runoff contribution into the rivers).

Another measurement of variability in the naturalized flow is the difference between minimum and maximum flows, in comparison to average flows (Table 2 and Figure 6). The average annual flow volume on the Red Deer ranges from 0.79 to 3.58 million dam^3/year. The range is 2.79 million dam^3/year, which is over 170% or nearly twice the average annual flow. The Oldman River ranges from 1.55 million dam^3/year to 5.10 million dam^3/year, a difference of 3.55 million dam^3 (108% of the average annual flow). The annual flow volume on the Bow River is the least variable, ranging from 2.68 to 5.17 million dam^3/year, a difference of 2.49 million dam^3 (66% of the average annual flow) and represents the most predictable water source in the basin. The flow into the South Saskatchewan River at Lake Diefenbaker ranges from 5.11 to 13.00 million dam^3/year, a difference of 7.89 million dam^3 (92% of the average annual flow). Such a large range of flow means that in any given year water may be scarce or abundant, which clearly impacts both human and ecological needs. These large natural variabilities in river flows affect the aquatic and shoreline ecosystems, and also represent a significant challenge to water managers within the SSRB.

An analysis of the flow contribution and HYDAT areas from the Montane, Boreal, and Prairie Ecozones was performed. The Oldman, Bow, and Red Deer River flows originating from the Montane and Boreal Ecozone portions of the watersheds are critical for surface water flow, supplying on average about 77% of the total flow of the South Saskatchewan River into Diefenbaker Lake (Table 3). These flows originate from 25,000 km^2 (Table 3). By contrast the prairie portion (111,000 km^2) of the SSRB supplies only 23% of the flow (Table 3), although its area is four times as large. In the prairie region most of the moisture from precipitation is absorbed by the soil, returns to the atmosphere at the same location by evapotranspiration or recharges groundwater supplies. The very different hydrology of these two portions of the SSRB can also be described through the contrast in *mean annual runoff.* Mean annual runoff is calculated as a depth of water over an area; for the humid Montane-Boreal portion the mean annual runoff is 267 mm while the dry prairie zone has only 18 mm (Table 3). The concept of mountains as water towers supplying most of the water in the basin's river system is certainly valid for the SSRB.

The glaciers also have a role to play in water flows for the SSRB watershed, particularly in the fall months, the transition time when river flows are at their minimum. Glaciers can be viewed as large storage reservoirs which capture snow and store it in the form of ice, eventually releasing it as meltwater at the glacier toe. The flows originating from the glaciers represent the annual input of snow and rain to the glaciers (i.e., a portion of the annual precipitation input into the mountain region) combined with the wastage of ice from the glaciers (i.e., the long-term melting back of the shrinking glaciers). Recent research shows the largest contribution of the glacier wastage to the river flow occurs during the lower flow period from July to September inclusive, contributing an average of 4.8% of the Bow River flow at Banff for the period 1975 to 1998.[14] When weighted as a percentage of the annual natural flow, the contribution from the glaciers for the flow at Banff is 2.2% annually. The Bow River is typical for glacier impacted rivers in the SSRB; glacier wastage contribution to the SSRB ranged from below 1% to 4% annually and from 1% to 12% in the July to September period. While the glaciers wastage do not supply a large annual percentage of flow in the main rivers, they are important for late-summer flows in some headwater streams. Melt is the larger component of glacier runoff and contribution to streamflow. Of the total average annual 1975–1998 flows at Calgary, 6.2% is from the melt of accumulated snow and precipitation on glaciers while 2.8% is from wastage of glaciers for a total of 9% glacier contribution to flow.

Historic Trends in Stream Flow

Water supplies in any semi-arid region attract much attention, and there has been significant research to understand the trends in stream flow over the past century for the South Saskatchewan River Basin. Data collected over roughly the last 100 years reveals a negative trend, or declining average annual flows for the SSRB caused by increases in annual air temperature.[15] Declining flow trends for the SSRB are consistent for stream flow trends in western Canada below the 60° parallel.[16] For instance, over the 88 years (1911–2002) of recorded flow of the Bow River at Banff, the mean annual flow has declined an average of 0.106% per year.[17] The declining trends in stream flow are echoed by the water level records for closed-basin prairie lakes, which show a long-term declining trend throughout most of the 20th century.[18]

Seasonal changes in flow trends are also occurring in prairie streams. For example, the spring snowmelt is occurring earlier with decreasing peak flow measurements. These trends are attributed to a combination of reductions in snowfall and increases in temperatures during the winter months.[19] The March 1 to October 31 runoff volumes are decreasing, as are the annual and minimum daily flows. A decrease in minimum daily flows will have an effect on water quality. For example, waste water discharged into rivers will form a larger percentage of the flow when a river is under low flow conditions. River water temperatures have also been shown to be higher when the rivers are under low flow conditions, and are likely to have some effects on the aquatic ecosystems.

Surface Water Hydrology of the Prairie Ecozone Region

Runoff from the Prairie Ecozone is small because most of the precipitation in this region is absorbed by the soil and returned to the atmosphere by evaporation and transpiration from vegetation or recharged ground water supplies. The contribution of the area to flow in the main rivers is further reduced because much of the area drains into closed basins. The hydrological concept of a landscape that does not contribute to flow in the rivers was originally presented in 1957 by Stichling and Blackwell[20] who noted that the *effective drainage area* of a basin on the Canadian Prairies fluctuated by year, season, and hydrologic event due to the glacier-formed landscape and the climate. In their work with the Prairie Farm Rehabilitation Administration (PFRA) in 1975, Godwin and Martin[21] further refined these concepts to delineate drainage basin areas by defining the *gross drainage area* of a stream to be enclosed by the height of land defining the drainage. They defined the *effective drainage area* of a stream to be that portion of a drainage basin which might be expected to contribute runoff to the stream during a runoff event with a return period of two years or more, i.e., every second year on average. The *non-contributing area* of any basin is the area between the gross drainage area and the effective drainage area. Two sources of areal data, the PFRA and the EcoAtlas of Canada, quantify the *non-contributing area* of the SSRB at 49%[22] and 43%[23] respectively.

The Water Survey of Canada HYDAT database gives the effective drainage area of the South Saskatchewan drainage basin above Gardiner Dam to be 87,000 km² or 64% of the 136,000 km² gross drainage basin area above the dam. The non-contributing HYDAT drainage area is therefore about 49,000 km² (136,000 km² less 87,000 km² or 36% of the gross drainage basin area). As indicated in Figure 7, the non-contributing areas primarily lie within the 111,000 km² Prairie Ecozone portion of the SSRB, where the potential evapotranspiration exceeds the precipitation. Much of this non-contributing portion of the SSRB rarely or never contributes flow to the main rivers, and the area is marked by the occurrence of highly saline lakes. The average annual runoff to the main rivers from the entire 111,000 km² Prairie Ecozone portion of the SSRB is about 18 mm (Table 3), which translates to 32 mm from the effective or contributing area of 62,000 km² (Table 3).

A generic water balance for the Prairie Ecozone portion of the SSRB can be estimated. The average annual rainfall and snowfall are about 280 and 105 mm respectively (calculated from Table 1), for a total precipitation of 385 mm. Most surface runoff from uplands (i.e., the dry land outside wetlands and lakes) occurs during the snowmelt season when the ground is still frozen (limiting infiltration), but localized surface runoff occurs occasionally in summer after high-intensity rainfall events. The mean annual runoff from the uplands, as indicated by the 32 mm (Table 3) average runoff from the effective drainage area or contributing area, is about 30 to 40 mm, including a small proportion of groundwater flow. About 75% of the surface runoff occurs during snowmelt, the other 25% from summer rainstorms. Most of the moisture available from snowmelt and rain events infiltrates and is temporarily stored in the soil but is subsequently lost to evaporation and transpiration. A simple water balance consideration implies that actual evapotranspiration from the uplands is equal to precipitation minus runoff, and amounts to about 385—40 = 345 mm per year on average. In other words, evapotranspiration is the dominant water loss at 345 mm in the prairie zone, approximately ten times bigger than the 32 mm of runoff. However, compared to the potential evapotranspiration of 600 to 1,000 mm per year, the actual evapotranspiration of 345 mm per year is still much smaller because it is limited by the available moisture. In the non-contributing areas of the SSRB the runoff does not flow to the main rivers, but collects in wetlands and lakes. These remain wet all summer and therefore lose water at potential evapotranspiration rates, thus consuming the runoff from the uplands. In wet years some of these wetlands and lakes overflow, thus enlarging the effective contributing area in such years.

The amount of runoff is very sensitive to the conditions of the soil, vegetation, and climate at any site and can vary drastically from year to year, whereas the potential evapotranspiration is fairly constant. This is why the annual spring pond counts for the prairie region, carried out since 1955 by the Canadian Wildlife Service and the U.S. Fish and Wildlife Service, show large fluctuations in pond numbers from less than 2 million in dry periods to over 6 million in wet periods.[24] Changes in agricultural practices, such as the ongoing change from conventional tillage (plowing land with tills) to conservation tillage (reduced or no-tillage practices), coupled with continuous cropping, can result in increased infiltration and reduced surface runoff and therefore may have large impacts on soil moisture, wetlands, and lakes. As the total area of wetlands and lakes is very sensitive to changes in runoff, whether due to changes in land-use or climatic variability, the conversion from cultivated land to permanent grassland can also result in a dramatic decrease of water input to wetlands.[25]

Groundwater of the Prairie Ecozone Region

In the Prairie Ecozone region of the SSRB, groundwater recharge occurs mostly beneath upland depressions where enough runoff water collects to maintain a water level above ground surface in low lying ponds or depressional areas for most or all of the summer season. Groundwater recharge beneath non-depressional uplands is very small because the water-holding capacity of the clay-rich soils means virtually all infiltrated water is held in the soil and returned to the atmosphere by evapotranspiration. Significant recharge beneath uplands occurs only in areas with surficial sands, where up to 50 mm of annual recharge may occur. Recharge to deeper aquifers, which are confined by layers of clay and clay-rich glacial till, is limited by the low permeability of the confining layers. Average annual recharge to such aquifers can vary from practically 0 mm for very deeply confined aquifers, to as much as 40 mm for shallow aquifers.[26] Groundwater discharge occurs to streams, lakes, wetlands, and springs. Where the water table is within 3 or 4 m of the ground surface, the groundwater is available to deep-rooted vegetation and can be lost to the atmosphere by transpiration from the vegetation.

Most of the groundwater supply wells in the prairies extract water from groundwater reserves less than 200 m depth below the ground surface.[27] Deeper permeable zones are generally isolated from the hydrologic cycle and generally contain saline waters. Most groundwater that supplies water wells in the SSRB is recharged within at most a few tens of kilometers of distance from the wells. Total groundwater use in the basin, about 50 million m^3 per year,[28] is equivalent to an extraction rate of about 0.4 mm per year over the area of the Prairie Ecozone. This low usage rate implies that groundwater use could be significantly expanded; however, groundwater discharge is also important to maintain flow from springs and water levels in wetlands and streams. Locally, heavy groundwater usage from deep confined aquifers can exceed the recharge rate and be unsustainable in the longer term. Sustainable groundwater withdrawal rates are generally too low to support any potentially significant irrigation demands, even in the rare cases where the groundwater quality is suitable for irrigation, as described in the next section.

Deep confined aquifers can yield large volumes of water for periods of up to several years. However, recharge rates are very low, and depleted aquifers may require tens to hundreds of years or even more to recover after pumping ceases. Deep aquifers can be considered as large storage reservoirs; as an adaptation strategy, these deep reservoirs could be used during droughts to augment surface water supplies for short-term or emergency uses, to increase resiliency and reduce the risks caused by drought. A prime example of such use of a deep aquifer is the case of the Estevan buried valley aquifer in south-east Saskatchewan which was pumped at a high rate from 1988 to 1994 to supply cooling water for thermal power plants. This aquifer is only slowly recovering, and will take decades to fully recover.[29] The use of deep groundwater aquifers as a reserve for mitigating drought effects will obviously depend on access (i.e., appropriately sized wells would need to be drilled), well yields, and water quality (treatment may be required to improve the natural chemical composition). While saline groundwater may not be economically treated for crops, short-term treatment strategies may very well be viable for industrial needs, domestic needs and select agricultural needs (e.g., livestock watering).

Water Quality in the SSRB

Surface Water Quality

Surface water in the main streams and rivers, and in most of the reservoirs of the SSRB, is derived from overland runoff and shallow groundwater and, therefore, generally originates as good quality water, with some contaminant characteristics from airborne or land use sources. In the Prairie Ecozones, the effects of evapotranspiration and groundwater discharge can increase the salt concentrations of some surface water bodies, depending on the geologic and hydrologic influences. In the non-contributing portions of the SSRB, bodies of water with no discharge are referred to as terminal lakes and wetlands; these are closed saline systems with very high concentrations of salts that have accumulated and concentrated over time by evaporation.

Water is a natural solvent that acquires chemical, biological, and physical characteristics from the natural geology, influences from environmental exposure and any pollutants or land-use contaminants. Water quality varies for each source by acquiring natural and anthropogenic characteristics. For example, surface water acquires characteristics from:

- land runoff: from both natural and developed land, including rural and urban sources of runoff, with varying sources of contaminants;
- wastewater return flows from human waste and industrial discharges;
- domestic and wildlife animal activities and waste impacts;
- internal aquatic life organisms;
- industrial and chemical sources of pollution by a variety of land, air and water mechanisms of movement across the landscape and within the water systems.

The surface water quality of the main streams and rivers in the SSRB can be characterized as fresh, soft water (low in calcium and magnesium hardness), containing a variety of constituents including organic matter, micro-organisms from natural and human sources (some of which are disease-causing), and low levels of dissolved minerals. An indicator of mineral content is the measurement of Total Dissolved Solids (TDS). Most good quality surface water sources (rivers and freshwater lakes) in the SSRB have concentrations ranging from 200 to 400 mg/ L TDS.[30] Poorer quality saline surface water bodies can have much higher concentrations.

Both Alberta[31] and Saskatchewan[32] have developed comprehensive surface water quality objectives. The Canadian Environmental Quality Guidelines are established by the federal and provincial Canadian Council of the Ministers of the Environment to protect aquatic ecosystems.[33] Alberta has applied a River Water Quality Index to assess the water quality in river systems.[34] This water quality assessment is based on concentrations of metals and ions, pesticides, nutrients, and bacteria. Portions of the Bow, Oldman and Red Deer rivers are classified as "fair," depending on location, while most of the major river systems in the SSRB are classified as "good."[35]

Watershed health and water quality in the Saskatchewan portion of the South Saskatchewan River has been classified as "*stressed,*"[36] using the Canadian Water Quality Index indicator developed by Environment Canada's National Water Research Institute. This indicator rates water quality considering concentrations of nutrients, dissolved oxygen, dissolved ions, heavy metals, pesticides, and bacteria.

Even though water quality in the South Saskatchewan River is under stress, surface water quality within the SSR remains a good source of water for many uses, including drinking water (providing proper water treatment is used), recreation, sustaining aquatic ecosystems, industrial needs, power generation, and many other uses. Other surface water sources in the SSRB (streams, tributaries, ponds, lakes, and reservoirs) have a wide range of quality, varying from good quality water in water bodies that are regularly flushed with fresh runoff water, to poor quality (e.g., nutrient and organic-rich water in shallow, stagnant reservoirs, and highly saline waters in terminal lakes and wetlands). Such saline water bodies, while part of the natural ecosystem, are not desirable water sources for most water uses, including livestock watering or other agricultural needs, because water quality is too poor and water treatment is simply too costly.

Groundwater Quality

Groundwater aquifers are recharged by rain and snowmelt as water infiltrates into the subsurface. In the Prairie Ecozones, most groundwater recharge occurs in the spring season beneath depressions where snowmelt water collects and has sufficient time to saturate the soil and infiltrate into the ground. Groundwater moves very slowly and may be in contact with the geologic formations for extended time periods, from days to hundreds or thousands of years, and its characteristics naturally change as water acquires the characteristics of the geology with which it is in contact.

Whether shallow or deep, groundwater acquires the biological, chemical and physical characteristics of the constituents the water is exposed to in the sub-surface geology. Groundwater chemistry in the glacial deposits that cover most of the SSRB tends to be hard (high in calcium, magnesium) and high in sulphate, reflecting the high sulphur and carbonate content of the weathered and oxidized glacial tills through which most of the groundwater flows on its way to recharging the deeper aquifers. In surficial sand aquifers water quality can be very good if the sand contains little reactive material. Groundwater in the deeper bedrock aquifers beneath the glacial deposits tends to be soft (low in magnesium) but high in sodium, bicarbonate, and chloride.

For the SSRB, most ground water tends to be highly mineralized. The total dissolved solids (TDS) may range from less than 100 to over 100,000 mg/L. As a reference, ocean seawater is about 33,000 mg/L TDS. In Saskatchewan, the majority of wells are less than 200 m deep and have total concentrations that range from 500 to 3,000 mg/L TDS; about 52% of wells have TDS in excess of 1,500 mg/L, the provincial aesthetic objective.[37] Common groundwater quality problems may include a variety of inorganic and organic matter (e.g., calcium/magnesium hardness, iron, manganese, arsenic, ammonia, sodium, sulphate, radon, uranium, dissolved organic carbon).

Groundwater in the prairie portion of the SSRB is generally not suitable for use on irrigated crops, because the water is too salty for plants, and could deteriorate soil quality. Likewise, the mineralized nature of the groundwater in the SSRB makes this water challenging to use for drinking or household water as it often requires extensive treatment processes.

In a Saskatchewan study, 99.6% of private rural wells exceeded one or more health or aesthetic guidelines,[38] and 35% exceeded health-related guidelines.[39] In a similar Alberta study, 93% of rural supplies exceeded one or more aesthetic guidelines, with 35% exceeding health guidelines.[40] Where health-related guidelines are exceeded (e.g., arsenic, nitrates, disease-causing micro-organisms, and other contaminants from both natural and anthropogenic influences), treatment is essential to ensure water is safe for drinking. As noted above, the exceedance of drinking water quality health guidelines tends to be in the 35% range. The groundwater quality in the SSRB is consistent with national estimates, which show that an estimated 20% to 40% of rural wells across Canada have nitrate or coliform bacteria concentrations exceeding drinking water guidelines.[41]

Threats to Water Quality

In an assessment by Environment Canada, common threats to water sources are identified as pathogens, algal toxins, pesticides, persistent organic pollutants and mercury, endocrine disrupting substances, nutrients, acidification, genetically modified organisms, municipal wastewater, industrial discharges, urban runoff, landfills and waste, agriculture and forestry, trace elements from natural sources and impacts from dams, diversions, and climate change.[42] All these threats to water quality exist in the SSRB and are related to land use concerns and human activities. Human activities (e.g., land use, industrial and urban development, waste management) must be properly managed in any watershed to sustain water quality and quantity for ecological and human needs.

Surface water quality in the SSRB is good from the perspective that most water sources are not impaired by pollution; however, selected locations do show signs of stress as reported in the Alberta River Water Quality Index;[43] for example, bacteria concentrations are rated from poor to fair on the Oldman River, fair on the Red Deer River and Bow River at Ronatane (30 km east of Calgary), but are good on the South Saskatchewan River upstream of Medicine Hat. Pesticide concentrations are fair on the Bow River at Ronatane, but are rated good to excellent in other locations in Alberta. Nutrient concentrations are rated fair on most recorded locations of the Bow, Red Deer, Oldman and South Saskatchewan Rivers in Alberta, while metal concentrations are rated good to excellent.

Climate Variability and Drought—the Past as a Guide for Future Adaptation in the SSRB

In Canada, drought has been defined as "disruptions to an expected precipitation pattern and can be intensified by anomalously high temperatures that increase evaporation,"[44] In the United States drought has been defined as "a deficiency of precipitation over an extended period of time resulting in a water shortage for some activity, group, or environmental sector."[45] Droughts are classified[46] according to the types of impacts:

- *Meteorological:* precipitation departure from normal;
- *Agricultural:* soil and irrigation water deficits (for specific crops);
- *Hydrological:* effects on surface or subsurface water supply; and
- *Socio-economic:* supply and demand of some economic good or service (e.g., water, hydroelectric power) is affected by precipitation shortages.

Drought has a continuum, beginning with meteorological drought—a short-term precipitation deficit. When meteorological drought extends for more than one growing season, agricultural drought follows as plant growth draws the remaining water storage reserves from the soil profile. When drought extends beyond one or two years, hydrological drought develops, as surface water bodies and subsurface groundwater reservoirs are depleted. Socio-economic droughts may occur at any stage of the drought continuum, with varying degrees of impact to people and the economy.

Droughts are also classified by specific thresholds of:

- *Intensity:* degree of precipitation shortfall and/or severity of impacts;
- *Duration:* usually from a minimum of three months to years;
- *Spatial extent:* synchronicity among basins.

There is an important distinction between drought, a recurring feature of climate variability not limited to regions of low precipitation areas, and aridity, the dominant climatic characteristic of regions of low rainfall.[47] Seasonal water deficits occur throughout Canada, particularly in the southern prairies where drought can persist for seasons to years with ecological and socio-economic consequences. The natural grassland ecoregions that characterize much of the SSRB evolved in response to a semi-arid climate, but also to frequent, and occasionally severe and prolonged drought. The most notorious historical drought occurred over an extended period of time (years) during the 1920s and 30s, and caused the dust bowl of the "Dirty Thirties." Large areas of the SSRB were depopulated and the rural population has still never recovered to pre-1930s levels.[48] Research shows prairie drought is Canada's most costly natural hazard.[49] For example, recent droughts did not have the social or ecological consequences of the 1930s, but there was a significant economic impact. The cost of the drought in 1988–89 was estimated at $2.5 billion.[50] The Canadian drought of 2001–02 affected more land than the 1931 drought and caused a $3.6 billion drop in Canadian agriculture, a $5.8 billion drop in Canada's Gross Domestic Product, and 41,000 job losses.[51] In spite of these losses, the impact of the 2001–02 drought was not as severe on the landscape, largely because the drought did not last longer than two years, and many adaptations implemented since the 1930s helped safeguard the prairie and its population from more severe long-term impacts (e.g., improved agricultural practices such as conservation tillage reduced the effect of soil loss from wind; water development, storage reservoirs and water distribution infrastructure also ensured dependable supplies for most people on farms and in communities).

Prolonged drought occurs when large-scale anomalies in atmospheric circulation patterns persist for months, seasons, or longer periods of time. There is a relationship between climate, the earth's surface and the influence of the ocean. *Teleconnections* have been identified between western Canadian climate to large-scale oscillations of ocean temperature and pressure. The term *teleconnection* refers to a recurring and persistent, large-scale pattern of pressure and circulation anomalies that span vast geographical areas. Examples of these include: the ENSO or El Niño Southern Oscillation Index, the PDO or Pacific Decadal Oscillation, the NAO (North Atlantic Oscillation), and the AMO (Atlantic Multidecadal Oscillation).[52] These teleconnections tend to be stronger in winter. Droughts can start anytime and persist through both warm and cold seasons, particularly with feedback interconnections between the earth's surface (e.g., low soil moisture) and the lower atmosphere. Anomalies in the atmospheric circulation patterns and the feedback process that produces drought tend to be especially persistent in the North American Great Plains.[53]

The Palmer Drought Severity Index (PDSI) uses temperature and rainfall data in a formula to determine the severity of drought. The Palmer Index is most effective in determining long-term drought—a matter of several months—and is not as good with short-term forecasts (a matter of weeks). It uses a '0' as a normal baseline, and drought is shown in terms of negative values: for example, minus 2 is moderate drought, minus 3 is severe drought, and minus 4 is extreme drought.

Table 4 lists the most severe and sustained prairie droughts of the past millennium.[54] These droughts are recorded in weather and wheat yield data, archived in historical documents, and inferred from climate proxy data. Records of lake sediment biology have revealed shifts in hydrologic regimes at centennial scales,[55] and shorter cycles are evident in moisture-sensitive tree-ring chronologies.[56] It is evident from Table 4 that drought is a naturally recurring phenomenon within the SSRB. Over the extended temporal record, it is also evident that severe and extreme drought is also recurrent. Yet, current water resource management practices and policies have assumed a stationary hydrological regime. Western water use, policy and management were established during a period of fairly stable and reliable water supplies as compared to the recent past and projected climate of the near future.[57] The proxy climate reconstructions indicate the last 100 years of instrumental records do not capture the full range of regional hydroclimate, and specifically do not capture droughts approaching or exceeding a decade in duration.[58] Thus the hydrologic flows recorded at stream gauges in recent decades cannot necessarily be regarded as representative of the long-term hydrology of the SSRB when considering both the pre-instrumental hydroclimate and the forecasted scenarios of global warming.

Global warming and climate change influences may very well affect the SSRB in the future. What may even be more significant is acquiring an understanding that the region is naturally exposed to wider ranges of climate and drought impacts, as shown by the extensive period of inferred past droughts identified in Table 4. If this past inference is accurate, the hydrologic record used to guide water management today may very well overestimate the water availability within the SSRB into the long-range future. Any severe or extreme drought will obviously have consequences to available water supplies, affecting both the quantity and quality of water within the basin.

Concluding Remarks

The South Saskatchewan River Basin is remarkable due to the wide variability it experiences in climate, geography and hydrology. The precipitation falling over the spatial extent of the watershed ranges from 1,200 mm in the western mountainous portion, to less than 300 mm in the central Prairie environment. Inter-annual variability is pronounced. When human intervention is removed from the recorded flows on the river system it is evident that flows experience a wide range from minimum to maximum annual flows. Over 75% of the naturalized surface water flow in the main rivers of the SSRB comes from the Rocky Mountains and foothills while the Prairie Ecozone contributes less than 25% of the naturalized streamflow into Lake Diefenbaker on the South Saskatchewan River. Both humans and ecosystems have and must continue to adapt to this wide ranging variability, the decreasing trends in annual streamflow and the earlier and decreasing volumes of the spring snowmelt event.

An analysis of the stream flow within the rivers of the SSRB does not tell the entire hydrological story. By analyzing the SSRB's effective drainage areas, it is estimated that about 40% to 50% of the basin does not contribute flow to the basin's rivers in average years. In these non-contributing areas of low relief and high aridity, the runoff water resulting from precipitation is stored locally in wetlands and lakes or in the ground as soil moisture or groundwater. The remaining 50% to 60% of the area of the SSRB contributes surface runoff flows to the main rivers. Across the prairie portion of the SSRB most of the water from precipitation is stored in the soil and thus supports crops and other vegetation. Understanding and managing the vertical water balance in the prairie portion of the basin is critical for agricultural and other users of water who are not able to depend on the main rivers for water supply.

In areas that are a remote distance from rivers, the variability in precipitation has a heightened impact ranging from drought to floods. The SSRB has a large reliance on groundwater, particularly in rural areas, to supply water for human and livestock needs. The groundwater can be highly mineralized, and may have some natural dissolved constituents, such as sodium or arsenic, that affect human health. Surface water supplies in the SSRB are affected by natural environmental exposure and land use activities. Climate-induced stress on surface supplies can be expected to affect surface water quality as well as water availability.

Drought and flooding are natural characteristics of the SSRB, and are a direct result of the natural climate variability and geography. By inference from the proxy hydrologic records of the last 1,000 years, the South Saskatchewan River Basin appears to be naturally and regularly exposed to drought of varying intensity (moderate to extreme) and duration (months to years). Water managers and agricultural producers in the SSRB have been relatively successful in adapting to the impact of drought on agriculture. Water management decisions are made with an understanding of assuming a certain degree of economic risk. Modern adaptations have no doubt equipped the region to cope with most of the vulnerabilities caused by the natural climate variability, except for the more extreme events. What may be problematic, is the possibility that current water management is based on a wetter than normal period of record in comparison to the last millennia. The true historic variability of the SSRB may be wider than the current understanding. When coupled with global warming scenarios, society and water managers in the SSRB will need to reassess whether or not to adopt a new degree of risk to cope with future vulnerabilities—higher temperatures, greater precipitation variability, and changing temporal occurrence of hydrologic events. Designing water management policies to address a wider range of climate and hydrologic variability in the basin will place decision-makers in a different position. Droughts and floods typically tend to be viewed as short-term catastrophic events with short-term responses. If the past is an accurate guide, more severe and longer duration droughts within the basin may be repeated into the future with greater social and economic impacts. In the event of future long-term drought, society may very well be challenged to consider new approaches and longer-term responses. The ability to address greater climate variability will require longer-term planning and even more significant adaptations to climate impacts on water availability in the South Saskatchewan River Basin.

Acknowledgements

The authors of this paper wish to thank the editors of this journal for the opportunity to participate in documenting the South Saskatchewan River Basin. Two independent and anonymous reviewers are acknowledged for their contributions. Others that provided significant feedback are Brent Paterson (Alberta Irrigation), Sal Figluzzi (Alberta Environment), Gord Bell (PFRA, Saskatoon), Elwood Scott (PFRA, Saskatoon) and Brian Bell (PFRA, Calgary); their comments and discussions were instrumental in strengthening this document.

Figures and Tables

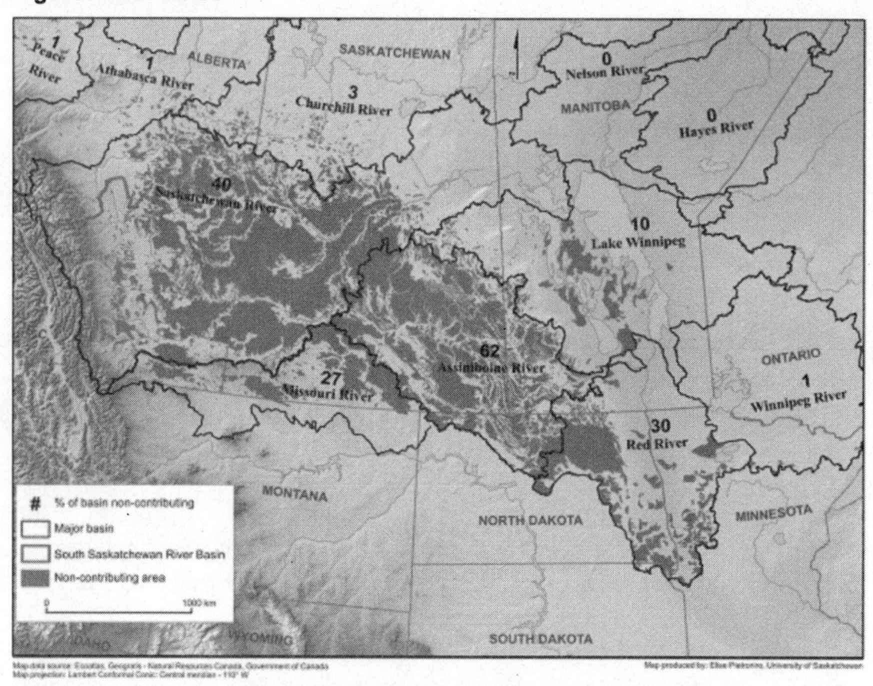

Figure 1. The drainage basins of the Prairie Provinces (source: PFRA). The brown area is non-contributing; that is, it is poorly drained and consists of dry areas contributing runoff to the main rivers only in wet years. The numerical values are the percentage of poorly drained area: for example, 40% of the entire Saskatchewan River Basin is poorly drained.

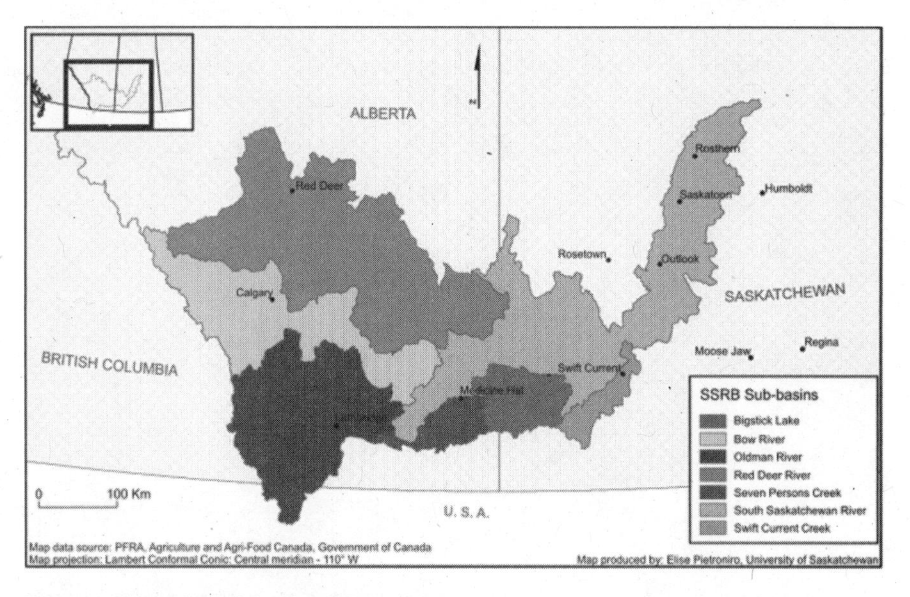

Figure 2. The major sub-basins of the SSRB.

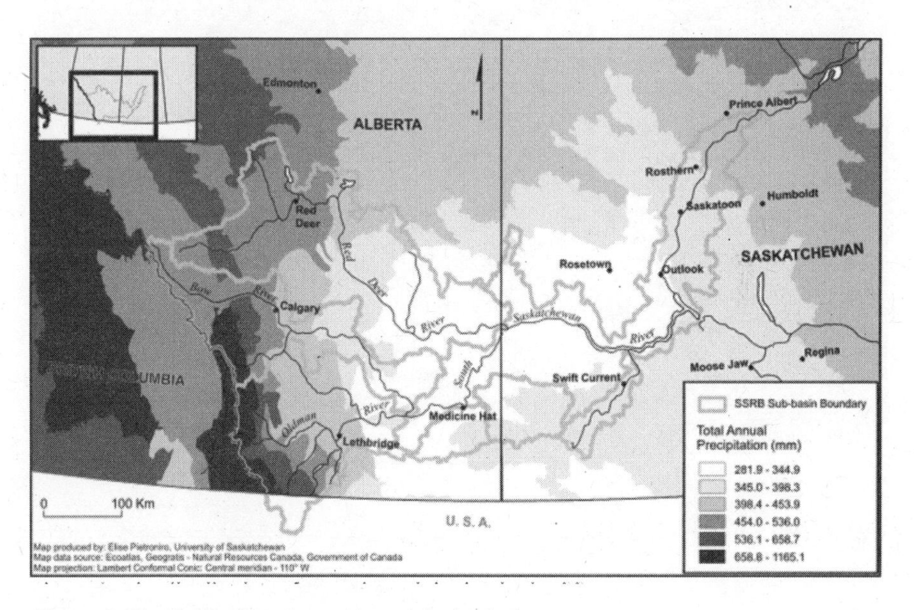

Figure 3. The distribution of annual precipitation in the SSRB.

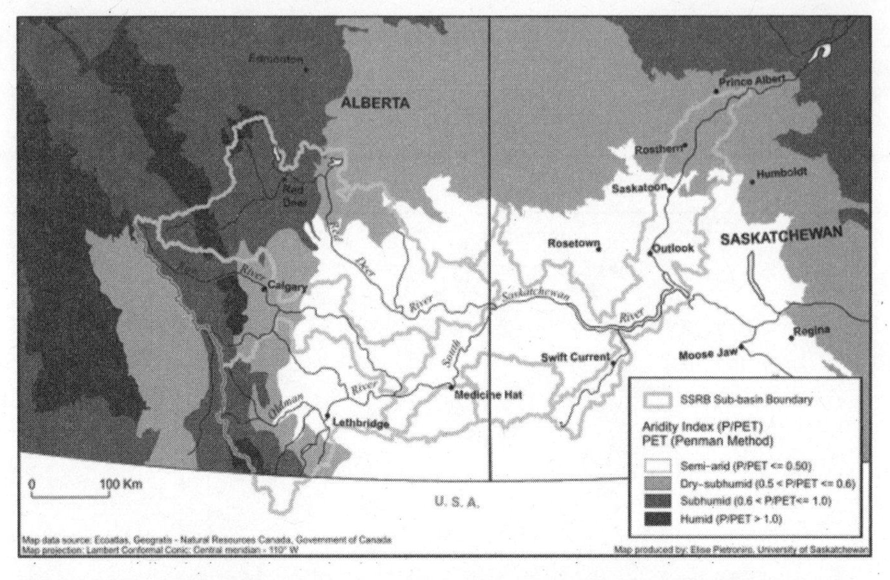

Figure 4. Aridity index (P/PET) in the SSRB. PET was estimated using the Penman method.

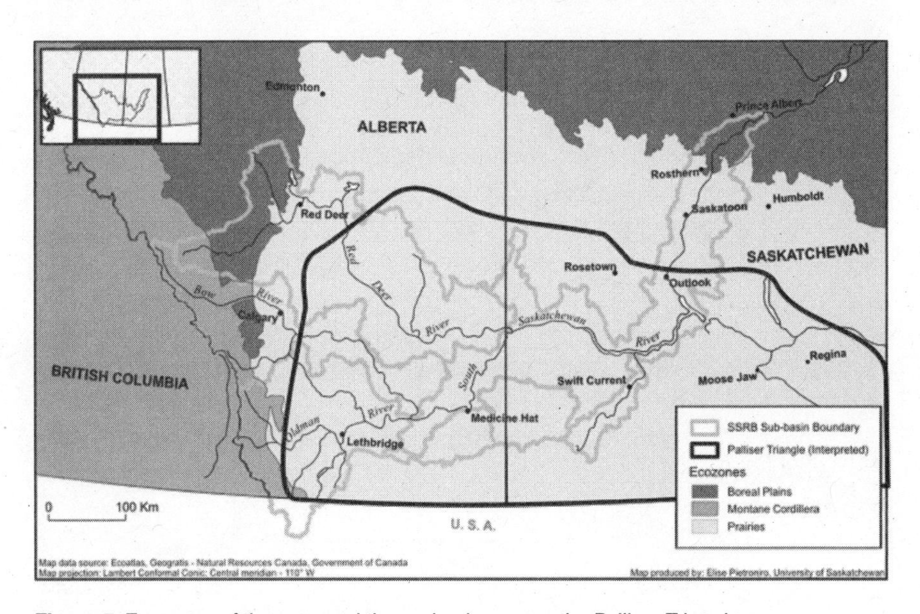

Figure 5. Ecozones of the SSRB and the region known as the Palliser Triangle.

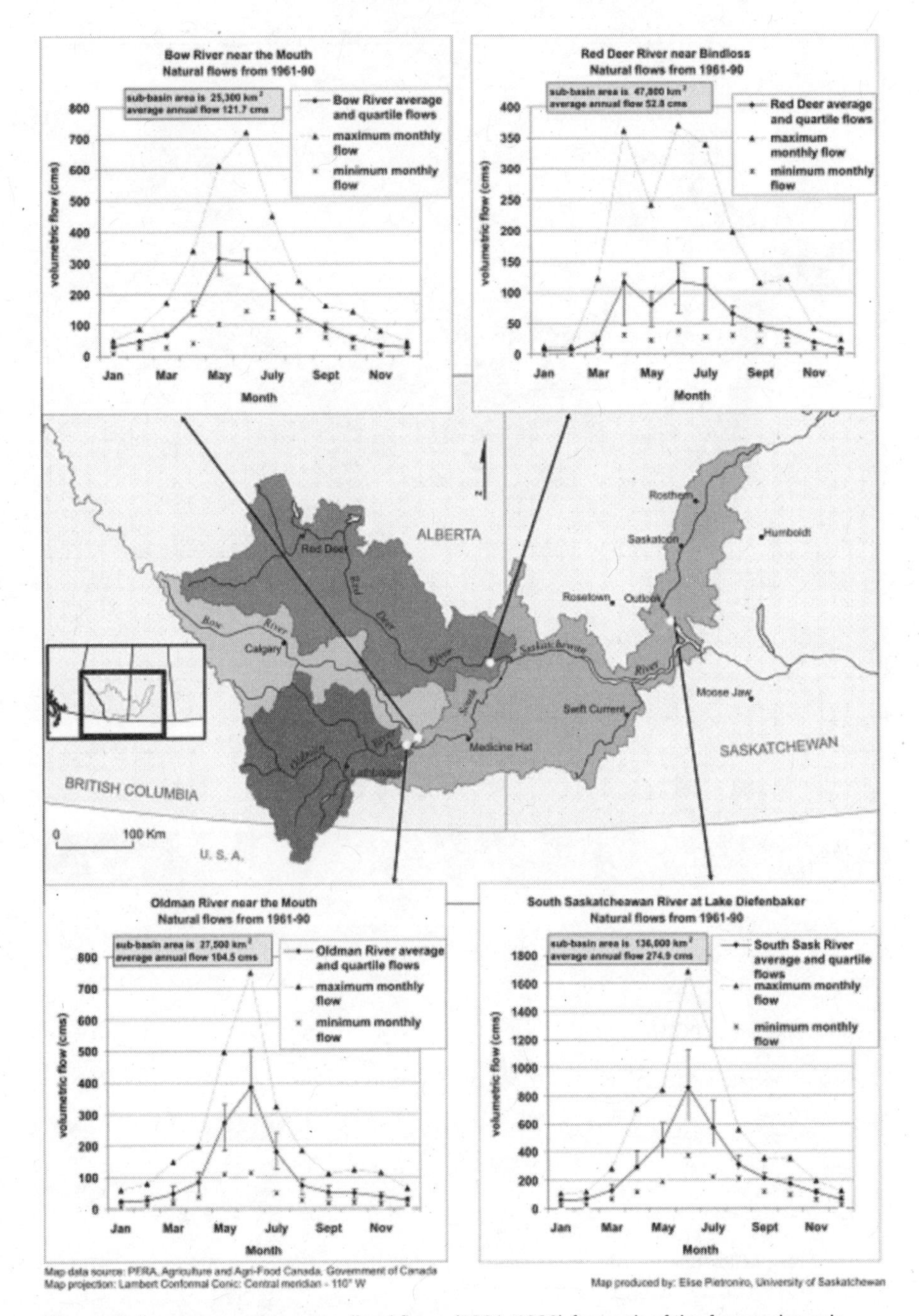

Figure 6. Average monthly naturalized flows (1961–1990) for each of the four major sub-basins of the SSRB.

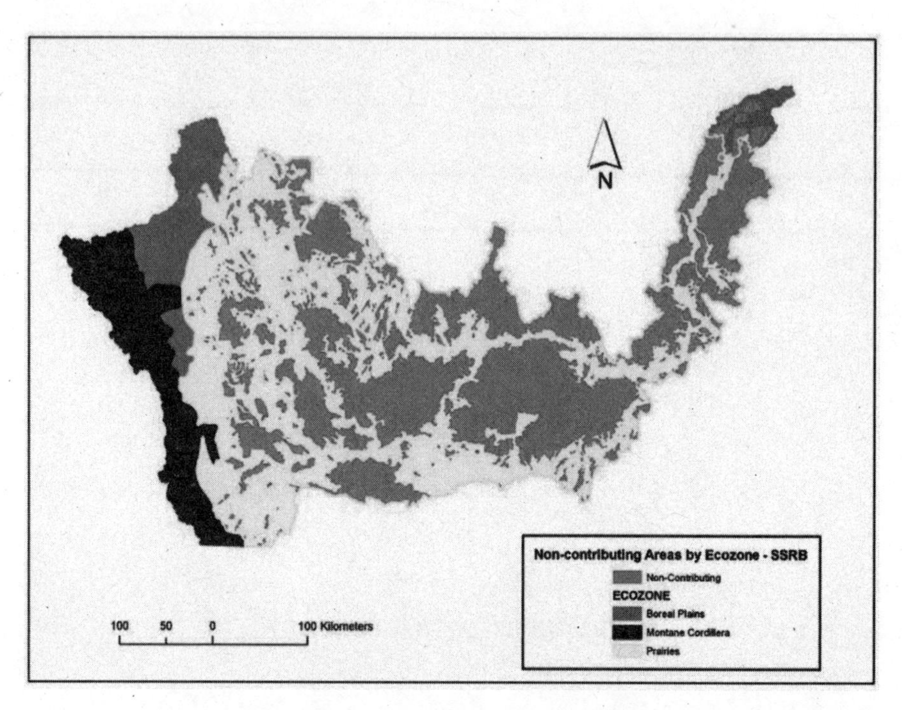

Figure 7. Non-contributing areas within the Ecozones of the SSRB. The non-contributing areal data is sourced from PFRA and the Ecozone data originates from the EcoAtlas of Canada.

	BANFF	CALGARY	LETHBRIDGE	RED DEER	MEDICINE HAT	SASKATOON
Elevation (MASL)	1397	1077	929	905	717	501
TEMPERATURE						
Annual Daily Average (°C)	2.9	3.9	5.6	2.3	5.5	2.0
Extreme Maximum (°C)	34.4	36.1	39.4	36.1	42.2	40.6
Extreme Minimum (°C)	-51.2	-45.0	-42.8	-46.1	-46.1	-50.0
PRECIPITATION						
Annual Rainfall (mm)	281.2	300.1	262.9	358.9	230	253.7
Annual Snowfall (mm SWE)	186.7	98.6	134.7	111.3	92.6	93.5
Annual Precipitation (mm)	467.9	398.7	397.6	470.2	322.6	347.2
% as Snow	39.9	24.7	33.9	23.7	28.7	26.9

Source: http://climate.weatheroffice.ec.gc.ca/climate_normals/index_1961_1990_e.html

Table 1. Selected climate normals (1961–1990) for selected cities in the South Saskatchewan River Basin. [59]

DRAINAGE SUB-BASIN	ANNUAL FLOW VOLUME (MILLION CUBIC DECAMETER*)			RANGE FROM MIN TO MAX AS A PERCENTAGE OF AVERAGE
	MINIMUM	AVERAGE	MAXIMUM	
Red Deer River at Blindloss	0.79	1.63	3.58	171.3
Oldman River at the mouth	1.55	3.29	5.10	108.1
Bow River at the mouth	2.68	3.79	5.17	65.6
South Saskatchewan River into Lake Diefenbaker	5.11	8.60	13.00	91.6

*A cubic decameter (dam3) is a volume of 10 metre × 10 metre 10 metre. 1 dam³ = 1000 m³.

Table 2. Annual volume of naturalized flows for each of the SSRB sub-basins, 1961–1990.

ECO-ZONE	TOTAL AREA (KM²)	AVERAGE FLOW (M³/S)	AVERAGE ANNUAL FLOW (MILLION DAM³)	PERCENTAGE OF INFLOW INTO LAKE DIEFENBAKER	AVERAGE ANNUAL RUNOFF (MM)	EFFECTIVE OR CONTRIBUTING BASIN AREA (KM²)	RUNOFF FROM CONTRIBUTING AREA (MM/YEAR)
Montane-Boreal	25,000	211	6.7	77	267	25,000	267
Prairie	111,000	64	2.0	23	18	62,000	32
Total (above Lake Diefenbaker)	136,000	275	8.7	100		87,000	

Areas and flows based on natural or naturalized HYDAT data for the Red Deer River at Red Deer, Bow River at Calgary, Sheep River at Black Diamond, Three Point Creek near Millarville, Highwood River at Diebel's Ranch, Oldman River at Waldron's Corner, Crowsnest River at Frank, Castle River near Beaver Mines, Waterton River near Waterton Park, and the Belly River near Mountainview.

Table 3. Summary of runoff and flow from Montane-Boreal and Prairie Ecozones.

YEAR(S)	OBSERVATION
1999–2002	Severe drought with vast spatial extent in 2001–2002 including most southern Canada
1980s	Extensive agricultural and hydrological drought in the Prairies in 1980–1981, 1984, 1988–1989
1977–1978	Extensive hydrological drought in 1977–1978; steady decline in spring snowmelt runoff time since ~1977 in west-central Canada
1960s	Most extensive agricultural drought in 1961, other extensive agricultural droughts in 1967–1968, and 1964
Late 1910s–early 1940s	April-June rainfall at Medicine Hat <100 mm/yr (1917–1926); severe drought in the Palliser Triangle (1929–1937); prolonged drought in the southern cordillera, Alberta (1917–1941); agricultural drought extensive in the Prairies (1936–1938); steady decline in lake levels
1880s–1890s	Multi-year drought in southern the Prairies (late 1880s and early 1890s)
1850s–1860s	Severe drought described by John Palliser exploration (1857–1859); negative July PDSI reconstruction from tree-rings
1820s	Negative July PDSI from tree rings from Cypress Hills
1700s	Suppressed tree growth in the Cypress Hills and eastern slopes in the 1690s, 1720s, 1750s, 1760s, and 1790s; reactivation of sand dunes in the Great Sand Hills
1560s–1570s	Major period of low flow of the South Saskatchewan River inferred from tree rings
1300–1500	Medieval Climatic Anomaly: high frequency of extreme droughts in the northern Great Plains from tree-ring and lake sediment records; major shift from wet to dry in the northern plains at ~500–700 a.d.; peak aridity in the southern Canadian Prairies at 900–1200 a.d.

Table 4. Recorded and inferred prairie droughts of the past millennium.

Endnotes:

1. Natural Resources Canada, *EcoAtlas* from Geogratis. Available at: http://geogratis.cgdi.gc.ca/geogratis/en/index.html (cited 17 December 2008); see also Agriculture and Agri-Food Canada-Prairie Farm Rehabilitation Administration (AAFCPFRA), *PFRA Watershed Project-Version* 7 (2007).
2. AAFCPFRA, *PFRA Watershed Project-Version 7.*
3. Natural Resources Canada, *EcoAtlas.*
4. Longley, R.W., "Variability of Annual Precipitation in Canada," *Monthly Weather Review* 81 (1953): 131–134.
5. Natural Resources Canada, *EcoAtlas.*
6. Canadian Council on Ecological Areas, 2004 *Ecozones of Canada. Canadian Council on Ecological Areas,* Environment Canada (http://www.ccea.org/ecozones/index.html).
7. Natural Resources Canada, *EcoAtlas.*
8. W.L. Strong and K.R. Leggat, *Ecoregions of Alberta* (Edmonton: Alberta Forestry, Lands and Wildlife, Land Information Services Division, Resource Information Branch, 1992).
9. Natural Resources Canada, *EcoAtlas.*
10. Natural Resources Canada, *EcoAtlas.*
11. D.F. Acton, G.A. Padbury, and C.T. Stushnoff, *The Ecoregions of Saskatchewan* (Regina: Canadian Plains Research Centre, 1998).
12. W. Nemanishen, *Drought in the Palliser Triangle: A Provisional Primer* (Calgary: Prairie Farm Rehabilitation Administration Agriculture Drought Committee, 1998); see also D.C. Jones, *Empire of Dust: Settling and Abandoning the Prairie Dry Belt* (Edmonton: University of Alberta Press, 1987).
13. Saskatchewan Environment and Resource Management, *Saskatchewan's State of the Environment Report 1997—The Prairie Ecozone: Our Agricultural Heartland* (Regina: Environment and Resource Management, 1997); for a more complete description of dryland agricultural adaptations, including conservation tillage and irrigation practices, see J. Bruneau, D. R. Corkal, E. Pietroniro, B. Toth, and G. van der Kamp, "Human Activities and Water use in the South Saskatchewan River Basin," in this issue of *Prairie Forum.*
14. A. Pietroniro, M.N. Demuth, L.E.L Comeau, C. Hopkinson, P. Dornes, N. Kouwen, B. Brua, J. Toyra, and A. Bingeman, "Streamflow shifts resulting from past and future glacier fluctuations in the eastern flowing basins of the Rocky Mountains." Environment Canada internal report to the Climate Change Resources User's Group of Alberta Environment, 2007.
15. S.B. Rood, G.M. Samuelson, J.K. Webber, and K.A. Wywrot,. "Twentieth-Century Decline in Streamflows from the Hydrological Apex of North America," *Journal of Hydrology* 306 (2005): 215–233; see also J.R. Westmacott and D.H. Burn, "Climate Change Effects on the Hydrologic Regime within the Churchill-Nelson River Basin," *Journal of Hydrology* 202 (1997): 263–279; and J. Yulianti and D.H. Burn,

"Investigating Links between Climatic Warming and Low Stream Flow in the Prairies Regions of Canada," *Canadian Water Resources Journal* 23 (1998): 45–60.

16. X. Zhang, K.D. Harvey, W.D Hogg, and T.R. Yuzyk, "Trends in Canadian Stream Flow," *Water Resources Research* 37–4 (2001): 987–998.
17. Rood, et al., "Twentieth-Century Decline in Streamflows," 215–233.
18. G. van der Kamp, D. Keir, and M. S. Evans, "Long-term Water Level Changes in Closed-Basin Lakes of the Canadian prairies," *Canadian Water Resources Journal,* 33–1 (2008): 23–38.
19. D.H. Burn, L. Fan, and G. Bell, "Identification and Quantification of Streamflow Trends on the Canadian Prairies," submitted to *Hydrological Science* (2008).
20. W. Stichling and S.R. Blackwell, "Drainage Area as a Hydrologic Factor on the Glaciated Canadian Prairies," International Union of Geodesy and Geophysics Proceedings, Volume 111, Toronto, 1957.
21. R.B. Godwin and F.R.J. Martin, "Calculation of Gross and Effective Drainage Areas for the Prairie Provinces," Canadian Hydrology Symposium—1975 Proceedings, Winnipeg.
22. AAFCPFRA, *PFRA Watershed Project-Version 7.*
23. Natural Resources Canada, *EcoAtlas.*
24. F.M. Conly and G. van der Kamp, "Monitoring the Hydrology of Canadian Prairie Wetlands to Detect the Effects of Climate Change and Land Use Changes," *Journal of Environmental Monitoring and Assessment* 67 (2001): 195–215.
25. G. van der Kamp, W.J. Stolte, and R.G. Clark, "Drying Out of Small Prairie Wetlands after Conversion of their Catchments from Cultivation to Permanent Brome Grass," *Hydrological Sciences Journal* Vol. 44, No. 3 (1999): 387–397.
26. G. van der Kamp and M. Hayashi, "The Groundwater Recharge Function of Small Wetlands in the Semi-arid Northern Prairies," *Great Plains Research* Vol. 8, No.1 (1998): 39–56.
27. H. Maathuis, *The Quality of Natural Groundwaters in Saskatchewan,* Saskatchewan Research Council Publication No. 12012–1E08, 2008.
28. Bruneau et al., "Human Activities and Water use in the South Saskatchewan River Basin."
29. G. van der Kamp and H. Maathuis, "Groundwater Sustainability in the Prairies—Some Examples." Proceedings (CD) of the 59th Canadian Geotechnical and 7th Joint IAH-CNC and CGS Conferences, Vancouver, b.c., October 1–4, 2006, paper No. 500, 2006.
30. Saskatchewan-Nelson Basin Board, *Water Supply for the Saskatchewan Nelson Basin: Appendix 7 Environmental Considerations* (1972): 423. Conductivity for the South Saskatchewan River at Saskatoon is listed as ranging from 268 to 632 _mhos. TDS was estimated by multiplying conductivity by a factor of 0.7.
31. Government of Alberta, *Surface Water Quality Guidelines for Use in Alberta* (Edmonton: Alberta Environment, 1999), 20.

32. Government of Saskatchewan, *Surface Water Quality Objectives,* Interim Edition, EPB 356 (Regina: Saskatchewan Environment, 2006), 9.
33. Canadian Council of the Ministers of Environment (CCME), *Canadian Water Quality Guidelines for the Protection of Aquatic Life* (1999). Available at: http://ceqg-rcqe.ccme.ca/ (cited 17 December 2008).
34. Alberta Environment, *Alberta River Water Quality Index.* Available at: http://www3.gov.ab.ca/env/water/SWQ/resources01.cfm (Cited 17 December 2008).
35. Alberta Environment, *Alberta River Water Quality Index 2005–2006.* Available at: http://environment.gov.ab.ca/info/library/7681.pdf, http://environment.gov.ab.ca/info/library/7680.pdf, and http://environment.gov.ab.ca/info/library/7683 (cited 24 December 2007).
36. Government of Saskatchewan, *State of the Watershed Report* (Regina: Saskatchewan Watershed Authority, 2007): pp. ii and 29.
37. Maathuis, *The Quality of Natural Groundwaters in Saskatchewan.*
38. Government of Canada, *Guidelines for Canadian Drinking Water Quality,* 6th edition (Ottawa: Health Canada, 1996).
39. J. Sketchell and N. Shaheen, "Groundwater Quality in Rural Saskatchewan: Emerging Issues for Drinking Water," in W. Robertson (ed.), *Proceedings of the 9th Conference on Drinking Water: Maintaining Drinking Water Quality—Lessons from the Prairies and Beyond* (Ottawa: Canadian Water and Wastewater Association, 2001), 242–258.
40. D. Fitzgerald, D. Kiely, D. Neilson, R. Shaw, S. Audette, R. Prior, M. Ashton, and E. Allison, "Alberta Farmstead Water Quality Survey," in the Technical Report submitted to the Canada-Alberta Environmentally Sustainable Agriculture Agreement, found in *Agricultural Impacts on Water Quality in Alberta,* vol. 1 (Edmonton: 1997).
41. G. van der Kamp and G. Grove, "Well Water Quality in Canada: An Overview," in M. Mahmoud and R. van Everdingen (eds.), *2001 An Earth Odyssey.* Proceedings of the 54th Geotechnical Conference, Calgary, Alberta, Canada, September 16–19, 2001 (Richmond, B.C.: Bitech Publishers Ltd., 2001), 45–49.
42. Government of Canada, "*Threats to Sources of Drinking Water and Aquatic Ecosystem Health in Canada,*" National Water Research Institute (NWRI) Scientific Assessment Report Series No. 1 (Burlington, Ontario: Environment Canada, 2001), 72.
43. Alberta Environment, *Alberta River Water Quality Index 2005–2006.*
44. Government of Canada, *Threats to Water Availability in Canada.* National Water Research Institute (NWRI) Scientific Assessment Report Series No. 3 and ACSD Science Assessment Series No. 1 (Burlington, Ontario: Environment Canada, 2004), 128.
45. National Drought Mitigation Center, "What's drought? Understanding and Defining Drought," (Lincoln, Nebraska: University of Nebraska-Lincoln, 2003). Available at: http://www.drought.unl.edu/whatis/concept.htm (cited 17 December 2008).

46. D.A. Wilhite, "Drought," in J.R. Holton, J.A. Pyle, and J.A. Curry (eds.), *Encyclopedia of Atmospheric Sciences* (New York: Elsevier Science Ltd., 2003).
47. D.A. Wilhite, "Drought," in *Global Assessment. Hazards and Disasters Series,* vol. 1 (New York: Taylor and Francis Group, Routledge, 2000).
48. G.P. Marchildon, S. Kulshreshtha, E. Wheaton, and D. Sauchyn, "Drought and institutional adaptation in the Great Plains of Alberta and Saskatchewan, 1914–1939," *Natural Hazards* 45 (2007): 391–411.
49. B. Bonsal, G. Koshida, E.G. O'Brien, and E. Wheaton, "Droughts," Chapter 3 in *Threats to Sources of Drinking Water and Aquatic Ecosystem Health in Canada,* (Burlington, Ontario: Environment Canada, 2001), 19–26; see also A. Shabbar and B.R. Bonsal, "Association between Low Frequency Variability Modes and Winter Temperature Extremes in Canada," *Atmosphere-Ocean* 42 (2004), 127–140.
50. N. Mayer and W. Avis, "Canada Country Study: Climate Impacts and Adaptation," *National Cross-Cutting Issues,* vol 8. (Envirnment Canada, 1998). Available at: http://dsp-psd.communication.gc.ca/Collection/En56–119–7-1998E.pdf (cited 17 December 2008).
51. E. Wheaton, V. Wittrock, S. Kulshreshtha, G. Sishida, C. Grant, A. Chipanshi, and B. Bonsal, *Lessons Learned from the Canadian Drought Years of 2001 and 2002: Synthesis Report* (2005). Available at: http://www.agr.gc.ca/pfra/drought/info/11602–46E03.pdf (cited 17 December 2008).
52. B.R. Bonsal and R.G. Lawford, "Teleconnections between El Niño and La Niña Events and Summer Extended Dry Spells on the Canadian Prairies," *International Journal of Climatology* 19 (1999): 1445–1458.
53. K. R. Laird, S.C. Fritz, K.A. Maash, and B.F. Cumming, "Greater Drought Intensity and Frequency before a.d. 1200 in the northern Great Plains," *Nature* 384 (1996): 551–554.
54. Sylvia Lac and Camilla Colan, "South Saskatchewan River Basin Biogeography," University of Regina, Prairie Adaptation Research Collaborative, IACC Project Working Paper No. 20, 2004.
55. A. Michels, K.R. Laird, S.E. Wilson, D. Thomson, P.R. Leavitt, R.J. Oglesby, and B.F. Cumming, "Multidecadal to Millennial-Scale Shifts in Drought Conditions on the Canadian Prairies over the Past Six Millennia: Implications for Future Drought Assessment," *Global Change Biology* 13–7 (2007): 1295–1307; see also D.J. Sauchyn, J. Stroich, and A. Beriault, "A Paleoclimate Context for the Drought of 1999–2001 in the northern Great Plains of North America," *The Geographical Journal* 169–2 (2003): 158–167.
56. S. St. George and D. Sauchyn, "Paleoenvironmental Perspectives on Drought in Western Canada," *Canadian Water Resources Journal* 31–4 (2006): 197–204.
57. D. Sauchyn, A. Pietroniro, and M. Demuth, "Upland Watershed Management and Global Change," Fifth Biennial Rosenberg Forum on Water Policy, Banff, Alberta, 2006.

58. D. Sauchyn and W.R. Skinner, "A Proxy Record of Drought Severity for the Southwestern Canadian Plains," *Canadian Water Resources Journal* 26–2 (2001): 253–272; see also R.A. Case and G.M. MacDonald, "A Dendroclimatic Reconstruction of Annual Precipitation on the Western Canadian Prairies since a.d. 1505 from Pinus flexilis James," *Quaternary Research* 44 (1995): 267–275.
59. Environment Canada, *Canadian Climate Normals or Averages 1961–1990* (2006). Available at: http://climate.weatheroffice.ec.gc.ca/climate_normals/index_1961_1990_e.html (cited 17 December 2008).

HUMAN ACTIVITIES AND WATER USE IN THE SOUTH SASKATCHEWAN RIVER BASIN

JOEL BRUNEAU, DARRELL R. CORKAL, ELISE PIETRONIRO, BRENDA TOTH, AND GARTH VAN DER KAMP

Abstract:

The South Saskatchewan River Basin (SSRB) is one of the largest dryland watersheds (167,765 km²or 16.78 million Ha) with the most variable geography and hydrology in all of Canada. Currently, about 2.2 million people rely on the major river systems within the basin. Together, human demand diverts close to 35% of average natural flows and consumes 24% of it. The largest water user by far is irrigated agriculture, accounting for over 90% of consumptive water demands within the basin. The extensive systems of canals, reservoirs, and dams, in conjunction with human consumptive demands, have significantly altered river flows. This paper describes how water is currently utilized within the SSRB. The region has adapted well to water limitations and drought; however, increasing population growth and water demands, coupled with the predicted changes in future climate patterns, will force the region to consider new adaptations in the future.

Sommaire

Le bassin de la rivière Saskatchewan-Sud (BRSS) est l'un des plus grands bassins hydrographiques situés en terres sèches (167 765 km² ou 16,78 millions d'hectares) et présente la géographie et le cycle hydrologique le plus variable au Canada. Aujourd'hui, environ 2,2 millions de personnes dépendent des grands réseaux hydrographiques de l'intérieur du bassin. En tout, afin de satisfaire au besoin de l'homme, près de 35 % du débit naturel moyen est dérivé ; l'homme consomme 24 % du débit moyen. L'agriculture irriguée est responsable de la plus grande partie de la consommation humaine d'eau, soit 90 % de la consommation à l'intérieur du bassin. Des systèmes étendus de canaux, de réservoirs et de barrages ainsi que les besoins de l'homme ont contribué à des modifications significatives au débit des rivières. Cet article décrit la façon dont l'eau est utilisée à l'intérieur du basin de la rivière Saskatchewan-Sud. La région s'est bien adaptée à la sècheresse et aux limitations en eau. Cependant, l'augmentation de la population et de ses besoins en eau et les changements climatiques prévus forceront la région à envisager de nouvelles stratégies évolutives.

Introduction

The South Saskatchewan River Basin (SSRB) is characterized by large dryland farms growing mainly cereal and oilseed crops, intensive farming, livestock production, and substantial field crop irrigation activities. Virtually all lands in the basin have been altered by human activities, and its water resources, both surface and ground, play a critical role in supporting human communities and economic activities. The relatively small population is concentrated in a few large urban centers with the rural population distributed across the basin in farms and small communities. These characteristics, combined with the unique prairie ecosystem and hydrology, have placed water at the centre of economic development, social cohesion, and public policy. Climate change is expected to place increasing pressure on society's ability to adapt to scarce water supplies; however, climate variability and adaptation is not new. First Nations and farm communities have a long history of adapting to climate variability in the SSRB, a trend which will continue into the future.

This paper provides a brief overview of the human dimension in the basin by focusing on the role water plays in maintaining communities, as well as on the stress human water demands have on the aquatic ecosystem. The intent is to provide a summary of the relationship between humans and the water regime in the basin.

Methodology and Basic Terms

This paper compares current water use in human activities to the average annual *natural flows* for the period from 1961 to 1990, a commonly-used baseline hydrologic record. Natural water flows are the flows in streams generated by a watershed in the absence of water diversions for human uses. Flows vary with the season and weather patterns. *Water diversions* refer to water withdrawals for human uses. This includes water used by households, municipalities, industries, and agriculture. The rivers in the South Saskatchewan River Basin are developed with dams, reservoirs, and water withdrawal systems diverting water for human uses. Only some of the water diverted by human activities is *consumed*—this water is no longer available for human uses or the natural ecosystem. The remainder is returned back to the river system (e.g., hydroelectric plants return water to rivers, communities return treated wastewater to rivers). To determine what the natural flow would have been without human intervention, we 'naturalize' the flow.[1] This is a set of calculations that considers the loss from known water diversions, the influence of evaporation from any constructed reservoirs, and the flow that is returned to the system to determine the flow that would have naturally occurred. Water consumption by human activities was calculated by subtracting the return flow from the total diverted flow. Though evaporation losses from reservoirs are properly attributed to human use, we do not include them as human consumption in the calculations below. Water use for human activities was totalled from various data sets using the most current available data. Diversion and consumption data comes from Environment Canada's Municipal Water Use Database (MUD) and the Industrial Water Use Database. Additional data is from Hydroconsult (2002) which also forecasts future water diversions and consumption. Irrigation diversions come from the irrigation districts in Alberta and Saskatchewan. Land use data comes from Statistics Canada.

Human Population, Land Use, and Economic Activity

Prior to European settlement, human activity in the SSRB consisted of aboriginal peoples who relied on hunting and gathering food from the natural environment. Archeological evidence records continuous human habitation as far as 11,000 years before present.[2] European settlement was encouraged by the Government of Canada by offering land in the 1800s. To encourage irrigation and agricultural activities in support of human settlement in the Prairies, the Northwest Irrigation Act (1894–1895) was established.[3]

While approximately 2.2 million people presently reside in the SSRB,[4] over 80% of the population lives in Calgary, Lethbridge, Medicine Hat, Swift Current, and Saskatoon. The remainder lives in over 200 smaller rural communities and residences.[5] Pipelines and diversions also take water to Saskatchewan communities and industry outside the basin. For instance, water from the South Saskatchewan River (SSR) is diverted into Buffalo Pound Lake, which provides drinking water for an additional 220,000 residents around Moose Jaw and Regina. Water is piped from the SSRB to Humboldt and Lanigan, in east-central Saskatchewan, where it supports potash mines.

Rural population densities in the basin are generally low with the Saskatchewan portion of the basin having lower rural densities than Alberta (see Figure 1). The higher rural densities in Alberta reflect a greater reliance on intensive livestock operations, value-added agricultural processing, and more extensive irrigation infrastructure (see below).

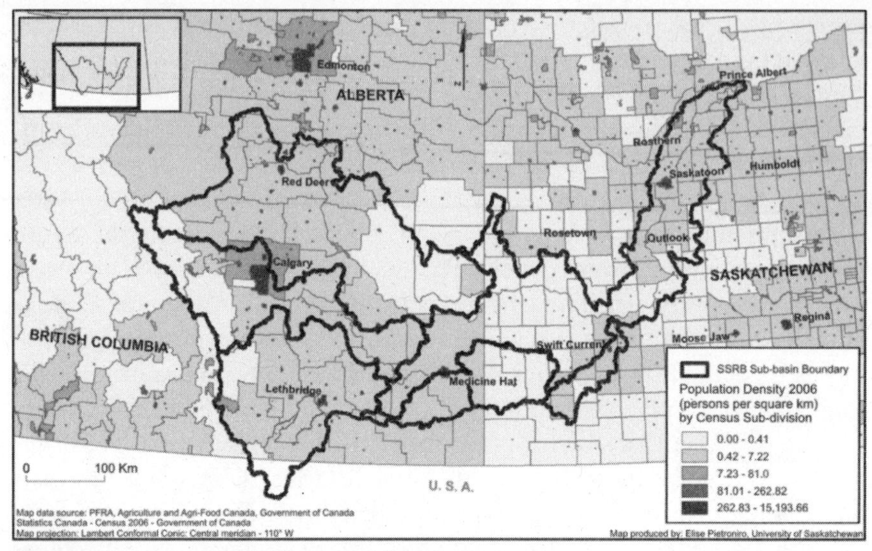

Figure 1. Population Density in the SSRB (2006).

Population in the basin has experienced strong growth.[6] The average annual growth rate was 2.3 % in the period 1991 to 2001 with the Alberta portion of the basin growing more quickly.[7] Since 1961, the SSRB population has tripled and this rate of growth is expected to continue into the future.[8] Most of the population growth will occur in the cities, and rural populations are expected to decline in both relative and absolute terms.[9]

Like other areas in Canada, the SSRB has a sophisticated and diverse economy that provided over a million jobs in 2001, and generates a Gross Domestic Product (GDP) that exceeds $69 billion.[10] Though services, government, retail and wholesale trade, and construction are the largest economic sectors, agriculture, irrigation, and agri-food manufacturing are also important in contributing over 7% of direct employment.[11] While irrigation is applied on only about 5% of the agricultural land, in Alberta alone it contributes about $830 million, or over 18%, of Alberta's agri-food Gross Domestic Product.[12] Estimates around Saskatchewan's Lake Diefenbaker show that economic value-added per hectare of irrigated lands are about five times that of dryland agriculture.[13]

Land use in the basin is predominantly dryland agriculture producing crops such as wheat and canola, and livestock such as cattle, pigs and poultry. Total land use allocated to farming in the SSRB is approximately 15 million hectares, an area that accounts for about 90% of all land in the basin. In 2001, about 46% of all farmland was used for cropping, while another 41% was used as pasture. The remainder of farmland was either summer fallow or under other uses.[14]

Table 1 shows the types of farms in the SSRB, their land use activities, as well as livestock production. About 44% of farms in the SSRB rely primarily on livestock operations raising cattle, hogs, poultry and other animals[15]; however, the fraction of farms in the Alberta portion of the SSRB relying primarily on livestock is much higher than in Saskatchewan which is more heavily focused on grains production. The Alberta portion of the SSRB raises over 80% of basin livestock with livestock density (based on hectares of available pasture) twice as high as in the Saskatchewan portion of the SSRB (and almost three times based on total farm lands). The production differentiation is also manifest in land use patterns with Alberta farms more heavily allocated to pasture.

	SSRB—AB	SSRB—SK	SSRB TOTAL
Number of Farms[1]	19,600	9,000	28,600
Primarily Livestock	53%	24%	44%
Primarily Grains	47%	76%	56%
Livestock (million head)			
Cattle	2.83	0.45	3.28
Density (head per ha)[2]	0.60	0.30	0.52
Hogs	6.06	1.12	7.18
Density (head per ha)	1.28	0.74	1.15
Poultry	1.27	0.23	1.50
Density (head per ha)	0.27	0.15	0.24
Land Use on Farms (million Ha)	9.89	5.34	15.23
Cropping[3]	42%	53%	46%
Pasture	48%	28%	41%
Fallow and Other Use	10%	19%	13%

1. Number of farms includes all operations with annual receipts greater than $2,499.
2. Density based on hectares of land allocated to pasture.
3. As a % of land allocated to farms (total SSRB area is 16.78 million Ha).

Source: Statistics Canada, *2001 Census of Agriculture, data by province, Census Agricultural Region (CAR) and Census Division (CD):* various tables.

Table 1: Farm Types, Land Use, and Livestock Production, 2001.

The high potential evaporation compared to precipitation in most of the SSRB means that irrigation offers significant benefits in terms of crop yields above what would arise under dryland practices. It is this high potential evaporation that has made the SSRB the most irrigation intensive area in Canada.[16] The SSRB has an extensive system of irrigation with numerous dams, reservoirs, water diversions and irrigation projects utilized to manage water resources (see Figure 2). For example, the Alberta portion of the SSRB has 13 irrigation districts and more than 2,700 individual projects outside of the dis-

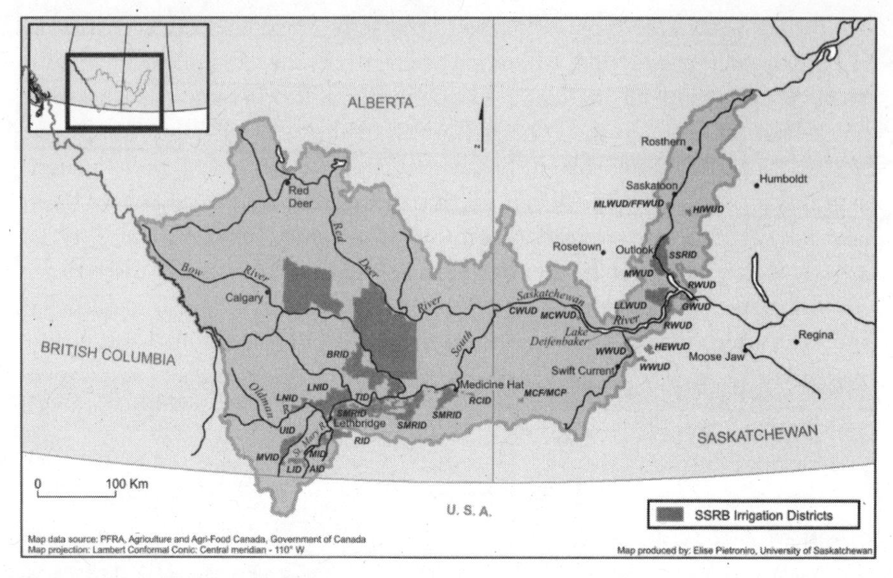

Figure 2. Irrigation Districts in the SSRB.

tricts,[17] while Saskatchewan has three major irrigation districts in the SSRB which manage individual and group irrigation projects.[18] Approximately 772,000 hectares in the SSRB are either under irrigation, or have irrigation structures in place. This distribution accounts for the majority of Canada's 1 million hectares of irrigated lands.[19] The Alberta portion of the SSRB is the most heavily irrigated and contains about 79% of irrigated SSRB lands. Irrigation acreage has been rising by over 1% per year in both provinces.

Water Management in the SSRB

The responsibility for water management in Canada essentially rests with the provincial governments.[20] In practice, water management involves many agencies with unique and, sometimes, overlapping mandates. The following is a brief summary of the key water institutions in the SSRB.[21]

Since the Saskatchewan River—the SSR is part of this larger system—lies within the jurisdiction of three provincial governments, Alberta, Saskatchewan and Manitoba, the regulation of its flow is administered with all three provinces' apportionment entitlements in mind. The Prairie Provinces Water Board (PPWB), created in 1948, administers, monitors and reports on the quantity and quality of inter-provincial water flowing from Alberta to Saskatchewan and then to Manitoba.[22] The Board is comprised of two federal representatives (Environment Canada and Agriculture and Agri-Food Canada—PFRA) and one representative from each of the three prairie provinces (Alberta Environment, the Saskatchewan Watershed Authority, and Manitoba Watershed Stewardship). The Board administers the 1969 *Master Agreement on Apportionment,* a legal agreement that ensures water in each river is equitably shared. The formula is simple: "Alberta and Saskatchewan may each take up to one half of the natural flow of water originating within its boundaries and one half of the flow entering the province. The remainder is left to flow into Manitoba."[23] For the major rivers in the SSRB (the Bow, Oldman, Red Deer, and South Saskatchewan), the apportionment obligations apply at the confluence of the Red Deer and South Saskatchewan rivers inside Saskatchewan, just east of the Alberta border. There are also minimum flow requirements that control how much water must remain in the river at any point in time, thus further limiting the extent to which Alberta and Saskatchewan can deplete the flow.

As long as provinces meet their annual apportionment obligations and minimum flow requirements, how and when they choose to use their share of water is left to their discretion. Figure 3 shows the share of natural flow that Alberta passes on to Saskatchewan as measured at the confluence of the Red Deer and South Saskatchewan Rivers. At no time since the agreement was formed has Alberta used more than 50% of the natural flow in any year. In 1988, a very dry year, flow in the Oldman basin dropped to a low of 41% of "apportionable flow" but the balance of flow to Saskatchewan was made up by water from the other sub-basins. The agreement permits such management flexibility allowing Alberta to meet Saskatchewan's entitlements by using any combination of flow from the Bow, Oldman, and Red Deer basins, which jointly contribute water to the South Saskatchewan River flowing into

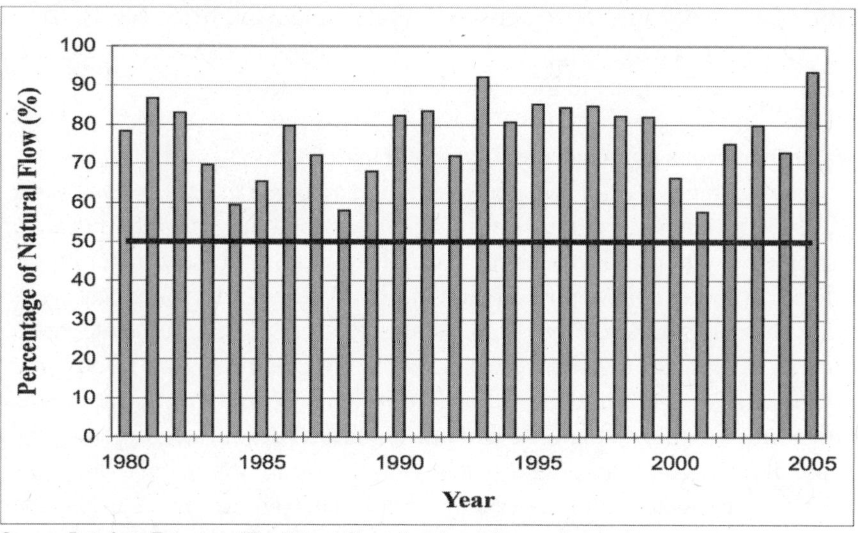

Source: Data from Toth et al., "The Natural Characteristics of the SSRB."

Figure 3. Ratio of Recorded Flow to Natural Flow at the Alberta-Saskatchewan Border.

Saskatchewan.[24] Water diversions in Saskatchewan are relatively small, so the province has always met its apportionment requirements to Manitoba.

A small fraction of the SSRB is in the United States (see Figure 2); for example, the St. Mary River, a tributary of the Oldman River, originates in the United States. A portion of the St. Mary flow is diverted into the Milk River system in the United States, with the remainder flowing into Canada and the SSRB. This international arrangement affects Montana, Alberta, Saskatchewan and Manitoba, and is managed by the *International Joint Commission,* based on the *1909* International Boundary Waters Treaty Act.[25] International water management involves Environment Canada and the federal Department of Foreign Affairs and International Trade.

Recent waterborne disease outbreaks causing death and illness from contaminated drinking water have convinced governments at all levels to place increasing emphasis on managing water as a whole system from the landscape watershed to the point of use.[26] Both Alberta and Saskatchewan do this by involving public, private, and community groups to participate in watershed planning. In 2002, Saskatchewan developed the *Long Term Safe Drinking Water Strategy.*[27] In 2003, Alberta developed a *Water for Life* strategy based on the recognition that water management involves many interests (social, economic, environmental), many players (government, private, and community), and requires managing water on a watershed basis.[28]

In Alberta, Alberta Environment is the principal agency responsible for issuing water allocations under a license, hydrologic flow monitoring and forecasting, and watershed management. In Saskatchewan, these responsibilities rest with the Saskatchewan Watershed Authority (SWA). Both agencies work in partnership with other provincial ministries, local watershed groups, communities, and local and federal governments.

Water Diversions and Consumption in the SSRB

Water availability is a critical constraint to life on the prairies. The SSRB is primarily semi-arid with annual precipitation less than half the potential water loss by evaporation and transpiration. On the prairies, average summer rainfall is less than potential evaporation and is highly variable; annual precipitation ranges from less than 300 mm on the plains, to over 800 mm in the mountains. Potential evaporation varies from approximately 450 mm to over 1,000 mm.[29]

As a result of these semi-arid conditions, many communities are situated on, or close to, the SSR and its tributaries, and utilize surface water sources. Though many communities rely exclusively on groundwater, only 7% of total municipal populations in Alberta and Saskatchewan do so.[30] On the other hand, about 85 to 90% of the rural populations in Saskatchewan and Alberta relied on groundwater in 1981.[31] This percentage may have been reduced somewhat since then by the expansion of regional piped water distribution systems reliant on surface water sources; nonetheless, groundwater is a still a primary supply source in most rural areas since it is both accessible and reliable.[32] Groundwater use for irrigation is very small, however, due to unsuitable water chemistry and limited well yields. Total annual groundwater use in Saskatchewan and Alberta was about 0.1 million cubic decametres[33] (dam³) in 1981.[34] About half of this usage was in the SSRB, so the present groundwater use in the SSRB is estimated to be 0.05 million dam³ per year, about 10% of the non-irrigation water use in the basin.

Water diversions from both surface and groundwater can be split into five major use sectors[35]: municipal, industrial, thermal and hydro power generation, livestock, and irrigation. By far, the largest water use sector in the basin is irrigation. Water diversions for the Alberta irrigation districts were 1.66 million dam³ in 2006. Imputing additional diversions from Saskatchewan and private irrigators in Alberta can bring total irrigation diversions to over 2.60 million dam³, though this depends on the year. Non-irrigation water demands, though substantial in themselves, are much smaller. Total water diversions in the SSRB in 1996—the last year in which comprehensive data was collected—for municipal, industrial, thermal power generation, and livestock watering amounted to only about one fifth of that used for irrigation[36] (see Table 2). Though irrigation demands were above average in 1996, even in a low demand wet year such as 2005, irrigation diversions are around three to four times the amount used for non-irrigation purposes.

	DIVERSIONS		RETURN FLOW		CONSUMPTION	
	MILLION DAM³	SHARE OF NATURAL FLOW	MILLION DAM³	SHARE OF TOTAL DIVERSIONS	MILLION DAM³	SHARE OF NATURAL FLOW
Non-irrigation Demand¹	0.51	5.9%	0.36	71%	0.16	1.8%
Municipal	0.31	3.5%	0.24	77%	0.07	0.8%
Livestock	0.07	0.8%	0.02	28%	0.05	0.5%
Industrial	0.05	0.6%	0.02	40%	0.03	0.3%
Thermal Hydro	0.09	1.0%	0.08	88%	0.01	0.1%
Irrigation Demand²	2.52	29.1%	0.59	23%	1.93	22.3%
Total Supply³	**8.67**				**8.67**	

1. Based on 1996 data which is the last year in which comprehensive demand data was collected.
2. The Alberta portion of the SSRB contains approximately 79% of the irrigated acreage in the SSRB (about 600,000 ha).
3. Total supply in million cubic decameters is based on estimated average natural flow from 1961–1990. (*Source:* Data from Toth et al., "The Natural Characteristics of the SSRB.")

Source: Data from Bruneau and Toth, "Dove-Tailed Physical and Socioeconomic Results in the SSRB."

Table 2. SSRB Diversions, Return flows, and Consumption.

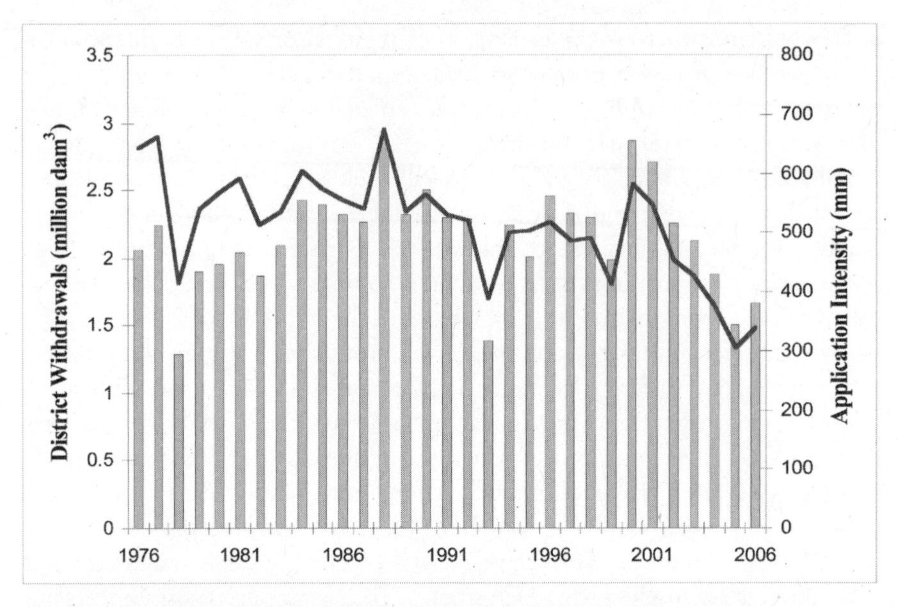

Source: Alberta Irrigation Information, *Facts and Figures for the year 2007* (Lethbridge, AB: Water Resources Branch Irrigation and Farm Water Division, 2008), Tables 7 and 9.

Figure 4. Water Diversions and Application Intensity in Alberta Irrigation Districts.

Due to extreme precipitation variability, irrigation demands can vary significantly over short periods of time. Figure 4 shows total diversions in the Alberta irrigation districts between 1976 and 2006.[37] Total diversions varied from a low of 1.29 million dam^3 in 1978, to a high of 2.88 million dam^3 in 1988, though it again reached 2.87 million dam^3 in 2000.[38]

Figure 4 also shows water application intensities. This is measured as the average amount of water per hectare, in millimeters, applied to available irrigated acreages in Alberta's irrigation districts. It has varied from a high of 700 mm to a low of about 300 mm per hectare. Total diversion of water depends on summer temperatures and rainfall, as well as irrigation infrastructure, acreage, and on-farm decisions. Figure 4 shows that total diversions have slowly trended upwards at about 0.1% per year. While irrigation application intensity follows annual precipitation closely, overall, irrigation intensity has decreased on average at about 1.2% per year. The reduction in average intensity can be due to investments in more water efficient infrastructure such as lining canals, complete replacement of canals with irrigation pipelines (to reduce water loss from evaporation or canal leakage), improvements to on-farm operations such as moving to low pressure pivot systems, and changes in crop mix. Also important is the increasing trend in land under irrigation (about 1.4% per year) allowing operators to use a greater proportion of their water licenses in any period.

To get a sense for how much of the water flowing in the SSR is being used in the basin, we can compare the total supply of water, based on hydrographic natural flows supplied by annual precipitation and calibrated to the 1961–90 period, with the total annual water diversions from all use sectors. Table 2 shows these calculations for 1996 since this was the last year comprehensive demand data were collected. As shown, despite the population of over 2 million that used the SSRB for water, the total volume of water diverted by municipalities for all uses constituted about 3.5% of total water available in an average year. In total, non-irrigation demand took up only 5.9% of average available water supplies, while irrigation diversions took in over 29%. The remaining 65% of the water was left to flow in the river.[39]

Diversions of water, however, can overstate the impact on downstream water flows since some of this water is returned to the river. The fraction of return flow differs significantly depending on the water user, so the actual consumption of water differs across activities. For instance, of the municipal water diverted, only about 23% is permanently removed from the river.[40] The remainder is returned as treated waste water or through storm sewers. The data shows that consumption by municipalities accounted for less than 1% of average annual water supply, while total water consumption from municipalities, industry, livestock watering, and thermal power generation, accounted for less than 2% of the average available water supply.

Irrigators consume a much higher share of their water diversions, though the consumption rate does vary depending on the irrigation system used, as well as climatic conditions. For instance, despite the increase in irrigation diversions and acreage, total return flows in the Alberta districts averaged about 0.6 million dam³ during the 1990s.[41] Improvements to on-farm irrigation efficiencies show return flow rates continue to decline. Since irrigation is such a large component of water demand, these changes to consumption rates are significant and can influence downstream water flows.[42] Table 2 shows irrigators in 1996 consumed about 22% of the average annual water supply.

Another way to identify the impact of human activities on water resources is to look at how the pattern of flow in the SSRB has been altered by diversions and water storage. The SSRB has 49 reservoirs in Alberta with live (usable) storage capacity of 2.8 million dam³.[43] Saskatchewan's Lake Diefenbaker was formed as a multi-use water project for the water-scarce region after completion of the Gardiner and Qu'Appelle Valley Dams in 1967;[44] this reservoir by itself stores an additional 9.4 million dam³ of water;[45] a storage capacity equivalent to 1.4 years of average natural river flow.

Most of the SSRB reservoirs were constructed or modified to allow human control and provide greater flexibility in water management. This allows water managers to store or divert water and alter river flows to balance the needs of society and the environment. Reservoir levels and river flows are adjusted to achieve aquatic and natural environment objectives, to allow for recreational water use, and to allow for navigable waterway transportation. Water managers also need to ensure municipalities can extract water for urban uses while also meeting peak irrigation demand in the summer. Hydroelectric power, though generated year round, is particularly important in the winter months. Properly managed, the SSRB water storage reservoirs also provide a degree of protection from flooding and drought, and allow for some water management from one year to the next. The management of reservoirs and river flows generally results in storing water during major runoff events (which reduces peak flows), and releasing stored water during periods of low flow (which increases minimum flows).

Figures 5a and 5b illustrate the effect of these diversions and storage on average monthly flows for the period 1961–1990.[46] The hydrograph figures show the recorded and natural monthly flows in dam³ for the Bow River at its mouth, and the South Saskatchewan River at Saskatoon. For the Bow River, the recorded and natural flows show the same overall monthly pattern with high summer flows. For the South Saskatchewan River, the recorded flow is quite different with the low flow periods much higher than the natural flow, and the peak flows much lower. The difference over the year between recorded and natural flows is the amount of water diverted from the system. This is primarily from water diverted for irrigation use or to replenish reservoirs and evaporation losses. For the Bow River and South Saskatchewan River, the annual recorded flows are reduced from natural flows 27% and 25% respectively.

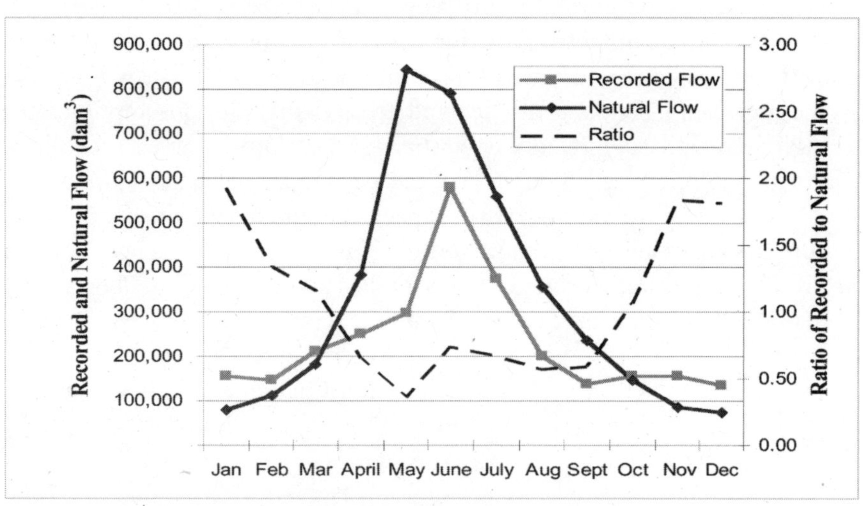

Source: Data from Toth et al., "The Natural Characteristics of the SSRB."

Figure 5a. Recorded and Natural Flows at the Mouth of the Bow River.

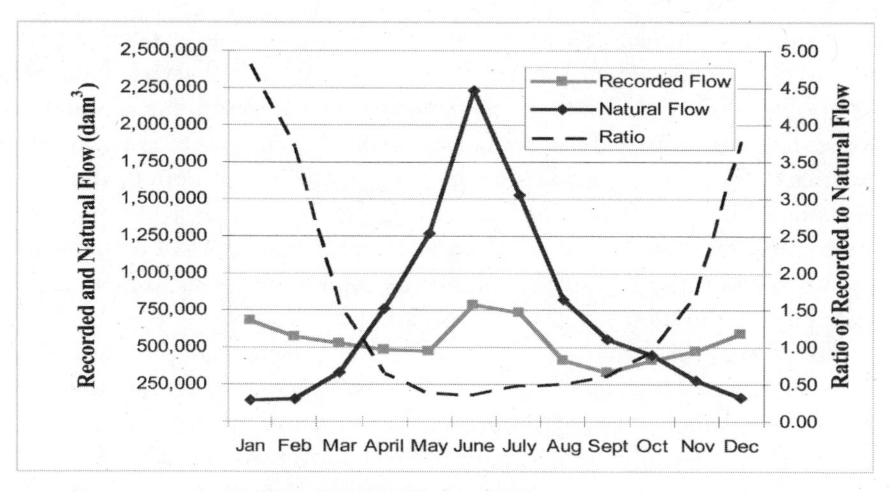

Source: Data from Toth et al., "The Natural Characteristics of the SSRB."

Figure 5b. Recorded and Natural Flows at the SSR at Saskatoon.

The hydrographs show the degree to which recorded flows differ from natural flows during the year. Changes in the pattern of these seasonal flows can impact water ecosystems, even if annual diversions are small. To highlight this we calculate the ratio of recorded to natural flows for each month and plot it as the dashed line (see figures 5a and 5b). A value less than one means the recorded flows are less than natural flows. For the Bow River, this ratio is almost always less than one, and falls below one half in the late summer. For the South Saskatchewan River, however, the ratio is sometimes higher than one. For instance, recorded flows between October and March are larger than natural flows, with January flows almost five times higher. In contrast, recorded flows in May, June, and July are less than half the natural flows.

Agricultural Water Use Adaptations in the SSRB

Since the beginning of human settlement in western Canada, people have adapted to the limitations and pressures caused by climate and water availability. After the retreat of the glaciers 11,000 years ago, human habitation was possible. For thousands of years before European settlement, First Nations communities relied on a plentiful natural food supply. First Nations nomadic pedestrian culture may have been an ideal adaptation to the natural climate and water variability of the SSRB. However, these adaptations were severely impacted by European-style agrarian settlement, the introduction of the horse and the gun, and the market-based fur trade, all of which led to the unsustainable harvesting of bison, their principal food source.[47]

Captain John Palliser surveyed the area in 1857–59, during a period that followed an extended dry cycle. Palliser mapped the 'Palliser Triangle' describing it as an extension of the Great American Desert and unsuitable for agriculture.[48] Despite Palliser's observation, European settlement of the prairies occurred in the late 1800s during a relatively wet period, and was encouraged by the federal North West Irrigation Act (1894–95) to promote economic growth in the West. Subsequently, prolonged decadal droughts in the 1920s and early 30s forced many rural people to relocate.[49] These long-term droughts in southern Alberta and Saskatchewan were so severe the two provinces were near bankrupt. The droughts forced significant provincial and federal responses, including administrative changes to land use and the creation of special agencies (e.g., Alberta's Special Areas Boards and Canada's Prairie Farm Rehabilitation Administration). Governments sought to promote more sustainable drought-tolerant agricultural practices, and undertook water development initiatives, the construction of irrigation projects, dams, reservoirs and canals, all in an effort to adapt the expanding settlement to the natural climate and water resource conditions of the prairies.

In the last 100 years, at least 40 droughts have occurred in the Prairie Provinces.[50] Despite the intensive irrigation used in the basin, irrigation expansion is limited by water availability (surface water supplies are limited or too far away, and the water quality of most groundwater supplies is unacceptable for irrigation). It is a reality, therefore, that the predominant land use in the SSRB is dryland agriculture, which is completely reliant on natural precipitation. Because the region is exposed to repeated droughts and has water supply limitations, dryland agricultural practices have evolved. Today, successful dryland farming relies on careful management of the *vertical water balance.* This is achieved by utilizing natural precipitation (rainfall, snowmelt runoff) with available soil moisture and conservation techniques, to grow the best-adapted drought-tolerant crops suited to the climate and soil type (typically cereal crops).

Agricultural adaptations for water management have evolved as agricultural producers and society have come to better understand the natural prairie environment. Adaptations were developed to cope with a limited water supply for dryland agriculture, to adapt to the natural climate variability, and preserve the natural ecosystem. Adaptation can take place at the farm-level, the municipal-level, or through provincial/federal strategies and initiatives. Several examples follow.

Reduced tillage, minimum tillage, or zero tillage are conservation practices. In essence, these practices attempt to ensure that a portion of the fall-harvested field crops are left standing and new crops are seeded directly into the ground, thereby preserving organic matter and soil structure, conserving moisture within the soil, and reducing soil loss caused by wind erosion. The use of zero tillage has expanded rapidly and now extends to 60% of crop land in Saskatchewan and 48% in Alberta. Conventional tillage is becoming less common, covering only 18% of crop land in Saskatchewan and 25% in Alberta.[51] Such changes in tillage practice require improved technologies, changes in cultural practices (e.g., increased herbicide use), and different farm equipment (e.g., seed drills). The cost and the effort to utilize such adaptations are clearly factors affecting their uptake.

Management of marginal lands is a practice designed to preserve sensitive lands in a more sustainable natural grassland state. When the prairies were settled, governments offered land for agricultural settlement. At the time, governments and the pioneers did not realize the limitations of the natural environment and the unique characteristics of the dryland prairies. Many marginal lands (soil with little or no organic matter) were broken for agricultural crop production.[52] These sensitive lands were soon discovered to be unsuitable for sustainable agriculture. Most marginal lands have now been converted back to more natural prairie grassland vegetation with the support of provincial and federal funding programs.[53] Such programs are established to manage land in a more sustainable fashion and to encourage biodiversity. While the conversion does not fully restore native prairie or avian communities, avian breeding habitat and diversity has improved.[54] In addition to permanent cover programs, federal and provincial crown pastures were established in Saskatchewan. These pastures are managed to benefit agriculture and the natural environment, thereby encouraging the biodiversity of flora and fauna native to the prairie grassland region. The management of marginal lands has been successful largely because of shared federal-provincial programming and the willing voluntary participation of affected private landowners. Community pastures have provided environmental benefits with a percentage of cost recovery from user-fees (commercial livestock operators are charged a rate to allow their cattle to graze on federal crown lands).[55]

Preservation of wetlands and lakes is another practice that has occurred as a result of learning more about the necessary benefits of natural ecosystems. Over time, settlement, economic growth and expansion, and increasing agricultural production in the prairie pothole landscape have led to the drainage of many natural wetlands. In the last two or three decades the change to minimum tillage and conversion to grasslands caused increased trapping of moisture on the uplands and reduced runoff to wetlands. Both these changes would be expected to lead to a decrease in the number of viable wetlands, with possible negative impacts to biodiversity and impairment of natural water recharge regimes. The proportion of natural wetlands that have been drained by human activities (e.g., agriculture, urban and industrial development) is not well documented. Drainage impacts on wetlands vary from slight impacts to complete disappearance. Estimates of prairie pothole loss by drainage range from 17 to 71%.[56] A comprehensive study of wetland drainage in the Canadian prairie region found that a 5% loss of wetlands occurred in the period from 1985 to 2001; the study indicated that the annual loss rate of wetland numbers has remained fairly constant for several decades.[57]

Potential deleterious impacts from land use developments and farming practices on wetlands and waterfowl continue to be a concern. Lakes, especially those located in closed drainage basins, are sensitive to cumulative changes of evaporation and of runoff due to changes of land-use and climate. Many closed-basin prairie lakes have experienced declining water levels and increasing levels of salinity over much of the 20th century.[58] As we have learned more about the natural benefits of wetlands, government, environmental agencies and agricultural producers are promoting the preservation of wetlands as a vital feature of the prairie landscape. The Saskatchewan Watershed Authority provides guidance on management of wetlands,[59] while Alberta is developing a new Wetland Policy as a part of its *Water for Life* strategy.[60]

The practice of irrigation itself is arguably the most significant modern-day adaptive reaction to climate variability in the SSRB. Even with successful adaptations, dryland farming remains at-risk from water shortages during extended dry periods. Irrigation can reduce this risk in addition to the increased value added per hectare. The value of irrigation water for this 'drought proofing' ranges from $37 to $42 per dam^3 in the Alberta portion of the SSRB.[61]

Adaptations within irrigation have also had significant implications for water management in the basin.[62] Because of the high water consumption from irrigated agriculture (irrigation diverts, on average, about 29% of the natural water flow and consumes about 22% of the natural water flow in the SSRB), improving water use efficiency is a major consideration. Over the years, irrigation conveyance works have been improved to conserve water loss (e.g., lined canals, the use of pipelines as distribution networks), flood irrigation has been converted to sprinkler irrigation, and sprinkler irrigation technologies have been converted to improve efficiency (e.g., converting sprinkler nozzles to low-drift, low-energy nozzles to conserve water and energy). Changes in water management have also occurred to improve irrigation efficiency, distribution, and productivity (e.g., development of water markets and transfers).[63]

The return flow water quality from irrigation return flows must also be monitored and managed to minimize the risk of contamination from unused nutrients or pesticides. This is accomplished by using agricultural best management practices that reduce transport of nutrients or farm chemicals (e.g., timing, methods, and application techniques).

Irrigated agriculture is also affected by drought years when river flows do not permit the desired full water allocations. For instance, in the drought years of 2001–02, irrigators in Alberta made difficult decisions as a specific coping strategy in which some irrigators shared, sold or traded their licensed allocation to allow others to receive sufficient water for their crops. This unique adaptive management strategy is seen by the local irrigators and communities as evidence of a strong sense of community and a willingness to work together to get through difficult years.[64]

In-stream Flow Needs: The Ecological Needs of the SSRB River System

Modern approaches in water management recognize that water allocations and water use are limited. As water becomes increasingly allocated in any watershed, the net result is that flow regimes are altered. It is clearly not desirable to extract more water than the amount that would be required to sustain healthy natural aquatic ecosystems as this could have a significant impact on the natural environment.

The Alberta government is currently conducting studies to determine in-stream flow needs for river systems within the SSRB.[65] In 2006, the *Approved Water Management Plan for the South Saskatchewan River Basin (Alberta)* established *Water Conservation Objectives* considering the natural rate of flow and in-stream objectives.[66] The conservation objectives stipulate that water flows must remain above 45% of natural flows, or 10% above the previously established in-stream objective, whichever is higher. As shown in figures 5a and 5b, however, irrigation demands and the diversion of water has altered rivers flows so that, at times, recorded flows fall below 50% of natural flows based on the 1961–90 data. This factor has been recognized by Alberta Environment such that current water usage and management of river flows are closely monitored to meet these conservation objectives.

Saskatchewan, on the other hand, has not yet identified specific objectives for in-stream flow needs. The *Background Report: South Saskatchewan River Watershed* notes no allocations exist for environmental or in-stream water use within the watershed, and Fisheries and Oceans Canada has initiated studies to investigate in-stream flow needs in the North Saskatchewan River.[67] The Saskatchewan Watershed Authority's River and Lake Water Quality Monitoring Program, however, does collect data to develop in-stream flow requirements.[68]

Concluding Remarks: Looking Forward

The hydrology and climate of the South Saskatchewan River Basin is highly variable; drought, floods, and variable precipitation are recurrent and natural features within the SSRB. Human populations have adapted to the hydrological realities over the last 11,000 years as First Nations cultures co-existed with the river systems and available food sources in a nomadic fashion with hunting and gathering.

Modern society within the SSRB still relies extensively on agricultural production; therefore, it is susceptible to climate vagaries. Nonetheless, dryland agriculture, though dependent on natural precipitation from summer rains and winter snowmelt, has adapted well to these conditions by relying on cereal crops throughout the basin and raising livestock in locations with suitable pasture land as well as widespread adoption of conservation tillage.

While the SSRB supports a human population of over 2.2 million people, the urban domestic consumptive water needs—only about 2% of available supply—are insignificant in relationship to the water consumed by irrigated agriculture. Though irrigated agriculture is practiced on only 5% of the land, it reduces risk, provides economic benefits, and consumes over 90% of the human consumptive water needs, permanently removing around 22% of available water in the South Saskatchewan River.

Water is essentially fully allocated in most of the Alberta sub-basins of the SSRB, and expansion of water use is now limited for the Bow, Oldman, and South Saskatchewan River basins. Water conservation and in-stream flow needs to sustain ecological integrity and aquatic ecosystems are two key areas that are receiving more attention by governments and water stakeholders. There is a need for targeted scientific research on in-stream flow needs to protect the aquatic environment and guide water management in the SSRB.

Water, on the other hand, is not fully allocated in the Saskatchewan portion of the SSRB. Expansion of water use, particularly by proponents of irrigation, is possible, but has not occurred to the extent it has in Alberta. Estimates suggest Saskatchewan has an irrigation growth potential of five times current usage, with acreage rising from 81,000 ha to around 400,000 ha. This potential largely exists in the SSRB, or would be serviced by the South Saskatchewan River.[69,70] Environmental groups express concern about dam and reservoir development associated with irrigation expansion, and advocate water conservation as a first step.[71] For irrigated agriculture to fully develop in Saskatchewan, the economic, social and environmental benefits will need to be realized, supported and driven by multiple stakeholders from agriculture, to environmental advocates, to citizens who see value and sustainability of land and water resources. Major development is not likely to occur solely for economic reasons.

Looking forward, climate variability and concerns over climate change can be expected to place greater pressure on the hydrologic resources of the SSRB. Climate change scenarios for the 2050s anticipate an increase in summer temperature, precipitation and evaporation, and historical records of climate variability also suggest the SSRB has experienced decadal droughts in the past lasting tens of years in duration.[72] Together, this can lead to a net loss of available water with the SSR experiencing reductions in flow of about 8% at Lake Diefenbaker, and the region will experience more severe droughts. Future human water uses within the SSRB will need to account for climate induced water variability. With population and economic growth continuing at a brisk pace, water allocations, and the distribution of water, will be affected by increased competition for water and a growing societal concern for safeguarding ecological needs under a changing Prairie climate. Undoubtedly, water management in the South Saskatchewan River Basin will require even more effective adaptations to balance economic, social, and environmental challenges.

Endnotes

1. Natural flow data comes from Brenda Toth, Darrell R. Corkal, David Sauchyn, Garth van der Kamp, and Elise Pietroniro, "The Natural Characteristics of the South Saskatchewan River Basin: Climate, Geography and Hydrology," in this issue.
2. David Meyer and Dale R. Russell, "Aboriginal Peoples from the Ice Age to 1870," in *Saskatchewan: Geographic Perspectives* (University of Regina: Canadian Plains Research Center, 2007), 101–128.
3. Agriculture and Agri-Food Canada, *The Health of Our Water: Toward Sustainable Agriculture in Canada* (Ottawa, Canada: Minister of Public Works and Government Services, 2000), 10.
4. Statistics Canada, *Table 051–0034: Total population, census divisions and census metropolitan areas.*
5. Suren Kulshreshtha and Wayne Thompson, "Basin Water Use in Regional Economic Perspective: Rationale for Input-Output Analysis," in *Climate Change and Water: SSRB Final Technical Report* (Ottawa, Canada: Natural Resources Canada, 2007), 51–92.
6. Suren Kulshreshtha, "Economic Value of Water in Alternative Uses in the South Saskatchewan River Basin," (University of Saskatchewan: Department of Agricultural Economics Research Paper, 2006).
7. Joel F. Bruneau, "South Saskatchewan River Basin: Water Withdrawal Forecasts and Scenarios," (University of Saskatchewan: Department of Economics Research Paper, 2006).
8. Hydroconsult EN3 Services Ltd., *South Saskatchewan River Basin Non-Irrigation Water Use Forecasts,* (Alberta Environment, 2002).
9. Bruneau, "SSRB: Water Withdrawal Forecasts."
10. Kulshreshtha, "Economic Value of Water," 12.
11. Kulshreshtha, "Economic Value of Water," 12.

12. Irrigation Water Management Study Committee, *South Saskatchewan River Basin: Irrigation in the 21st Century. Volume 1: Summary Report* (Lethbridge, Alberta: Alberta Irrigation Projects Associations, 2002), 140.
13. Saskatchewan Agriculture, Food and Rural Revitalization, *Irrigation in Saskatchewan* [cited Dec. 21, 2007]. Available from the World Wide Web: (http://www.agriculture.gov.sk.ca/Default.aspx?DN=59abfaba-401e-49fb-91d0-d357692288d4), 9.
14. Statistics Canada, *2001 Census of Agriculture, data by province, Census Agricultural Region (CAR) and Census Division (CD): Land Use.*
15. Census divisions (CD) do not align perfectly with the SSRB drainage area. Using GIS mapping, we can identify the fraction of a CD that lies within the SSRB. We use this share to identify human and livestock populations as well as land use activities in the SSRB. See: Lawrence Martz, Robert Armstrong, and Elise Pietroniro, "The South Saskatchewan River Basin: Physical Geography," in *Climate Change and Water: SSRB Final Technical Report* (Ottawa: Natural Resources Canada, 2007), 31–40.
16. Irrigation Water Management Study Committee (IWMSC), *Irrigation in the 21st Century* and Saskatchewan Agriculture, Food and Rural revitalization (SAFRR), *Irrigation in Saskatchewan.*
17. Alberta Agriculture, Food and Rural Development, *Irrigation in Alberta: Facts and Figures for the Year 2004* (Lethbridge, Alberta: Irrigation Branch, Alberta Agriculture, Food and Rural Development, 2005).
18. SAFRR, *Irrigation in Saskatchewan.*
19. Brace Centre for Water Resources, *Analysis of Issues Constraining Sustainable Irrigation in Canada and the Role of Agriculture and Agri-Food Canada* (Regina, Canada: Prairie Farm Rehabilitation Administration, 2005), 12.
20. For more details see Margot Hurlbert, Darrell R. Corkal, and Harry Diaz, "Government and Civil Society: Adaptive Water management in the South Saskatchewan River Basin"; as well as Margot Hurlbert, "Comparative Water Governance in the Four Western Provinces." Both in this edition.
21. For a more detailed overview of key water institutions in the SSRB, see Darrell R. Corkal, Bruce Inch, and Philip E. Adkins, *The Case of Canada—Institutions and Water in the South Saskatchewan River Basin* [cited Dec. 18, 2007]. Available from World Wide Web: (http://www.parc.ca/mcri/pdfs/papers/iacc045.pdf).
22. Prairie Provinces Water Board, *Overview* [cited Dec. 18, 2007]. Available from World Wide Web: (http://www.mb.ec.gc.ca/water/fa01/fa01s01.en.html).
23. Prairie Provinces Water Board, *Master Agreement on Apportionment* [cited Dec. 18, 2007]. Available from World Wide Web: (http://www.mb.ec.gc.ca/water/fb01/fb00s04.en.html).
24. Sal Figliuzzi, *South Saskatchewan River Sub-basin Contributions to International and Interprovincial Water-Sharing Agreements* (Alberta Environment, 2002).

25. International Joint Commission, *What is the Boundary Waters Treaty?* [cited Dec. 21, 2007]. Available from World Wide Web (http://www.ijc.org/rel/agree/water.html), and *Accredited Officers for the St. Mary—Milk Rivers* [cited Dec. 21, 2007]. Available from World Wide Web: (http://www.ijc.org/conseil_board/st_mary_milk_rivers/en/smmr_mandate_mandat.htm).
26. Health Canada, *From Source to Tap—The Multi-Barrier Approach to Safe Drinking Water,* (Ottawa, Canada: Health Canada, 2002); and Saskatchewan Watershed Authority and South Saskatchewan River Watershed Stewards Inc., *South Saskatchewan River Watershed Source Water Protection Plan* (Regina: Saskatchewan Watershed Authority, 2007).
27. Government of Saskatchewan, *About SaskH$_2$o, Saskatchewan's Safe Drinking Water Strategy.* [cited Dec. 21, 2007]. Available from World Wide Web: (http://www.saskh2o.ca/about.asp).
28. Alberta Environment, *Water for Life,* 2003 [cited Dec 21, 2007]. Available from World Wide Web: (http://www.waterforlife.gov.ab.ca/docs/strategyNov03.pdf), p. 32.
29. For a more detailed review of the physical characteristics of the SSRB see Toth et al., "The Natural Characteristics of the South Saskatchewan River Basin," in this issue.
30. Environment Canada, *Municipal Use Database (MUD).*
31. P.J. Hess, *Ground-water use in Canada, 1981* (National Hydrology Research Institute, Inland Waters Directorate, 1986), 43.
32. John Lebedin, Harry Rohde, Daryl Jaques, and Mike Hammer, *Rural Water Well Infrastructure Assessment on the Prairies: An Overview of Groundwater Development and Reliance Trends* (Prairie Farm Rehabilitation Administration, 2000).
33. One cubic decametre, or one dam^3, is 10 m by 10 m by 10 m or 1,000 m^3. One dam^3 of water is 1,000,000 Litres. In Imperial units, one dam^3 of water is equivalent to 0.81 acre-feet of water (one acre-foot of water is a volume of water one foot deep over one acre of land).
34. Hess, *Ground-water use in Canada,* 43.
35. Municipal data includes household, commercial, and industrial users. Industrial activities outside of municipal distribution networks include water for aggregate washing, oil and gas injection, gas and petroleum processing, and construction. For more details see Alberta Environment, *South Saskatchewan River Basin Non-Irrigation Water Use Forecasts* (Alberta Environment, 2002).
36. Joel F. Bruneau and Brenda Toth, "Dove-Tailed Physical and Socioeconomic Results in the SSRB" in *Climate Change and Water: SSRB Final Technical Report* (Ottawa: Natural Resources Canada, 2007), 207–236.
37. Alberta Agriculture, Food and Rural Development (AAFRD), Irrigation: Facts & Figures.
38. It is important to note that, at no time, did actual diversions in the Alberta districts come close to exceeding water license allocations at 3.341 million dam^3.

39. Table 2 shows the relative importance of human activity for the basin overall. However, the data hides both the seasonal variation in demands and supply as well as the geographic variation that takes place. For a more detailed analysis see Bruneau and Toth, "Dove-Tailed Physical and Socioeconomic Results in the SSRB."
40. Bruneau, "SSRB: Water Withdrawal Forecasts."
41. IWMSC, *Irrigation in the 21st Century. Volume 1,* 81.
42. Bruneau, "SSRB: Water Withdrawal Forecasts."
43. IWMSC, *Irrigation in the 21st Century. Volume 1,* 71.
44. Agriculture and Agri-Food Canada, *Lake Diefenbaker and how it operates,* 2007 [cited Mar. 6, 2008]. Available from World Wide Web: (http://www.agr.gc.ca/pfra/pub/lakedb_e.htm).
45. Saskatchewan Watershed Authority, *Dams and Reservoirs, Lake Diefenbaker* [cited Dec. 21, 2007]. Available from World Wide Web: (http://www.swa.ca/WaterManagement/DamsAndReservoirs.asp?type=LakeDiefenbaker).
46. Gardiner Dam construction began in 1959 and was completed in 1967. While natural flow records are calibrated to the period 1961–90, the effect of Gardiner Dam on river flow occurred after its completion in 1967. For a more complete description of natural river flow see Toth et al., "The Natural Characteristics of the SSRB."
47. Jim Daschuk and Gregory P. Marchildon, "Overview of Climate and Aboriginal Adaptation in the South Saskatchewan River Basin before the Settlement Period" (University of Regina, Institutional Adaptation to Climate Change (IACC) Project Working Paper No. 6., 2005).
48. Gregory P. Marchildon, Suren Kulshreshtha, Elaine Wheaton, and Dave Sauchyn, "Drought and Institutional Adaptation in the Great Plains of Alberta and Saskatchewan, 1914–1939," *Journal of Natural Hazards* 45 (2008): 391–411.
49. Marchildon et al., "Drought and Institutional Adaptation in the Great Plains."
50. Marchildon et al., "Drought and Institutional Adaptation in the Great Plains."
51. Statistics Canada, *2006 Census of Agriculture, farm data and farm operator data: Tenure, land use, land management.*
52. The historic record of how the Prairies were in ecological and economic ruin after the decadal droughts in the 1920s and 30s is well documented by James H. Gray who records strong institutional efforts focused on research and demonstration of new methods of agriculture, based on focused regional science. The success was largely achieved through the determination of committed researchers from the federal Dominion Experimental Farms which formed the Prairie Farm Rehabilitation Administration (PFRA) in 1935. Gray recognizes these individuals as agents which prevented the Canadian West from becoming a desert, and states, "the conquest of the desert in the Palliser Triangle in the 1930s is the greatest Canadian success story since the completion of the Canadian Pacific Railway . . . Canada could not have survived, economically or politically, if this vast area had been permitted to go back to the weed-covered wasteland and short grass cattle range." James H. Gray, *Men Against the Desert* (Saskatoon, Modern Press, 1978) VII.

53. *Jill S. Vaisey, Ted W. Weins, and Robert J. Wettlaufer*, "The Permanent Cover Program—Is Twice Enough?" *in Soil and Water Conservation Policies and Programs Success and Failures (CRC Press, Boca Raton, 2000), 211–224.*
54. D. Glen McMaster and Stephen K. Davis, "An Evaluation of Canada's Permanent Cover Program: Habitat for Grassland Birds?" *Journal of Field Ornithology* 72 (2001): 195–210.
55. Agriculture and Agri-Food Canada, *Community Pasture Program Business Plan—Optimizing Program Performance in a Working Landscape* (PFRA Land Management Division, 2007).
56. B.D.J. Batt, M.G. Anderson, C.D. Anderson, and F. D. Caswell, "The Use of Prairie Potholes by North American Ducks," in *Northern Prairie Wetlands* (Ames, Iowa: Iowa State University Press, 1989): 204–227.
57. M.D. Watmough and M.J. Schmoll, *Environment Canada's Prairie and Northern Regions Habitat Monitoring Program Phase 2: Recent Habitat Trends in the Prairie Habitat Joint Venture. Technical Report Series No. 493* (Edmonton, Alberta: Environment Canada, Canadian Wildlife Service, 2007), 135.
58. Garth van der Kamp, D. Keir, and M.S. Evans, "Long-term water level changes in closed-basin lakes of the Canadian prairies," *Canadian Water Resources Journal* 33 (2008): 23–38.
59. Denis Huel, *Managing Saskatchewan Wetlands* (Regina, SK: Saskatchewan Wetland Conservation Corporation, 2000).
60. Alberta Environment, *Recommendations for a New Alberta Wetland Policy* (Alberta Water Council, 2008).
61. A. Samarawickrema and S. Kulshreshtha, "Value of Irrigation Water for Drought Proofing in the South Saskatchewan River Basin (Alberta)," *Canadian Water Resources Journal* 33 (2008): 273–281.
62. IWMSC, *Irrigation in the 21st Century.*
63. Alberta Environment, *Approved Water Management Plan for the South Saskatchewan River Basin (Alberta)* (Calgary, Alberta: Alberta Environment, 2006).
64. Monica Hadarits, *Lethbridge Stakeholder Workshop on Water and Climate: Summary Report* (Regina, SK: Institutional Adaptation to Climate Change Project, 2007).
65. Alberta Environment, *South Saskatchewan River Basin Water Management Plan—Phase Two Background Studies, Aquatic Environment In-Stream Flow Needs Determination for the SSRB Report* (Alberta Environment, 2003).
66. Alberta Environment, *Approved Water Management Plan for the SSRB.*
67. South Saskatchewan River Watershed Stewards, *Background Report South Saskatchewan River Watershed* (South Saskatchewan River Watershed Stewards Inc., 2007), 57.
68. Saskatchewan Watershed Authority, *State of the Watershed Report, 2007* (Regina, SK: Monitoring and Assessment Branch Stewardship Division, 2007).
69. Brace Centre for Water Resources, *Analysis of Issues,* 16.

70. Saskatchewan Agrivision Corporation Inc., *Water Wealth: A Fifty Year Water Development Plan for Saskatchewan* (Saskatoon, SK: Saskatchewan Agrivision Corporation Inc., 2004), 219.
71. Saskatchewan Environmental Society, *Working for a Sustainable Future* [cited: Mar 7, 2008]. Available from the World Wide Web: (http://www.environmentalsociety.ca/issues/water/index.html).
72. See Suzan Lapp, David J. Sauchyn, and Brenda Toth, "Constructing Scenarios of Future Climate and Water Supply for the SSRB: Use and Limitations for Vulnerability Assessment," in this issue.

CONSTRUCTING SCENARIOS OF FUTURE CLIMATE AND WATER SUPPLY FOR THE SSRB: USE AND LIMITATIONS FOR VULNERABILITY ASSESSMENT

SUZAN LAPP, DAVID J. SAUCHYN, AND BRENDA TOTH

Abstract:

This paper describes scenarios of climate change and water supply constructed to assess vulnerability of communities to future conditions in the South Saskatchewan River Basin (SSRB). Output from five Global Climate Models (GCMs) forced with various future emission scenarios were used to construct a range of future scenarios of temperature and precipitation (i.e., median, warmest-wettest, warmest-driest, coolest-wettest, and coolest-driest) over the SSRB. Downscaling using the stochastic weather generator, LARS-WG, was also carried out at Lethbridge, Alberta, and Swift Current, Saskatchewan, and results were compared to future scenarios derived using the coarse resolution GCMs solely. The results between the GCM and LARS-WG scenarios were comparable; both showing increases in monthly temperatures, increases in winter precipitation, and typically decreasing summer precipitation, but with amplified variability. Scenarios of future flows of the South Saskatchewan River and its tributaries were derived by coupling the HadCM3 TAR model scenarios with the hydrological model WATFLOOD. Flow decreased and the dominant flow season shifted from summer to spring for some rivers.

Sommaire

Cet article établi des scénarios de changements climatiques et d'approvisionnement en eau afin d'évaluer la vulnérabilité des collectivités aux conditions futures dans le bassin de la rivière Sakatchewan-Sud (BRSS). Les résultats obtenus de cinq modèles du climat du globe (MCG) et de cinq différents scénarios concernant les futures émissions furent combinés pour établir une gamme de scénarios de températures et de précipitations (c.-à-d. médiane, plus chaud-plus humide, plus chaud-plus sec, plus froid-plus humide, plus froid-plus sec) pour le BRSS. Une réduction à l'échelle à l'aide d'un générateur stochastique de conditions météorologiques, le LARS-WG, a été effectuée à Lethbridge, en Alberta, et à Swift Current, en Saskatchewan, et les résultats furent comparés seulement à des scénarios futurs issus des MCG à résolutions grossières. Les résultats des scénarios des MCG et du LARS-WG étaient similaires. Les deux scénarios indiquaient une augmentation des températures mensuelles, une augmentation des précipitations l'hiver, et généralement une diminution des précipitations en été, mais avec une plus grande variabilité. Des scénarios projetant les futurs débits d'eau de la rivière Saskatchewan-Sud et de ses tributaires furent établis en couplant des scénarios issus des modèles HadCM3 TAR avec le modèle hydrologique WATFLOOD. Les scénarios prévoient une baisse du débit et que la saison principale d'écoulement passerait de l'été au printemps pour certaines rivières.

1. Introduction

A key component of the conceptual framework, and associated research methodology, of the IACC project is the assessment of future vulnerability to climate change. Achieving one of the project objectives "To examine the effects of climate change risks on the identified vulnerabilities"[1] requires that we develop future scenarios for climate risks identified by the studied communities. Those risks or current vulnerabilities, discussed in other articles in this volume, are mostly related to water. The river basins, the Elqui in Chile and South Saskatchewan in Canada, are arid and semi-arid, respectively, and dry environments have the most variable hydroclimate. In the southern part of Canada, the highest year-to-year variation in precipitation is in the prairies. The only region with a coefficient of variation (standard deviation/mean) above 25% nearly coincides with the South Saskatchewan River Basin (SSRB).

In the SSRB, the climate risks most often identified in the studied communities of Hanna, Taber, Outlook, Cabri, Stewart Valley and Blood Indian Reserve were drought, extreme weather events, such as intense thunderstorms and associated hail and flash floods, and low river flows affecting potable water. The vulnerability of these rural communities to climate change will depend on the extent to which these regional climate risks are affected by global warming. This paper presents scenarios of the future climate and hydrology of the SSRB using outputs from runs of Global Climate Models (GCMs) forced with anthropogenic greenhouse gases to simulate global warming.

A climate change scenario is a plausible representation of a future climate that is constructed from consistent assumptions about future emission of greenhouse gases (GHGs) and other pollutants, for explicit use in investigating the potential impacts of anthropogenic climate change.[2] Scenarios are not forecasts of future climate, but rather are intended to provide adequate quantitative measures of uncertainty represented with a range of plausible future paths.[3] Future greenhouse gas concentrations are an unknown because we cannot predict the kinds and extent of activities humans will engage in that will reduce or increase them.

Three types of climate scenarios provide input to hydrological, agricultural, socio-economic, and biophysical models for impact and sensitivity studies: (1) synthetic; (2) analogue (temporal and spatial); and, (3) derived from GCMs. Synthetic and analogue scenarios capture a wide range of possible future climates, and are useful for identifying thresholds or discontinuities of response beyond which effects are no longer beneficial or are detrimental.[4] [5] Synthetic scenarios apply an arbitrary change to a particular variable of an observed time series; for example, adding 2° C to the monthly average temperature. This new time series, however, maintains the variability of the original time series. Analogue scenarios represent potential future climate by using the observed climate regime from a typically warmer previous period or other location as an anticipated future climate. Temporal analogues are derived from either instrumental or paleoclimatic records. As these scenarios represent real historical climate states, they are physically possible and can be constructed for various climate variables. With spatial analogues one weakness is the lack of correspondence between climatic and non-climatic features between regions; therefore, these scenarios may not represent physically plausible scenarios for conditions in the study region. Also, most drivers of the analogue climates are likely natural variations, rather than a response to GHG-induced warming.

Global climate models "are the only credible tools currently available for simulating the response of the global climate system to increasing greenhouse gas concentrations"[6]; therefore, they were used in this study for the construction of future climate scenarios. GCMs are fully coupled mathematical representations of the complex physical laws and interactions between ocean/atmosphere/sea-ice/land-surface.[7] They simulate the behaviour of the climate system on a variety of temporal and spatial scales using a three-dimensional grid over the globe. A high level of confidence can be placed in climate models based on the fact they are[8]: (1) fundamentally based on established physical laws, such as conservation of mass, energy and momentum, along with numerous observations; (2) able to simulate important aspects of the current climate; and, (3) able to reproduce features of past climates and climate changes. Climate models have accurately simulated ancient climates, such as the warm mid-Holocene of 6000 years ago and trends over the past century, combining both human and natural factors that influence climate.

GCM experiments simulate future climate conditions based on estimated warming effects of carbon dioxide (CO_2), other GHGs, and the regional cooling effects of increasing sulphate aerosols beginning in the late 19th century or early 20th century using scenarios of future radiative forcing. The Intergovernmental Panel on Climate Change Third Assessment Report (IPCC-TAR)[9] published 40 different emission scenarios providing a range of future possible GHG emissions and atmospheric concentrations from socio-economic scenarios labelled SRES (Special Report on Emission Scenarios).[10] The recent IPCC Fourth Assessment Report (AR4)[11] describes the latest vintage of GCMs and experiments currently available. Most GCM experiments also consist of multiple (or ensemble) simulations for each of these experiments representing different initial boundary conditions of the GCM at the beginning of the experiment. Impact studies are adopting the combination of scenarios derived from ensemble simulations and scenarios reproducing multi-decadal natural climate variability from long GCM control simulations.[12]

2. Scenarios of the Future Climate of the SSRB Derived from GCMS

2.1. Constructing Climate Change Scenarios

Developing a climate scenario for an impact study requires that data for the relevant climate variable(s) be available from both the GCMs and the 'climatological' record, for two time periods typically each of 30 years: some future time period such as the 2020s, 2050s or the 2080s (i.e., 2010–2039, 2040–2069, and 2070–2099) and the baseline climate (1961–1990)*(see description below). A climate change scenario constructed using GCM output is typically expressed as a percentage change in precipitation or temperature change in degrees from a mean baseline of 1961–90 to a future 30-year period. These differences or ratios are then used to adjust the observed climatological baseline dataset to develop a future climate scenario. A 30-year time period is used to differentiate between the climate change signal and the inter-annual and inter-decal variability within the time series.

One of the limitations of GCMs for constructing climate change scenarios is the difference in climate sensitivity between models. Due to parameteriza-

---

* The baseline dataset or 'normal' period is representative of the observed, present-day meteorological conditions and describes the average conditions, spatial and temporal variability and anomalous events. The current 30-year normal period as identified by the WMO (World Meteorological Organization) is 1961–1990.[14]

tion and simplification of modeling processes and feedbacks, GCM simulations may respond quite differently to the same forcing.[13] While GCMs probably capture a large part of the uncertainty in modeling responses, they do not encapsulate the range of uncertainties in future emission scenarios. By choosing an array of GCMs and several future emission scenarios (SRES), a broad range of future climate scenarios (e.g., warmest-wettest, warmest-driest, coolest-wettest, and coolest-driest) can be generated to capture much of the uncertainty.

2.2. Data

The Fourth Assessment Report (AR4) of the IPCC[15] lists 24 climate models; only seven had the required variables of daily maximum/minimum temperature and precipitation for both the 1961–90 historical and future 2040–69 period (Table 1). The SRES[16] experiments A1B, A2 and B1 were available for the AR4 models. Output from the HadCM3 TAR (Third Assessment Report) also were used because the AR4 did not provide the required variables for this model; the A2 and B2 experiments were available for various runs of the HadCM3 TAR model, and this GCM has been used for previous climate change research in western Canada.[17] [18] [19] GCM output for the AR4 models (daily and monthly) was obtained from the World Climate Research Programme's (WCRP's) Coupled Model Intercomparison Project phase 3 (CMIP3) multi-model dataset.[20] Monthly and daily HadCM3 TAR output was available, respectively, from the IPCC Data Distribution Center[21] (IPCC-DDC) and the Climate Impacts LINK Project.[22] The 20th Century experiment (20CM) from each GCM provided the baseline period (1961–1990) output from the AR4 models and the SRES scenario experiments for the HadCM3 model.

Baseline observed historical gridded (0.5°) climate data (monthly precipitation, maximum and minimum temperature), covering North America from 1901–2000, were recently generated by the Canadian Forest Service.[23] The GCM output was interpolated to the historical data 0.5 degree grid, using the linear interpolation routine in Matlab 7.1, and mapped to illustrate future climate scenarios for the basin.

2.3. Future Climate Change Scenarios: SSRB

Figure 1 is scatter plot of climate change scenarios, the change in summer season average temperature (degrees C) and precipitation (%), for the 2050's (2040–69) relative to the 1961–1990 period for the seven GCMs and experiments. Five model experiments were chosen to represent the range of possible climates: MIROC3.2 MEDRES A2(1) (warm/dry: +3.3° C/-17.3%), HadCM3 TAR a2(1) (warm/wet: +2.9° C/+4%), CGCM3.1/T47 B1(1) (cool/dry: +2.1° C/-6.5%), CSIRO MK3.0 A1B(1) (cool/wet: +1.3° C/+10.3%), and CGCM3.1/T47 B1(2) (median: +2.2° C/+2.2%).

Figure 2 presents maps of the SSRB showing the annual average temperature (°C) change scenarios for the 2050s for the five climate models relative to the 1961–90 baseline period. Temperature increases range from +1 to +3.50C, with the greatest increase projected in the eastern part of the basin. Figure 3 maps the annual precipitation change scenario (%) throughout the basin for the 2050s period relative to the baseline. MIROC3.2 MEDRES A2(1) is the only GCM projecting a decrease (in the central area of the basin); all others show an increase in annual precipitation over the entire basin. These projections of increased annual temperatures, amplified during winter, and decreased summer and increased winter/spring precipitation, also extend over most of southwestern Canada.[24]

Spring/summer soil moisture is essential in the SSRB particularly for dryland farming. The Climate Moisture Index (CMI), the difference between annual precipitation (P) and annual potential evapotranspiration (PET), is a fairly simple indicator of soil moisture.[25] [26] PET values were calculated using the Thornthwaite equation because of its relatively simple routine; only mean monthly temperature and day length are required, and data for limited variables are available from the GCMs. Figure 4 maps the May-June-July P-PET (mm) for the 1961–90 period, 2020s, 2050s, and the 2080s for the median CGCM 3.1/T(47) B1(2) scenario. Overall, with decreased precipitation and increased temperatures in summer, the climate moisture index is decreasing (increasing soil moisture deficit) throughout the basin, particularly in the central and eastern portions which rely heavier on rainfall for soil moisture compared to irrigated areas. Areas that have access to irrigation or supplementary water may actually benefit from increases in summer temperatures; different crop type choices may be available with a longer growing season and more heat units.[27]

2.4. Future Climate Scenarios: Specific Sites

The GCM grid cells centred on Lethbridge, Alberta, and Swift Current, Saskatchewan, were selected for a more detailed analysis of future monthly minimum/maximum temperature and precipitation scenarios (Figure 5). Swift Current has a slightly cooler climate, particularly during the winter months, compared to Lethbridge; however, the two stations have similar seasonal precipitation distribution and annual amounts. GCMs project greater increases in minimum temperature (5° C at Swift Current and 4° C at Lethbridge) than maximum temperature during winter and spring. Increases in maximum monthly temperature of 1–4° C are more consistent among months.

Monthly scenarios of future precipitation (Figure 5) show that most models project changes in seasonal variability at both sites with higher winter and lower summer precipitation; annual total precipitation increases. Future monthly variability is highest during the late spring through summer, and into early fall, which is when most of the annual precipitation falls. This summer/fall variability relative to winter likely reflects the weaker ability of the models to simulate convective precipitation than frontal. Lethbridge tends to have an overall increase in all seasons except summer, with June precipitation decreased by nearly 10 mm. Most models also show decreases during July through September, leaving this season in a moisture deficit. Similarly, at Swift Current the majority of models show a drier summer and more variability with otherwise more precipitation particularly in winter and spring.

3. Scenarios of Future Flows of the SSRB Derived by Coupling GCM Scenarios and Hydrological Models

To assess the future trends in streamflow for the SSRB, GCM estimates of future temperature and precipitation were used in a simple hydrological model, WATFLOOD. This modeling exercise serves as an example of the applicability of GCM scenarios in future planning of water resources. WATFLOOD is a physically-based hydrologic model[28] representing the dominant vertical fluxes, precipitation, interception, infiltration. This model provides a horizontal utility that allows for overland, inter- and base flows to route water to the channel; the routing aspect demands that the model is fully distributed in space. The only distributed forcing time series data necessary for WATFLOOD are temperature and precipitation, making it suitable for coupling with GCM indicators of future climate. Other pieces of information used within WATFLOOD are elevations, the extent of the watershed, and other topographic features such as land cover characteristics, reservoir and channel properties.

Initially, the hydrologic model was run for the current climate; WATFLOOD was driven using current climatology in the form of station-observed temperature and precipitation data. The flows generated were compared to naturalized streamflow to ensure that the hydrologic model could replicate streamflow. In order to assess the influence of climate change within the South Saskatchewan River Basin, selected IPCC TAR temperature and precipitation change scenarios were used as input, and future scenario modeled flows were compared to the modeled current flows.

There are 21 stations with complete 1961–90 temperature and precipitation records. As high resolution spatial precipitation gauging enhances accuracy when modeling channel output, additional precipitation records were extracted to force WATFLOOD for the 1961–90 period.[29] There are 775 climate stations that have daily precipitation records for more than five years during the 30-year period.

The choice of climate scenarios was based on an analysis[30] of all publicly available GCM scenario outputs. The potential change in temperature was applied as offsets and precipitation was normalized and these values were applied during a data gridding process, resulting in anticipated distribution of future temperature and precipitation across the basin. Due to the coarse temporal and spatial scale of the available GCM data, a spatially weighted average change in temperature and precipitation was applied over the entire basin at a monthly time step. The significance is that the temperature and precipitation patterns of the current climate were simply replicated with a step change to provide the estimate of future climate temperature and precipitation patterns.

The example presented in this paper is the streamflow estimate yielded from the suggested climate change as represented by the A2(1) scenario of the Hadley HadCM3 TAR GCM. The projected HadCM3 changes to 1961–90 temperature and precipitation are presented in Figure 6, and are essentially a moderate view of future climate with a mean increase in temperature of less than 2.5°C and a slight increase in precipitation of about 6.4%. The GCM projects that all seasons will be warmer, and that fall (Sept-Nov), winter (Dec-Feb), and spring (Mar-May), will be wetter, while the summer months (Jun-Aug) will be drier.

The 29 years of simulated flows (October 1, 1961 to September 30, 1990) for each of the six scenarios were averaged for each month, and the A21 scenario results are shown in Figure 7. While the perturbations are applied uniformly across the basin, the response in annual future flows differs between the sub-basins (see Figure 7). The Oldman basin shows flows the least affected by the future change in climate with an average annual reduction in flow of 0.01%, with the Bow River showing changes of -7.6%, the Red Deer -12.6%, and the average decrease in flows of the South Saskatchewan into Lake Diefenbaker was 8.5%.

There are differences in the seasonality of current modeled flows across the basin (Figure 8). With current climatology, the flows in the Oldman basin are spring dominated, while the Bow River at Calgary exhibits summer dominated flows. Further downstream the current flows for the Bow River at the mouth are approximately equal in the spring and summer; similarly, current summer and spring flows are equally weighted for the South Saskatchewan at Lake Diefenbaker, although there is slightly more summer than spring flow to the reservoir.

Future flows for the Bow River at Calgary are also summer dominated, although the spring to summer flow ratio increases slightly, and future downstream flows at the mouth shift to a spring dominated flow system. The Oldman system essentially retains its spring flow dominated system. The integration of flows in the South Saskatchewan River shifts from a slightly summer dominated system in current flows, to a future flow regime that is slightly dominated by spring flows.

4. Limitations of GCMs as the Source of Climate Change Scenarios

While climate models are "the only credible tools"[31] currently available for simulating future climate scenarios, they have limitations that apply in general to climate impact studies and, specifically, to our attempt to link future climate to current climate risks in the SSRB. The coarse spatial resolution (100s km) is a commonly cited drawback of GCM derived climate scenarios. This problem is particularly acute for a study like ours where the aim is to evaluate the vulnerability of individual rural communities by providing future climate and water scenarios. Fortunately much of the SSRB, beyond the eastern slopes of the Rocky Mountains, has relatively low relief and homogeneous land cover. Even so, we are applying single values of climate variables for GCM grid boxes (thousands of km^2) to small rural communities. Various methods of downscaling have been developed to overcome this spatial resolution limitation downscaling using relationships between observed large-scale atmospheric processes and station-scale data. Downscaling provides information required for water resources management at scales much finer than the current resolution of any GCM for the interpretation of impacts related to climate change or climate variability.[32]

4.1. Downscaling: Two Approaches

The two common approaches to the downscaling of climate scenarios are dynamical and statistical. The confidence that may be placed in downscaled climate change information is foremost dependent on the validity of the large-scale fields from the GCM. Dynamical downscaling involves the use of high-resolution (regional) climate models (RCMs) to obtain finer resolution climate information from coarse-scale GCMs.[33] RCMs are nested within a GCM that provides the initial and lateral boundary driving conditions. The RCM incorporates better parameterization and more direct representation of some small, fine-scale processes and features such as topography and land cover inhomogeneity.[34] RCMs are more computationally demanding than global-scale models[35]; therefore, RCM data are typically available for only one run of a single model for limited time spans. The latest version of the CRCM.4.2.0 (Canadian Regional Climate Model) has output available for the CGCM3 SRES A2 and the time period of 2041–2070 with a 45-km horizontal grid-size[36]; previous versions of CRCMs provided data for a 20-year window (2046–2065). As illustrated in Figure 1, on the other hand, there are numerous runs of various GCMs providing a range of future climates or scenarios.

Statistical downscaling methods are much more popular than dynamical downscaling techniques for deriving future climate scenarios; they are a cheap way of obtaining climate change data at higher temporal or spatial resolution than can be provided by the GCM. The statistical downscaling of GCM data is based on a statistical model linking the climate simulated by the GCM and the current climate characterized by instrumental data. This technique has been widely applied to derive daily and monthly precipitation at higher spatial resolution for impact assessments.[37, 38] SDSM (Statistical Downscaling Model)[39] and LARS-WG (Weather Generator)[40] are two popular methods of statistical downscaling.

Here we provide downscaled future scenarios for Lethbridge and Swift Current using the LARS-Weather Generator and ask "Does downscaling provide better results than the climate scenarios derived from GCMs and presented in section 2.4?"

Future climate change scenarios were derived for the 2050s at Lethbridge and Swift Current using the five GCMs described in 2.3 above to adjust the LARS-WG parameters. Monthly-observed homogenized precipitation and minimum/maximum temperature datasets, used to calibrate the model, were obtained for the entire study area for 1961–90 from Environment Canada.[41, 42] Figure 5 compares the future monthly minimum/maximum and precipitation climate scenarios at the two stations between those derived using GCMs (section 2.4) and downscaled with LARS-WG. Overall, the results of the downscaled monthly station data are very similar to those derived from the GCMs. The principal difference between the two scales of climate scenarios is the monthly precipitation variability. At Lethbridge, the LARS-WG results show greater variability for the months of February, June, August, November and December. At Swift Current, March, April and August have greater precipitation variability than derived using GCMs. The scenarios from the coarse resolution GCMs are not substantially different from those derived using downscaling. Downscaling is more labour intensive, suggesting, in this case, that downscaling does not necessarily provide better results.

LARS-WG also generates a series of wet and dry days, and the agriculturally important extreme events of frost and high temperature. At both stations there is a decreased number of days below freezing (<0° C) and an increased number of hot days (>30° C) in the summer months (late spring into early fall; Figure 9). In July and August the models project that the number of days >30° C could double. Changes to the length of wet and dry spells are variable and fluctuate around the 1961–90 average monthly number of days; therefore, it is difficult to draw any conclusions. On average, the majority of the models favour increasing wet spell length for the winter months and increasing dry spell length for the late summer months. We are less concerned about changes in the winter months as compared to spring and summer when the impacts of drought are more severe.

4.2. Climate Variability

Because GCMs simulate atmospheric and oceanic states and processes at a global scale, the most robust projections are for the largest areas and for multi-decadal mean values; thus, the IPCC-TGICA[43] places highest confidence in the projections of global temperature trends expressed as the difference between a baseline of 30 years and a future 30-year time-slice. The GCM projections have decreasing reliability for progressively smaller areas and time periods, and for water-related variables versus temperature. Here we are interested in a single watershed and water-related events; thus, the coarse resolution is a constraint, although the SSRB spans 5–11 GCM cells (depending upon the grid cell size of the GCM). More problematic, however, is the greater sensitivity of the rural communities to climate variability, departures from mean conditions, than climate change, to shifts in the mean. Most of the climate risks identified by the communities are departures from mean conditions (e.g., droughts, floods, frost) and not a shift in mean conditions as projected by GCMs. Data on future extremes and variability have been extracted from GCMs for large regions; but, once again, the reliability declines as the region of interest decreases in size.

Despite these constraints, we derived, with caution, some scenarios on the degree of variability that might be expected under the climate change scenarios presented above. This analysis is a preliminary step to investigating long-term trends and variability of future climate scenarios using the Climate Moisture Index of P-PET, as modeled by the CGCM3.1/T67 covering the total Prairie Provinces area. An eight-year low pass digital filter[44] was used to smooth the annual P-PET for each grid cell to retain information that was coarser than the frequency of eight years. Principle Component (PC) analysis was used to analyze and compare trends between the 1961–90 observed P-PET and the 20th century modeled P-PET and the future SRES (A1B, A2, and B1) P-PET experiments. The B1 scenario is close to the median, and the A2/A1B scenarios tend to be warmer/drier scenarios (Figure 1) relative to all the models tested. Figure 10a shows the standardized PC1 for the change in P-PET over the Prairie Provinces; negative/positive values represent drier/wetter conditions. There is a strong Pacific Decadal Oscillation (PDO) pattern evident, switching from a positive phase in 1947, to negative and again positive in 1977. During negative PDO phases the sea surface temperature is warmer in the North Pacific and cooler along the west coast of North America, bringing generally cooler wet conditions to the Prairies[45]. The opposite pattern exists for positive PDO phases, and we expect drier conditions. The negative PDO pattern dominates during 1947–1976, but positive PDO indices occur in 1957–58 and 1969–1970, and are observed in the P-PET PC1 as negative values during these periods. During the 1977–78 "regime shift,"[46] when the

PDO shifted from a negative to positive phase (Figure 11), the P-PET for the PC1 values become negative (Figure 10a). This teleconnection between positive PDO phases and dry periods on the prairies has been well documented.[47 48 49] Our preliminarily results indicate that, at timescales great than eight years, the 20th century model reproduces a similar pattern to the observed data (see Figure 10a), placing increased confidence in the GCMs ability to produce future climate scenarios.

Future P-PET scenarios based on the standardized PC1 and SRES experiments also maintains similar natural variability as the observed and modeled P-PET (Figure 10b), but an increased moisture deficit (more negative P-PET) is projected by the three experiments extending to the end of this century. This decline in moisture is associated with decreasing precipitation and increased temperatures during the spring/summer period.

The variability of future climate will be a function of the natural climate cycles modulated by greenhouse gas warming. Research on the nature and degree of this modulation is in early stages, but one approach involves the analysis of natural climate variability that will underlay the variation in future climate. Historical weather data contain detailed information on the variability of climate at daily to decadal scales; however, these records are relatively short in western Canada, at most 120 years, and mostly considerably shorter. Longer records that pre-date the instrumental period are available from climatically sensitive geological and biological archives. Figure 12 is a plot of the annual flow of the South Saskatchewan River for the period 1402–2002. Axelson[50] developed a statistical relationship between streamflow and tree growth in the basin; both respond to the effective precipitation (P-PET) that recharges the soil moisture balance, and is discharged from the basin in the stream channel. This plot of departures from the mean flow illustrates that negative departures or drought occurs periodically and with greater duration and severity before the 20th century; thus, communities in the basin can expect severe and prolonged drought, simply because it is characteristic of the long-term hydroclimatic variability, with or without human-induced global warming.

5. Conclusions

Output is available from various GCMs to develop scenarios of the future climate of the SSRB, and model stream flows, and thereby assess future vulnerabilities and climate risks. Application of these scenarios to vulnerability assessment, however, requires an understanding the source, derivation and limitations of the scenarios because model projections are only simulations of possibilities. Some climate risks cannot be properly evaluated because current models and methods do not provide reliable information at the relevant spatial and temporal scale for the variables identified by stakeholders. For example, one important variable not examined is the change in wind frequency and speed, which can have dramatic effects on snow sublimation and evaporation of water from soil and storage ponds.

GCM scenarios suggest the SSRB will experience an increase in both temperature and precipitation by 2050. Less precipitation is expected in summer, and more in winter, increasingly in the form of rain with rising temperatures. Warmer temperatures will result in a longer growing season, but there also will tend to be less available soil moisture in mid to late summer. The projected changes in temperature will influence snow accumulation in the mountains which feed the rivers that communities depend on for their water supply. Decreased runoff, and a shift in the dominant flow season from summer to spring, will cause river flows to decrease throughout the summer and fall months. Increased temperatures will result in an increased number of days with net positive evaporation from soil, storage dugouts, rivers, lakes, and reservoirs. When a median GCM scenario is coupled with a hydrological model, mean flows for the 2050s are reduced for all the major streams in the SSRB. The flows of the South Saskatchewan River into Lake Diefenbaker are reduced by 8.5%. Also with earlier peak flow, there is an increase in the ratio of spring to summer flow.

These new scenarios of the future climate and surface water supplies of the SSRB represent the shift in average conditions that communities and institutions can expect. This is critical information in anticipation of the impacts of climate change, but not necessarily the most relevant information for the rural communities, or at least those studied by IACC project researchers.[51] [52] [53] The major climate risks (drought, flooding, storms) and vulnerabilities identified through the community assessments are departures from mean conditions and sensitivities to this climate variability and extreme events; therefore, this paper supplemented the conventional GCM scenarios of climate change with sources of hydroclimatic data at finer scales: downscaled GCM output, annual Climate Moisture Index (CMI) data from a GCM, and proxy climate (tree-ring) records that capture the natural hydroclimatic variability that underlies the trends imposed by global warming. These approaches and information are necessitated by the nature of vulnerability to climate change on the Canadian plains as revealed through the "bottom up" approach to vulnerability assessment undertaken in the IACC project. The "top down" approach of providing conventional GCM-based scenarios of shifts in mean conditions also is useful in terms of informing stakeholders about the directions of climate change, whether or not they perceive these trends as immediate climate risks.

Our scenarios of the mean conditions and variability of future climate and water resources have considerable implications for economic, environmental and social processes within the SSRB. The forces driving the Prairies' climate, its variability and its water resources, need to be understood in greater depth for society to be better prepared for the future. Planning and implementing adaptation to climate change requires communities and institutions to develop practices and policies that can be implemented when there is uncertainty about future conditions.

Tables and Figures

CLIMATE MODELING CENTRE	MODEL	SRES SIMULATION	GRID CELL SIZE (DEGREES)	DIMENSIONS
Canadian Centre for Climate Modelling and Analysis Canada	CGCM3 (T47)	A1B*, A2*, B1*	3.75° × 3.75°	238.125–256.875W 44.52–55.77N
	CGCM3 (T63)	A1B, A2, B1	2.81° × 2.810	241.8725–258.7475W 46.04–57.29N
Met Office Hadley Centre UK	HadCM3	A2*, B2 (TAR)	3.75° × 2.55°	241.885–260.625W 46.25–56.25N
National Institute for Environmental Studies Japan	MIROC3.2-MEDRES	A1B*, A2*, B1*	2.8125° × 2.8°	240.468–257.343 W 44.64–55.84N
Geophysical Fluid Dynamics Laboratory USA	GFDL 2.0	A1B, B1	2.5° × 2.00	241.875–259.375W 46–56N
Max-Planck-Institut for Meteorology Germany	ECHAM5-OM	A1B, A2, B1	1.875° × 1.87°	240.9375–257.8125W 46.629–55.979N
Australia's Commonwealth Scientific and Industrial Research Organization Australia	CSIRO-MK3.0	A1B, A2, B1	1.875° × 1.87°	240.9375–257.8125W 46.6312–55.637N

*More than one experiment was carried out for these emission scenarios.

Table 1. Information about the climate models chosen for this study: the country of origin, SRES simulations available, grid cell size and dimensions of the area for each model. The output is available from the IPCC Fourth Assessment Report (2007) for all models except the HadCM3 from the Third Assessment Report (2001) at the IPCC Data Distribution Centre (http://www.ipcc-data.org/) and Program for Climate Model Diagnosis and Intercomparison (http://www-pcmdi.llnl.gov).

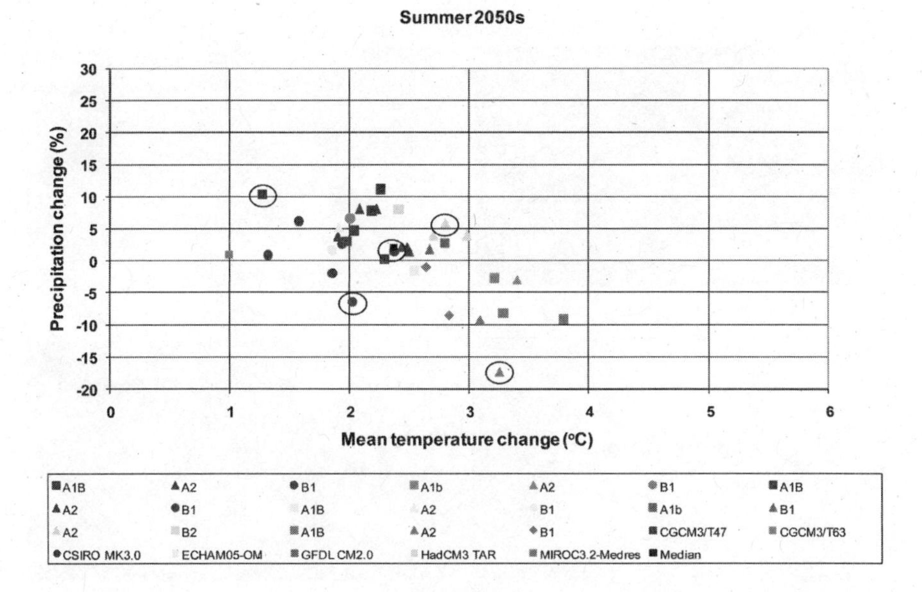

Figure 1. Scatter plot indicating mean temperature (°C) and precipitation (%) change for the SSRB for the 2050s summer. The models were chosen based on the availability of the required climate variables: daily minimum and maximum temperature and daily precipitation. The colours correspond to the GCM and the symbols identify the scenario. MIROC Medres A2(1) (warm/dry), HadCM3 TAR A2(a) (warm/wet), CGCM3.1/T47 B1(1) (cool/dry), CSIRO MK3.0 A1B(1) (cool/wet), CGCM3/T47 B1(2) (median) are all circled. (Value in brackets identifies the run number).

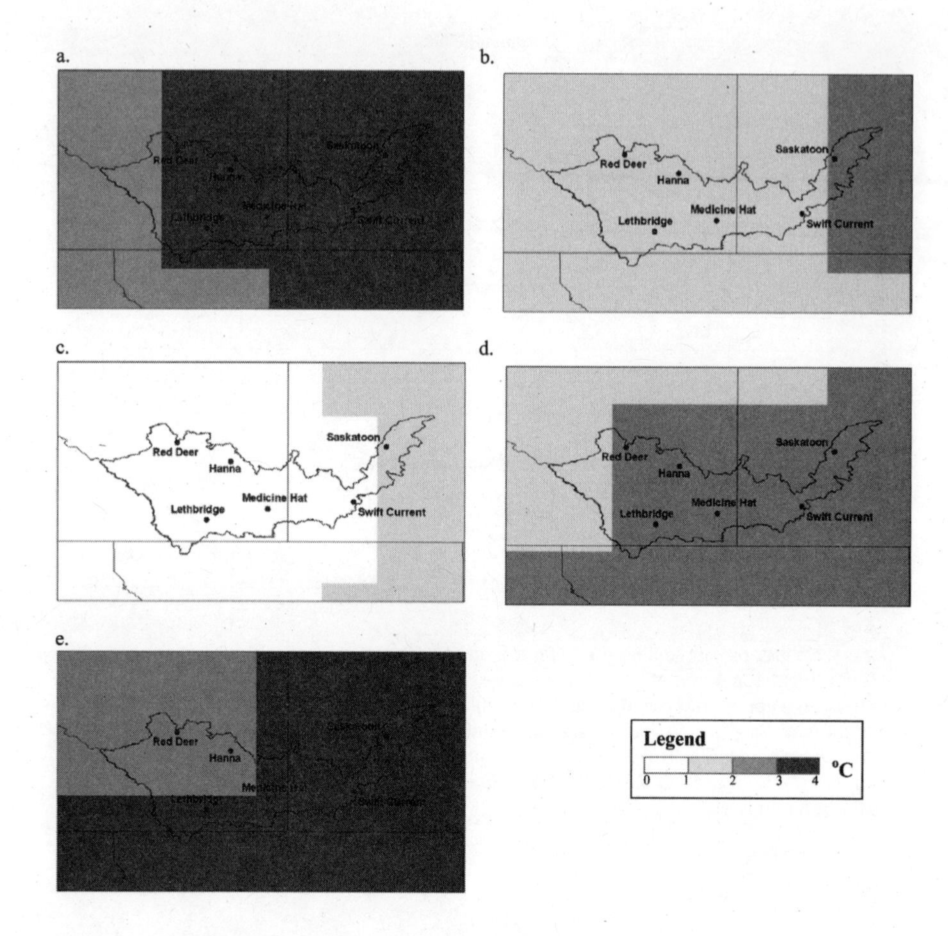

Figure 2. Maps of annual temperature change scenarios (°C) for the 2050s relative to 1961–90. a. CGCM3.1 T47 B1(1) (cool/dry) b. CGCM3.1 T47 B1(2) (median) c. CSIRO MK3.0 A1B(1) (cool/wet) d. HadCM3 TAR a2(1) (warm/wet) e. Miroc Medres a2(1) (warm/dry).

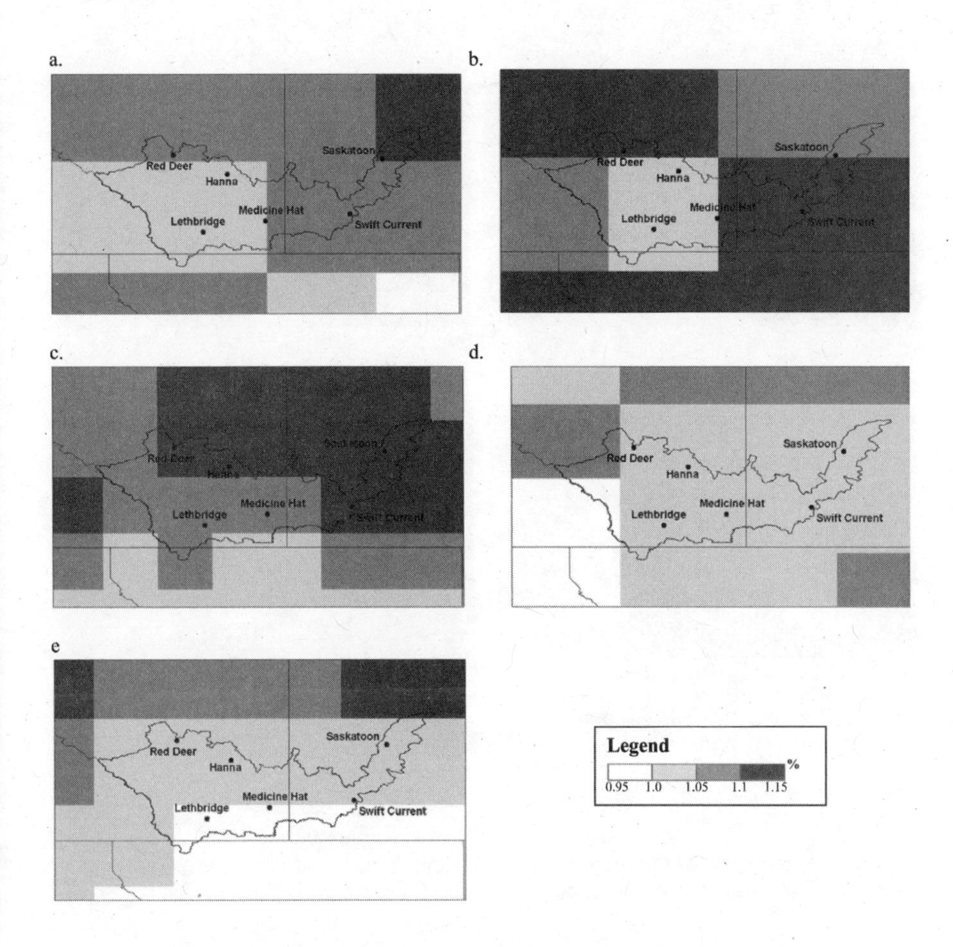

Figure 3. Maps of annual precipitation change scenarios (%) for the 2050s relative to 1961–90. a. CGCM3.1 T47 B1(1) (cool/dry) b. CGCM3.1 T47 B1(2) (median) c. CSIRO MK3.0 A1B(1) (cool/wet) d. HadCM3 TAR a2(1) (warm/wet) e. Miroc Medres a2(1) (warm/dry).

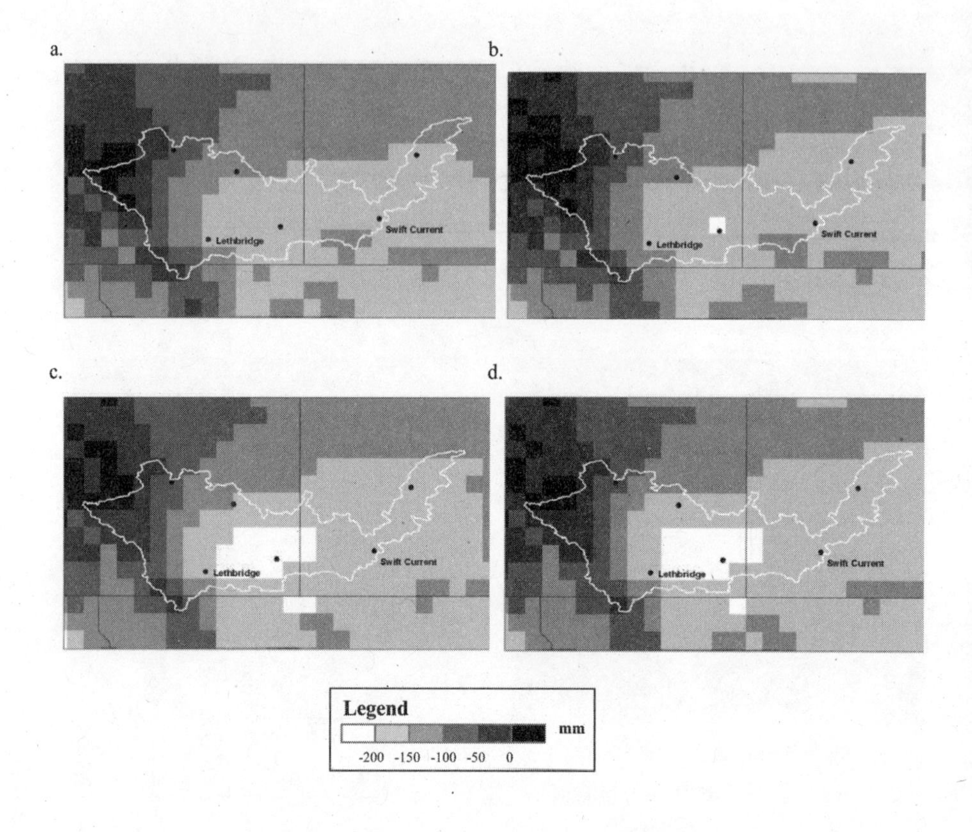

Figure 4. Future Climate Moisture Index (P-PET (mm)) maps for May-July using CGCM3.1/T47 B1(2) (Median) Scenario (a) 1961–90 (b) 2020s (c) 2050s (d) 2080s.

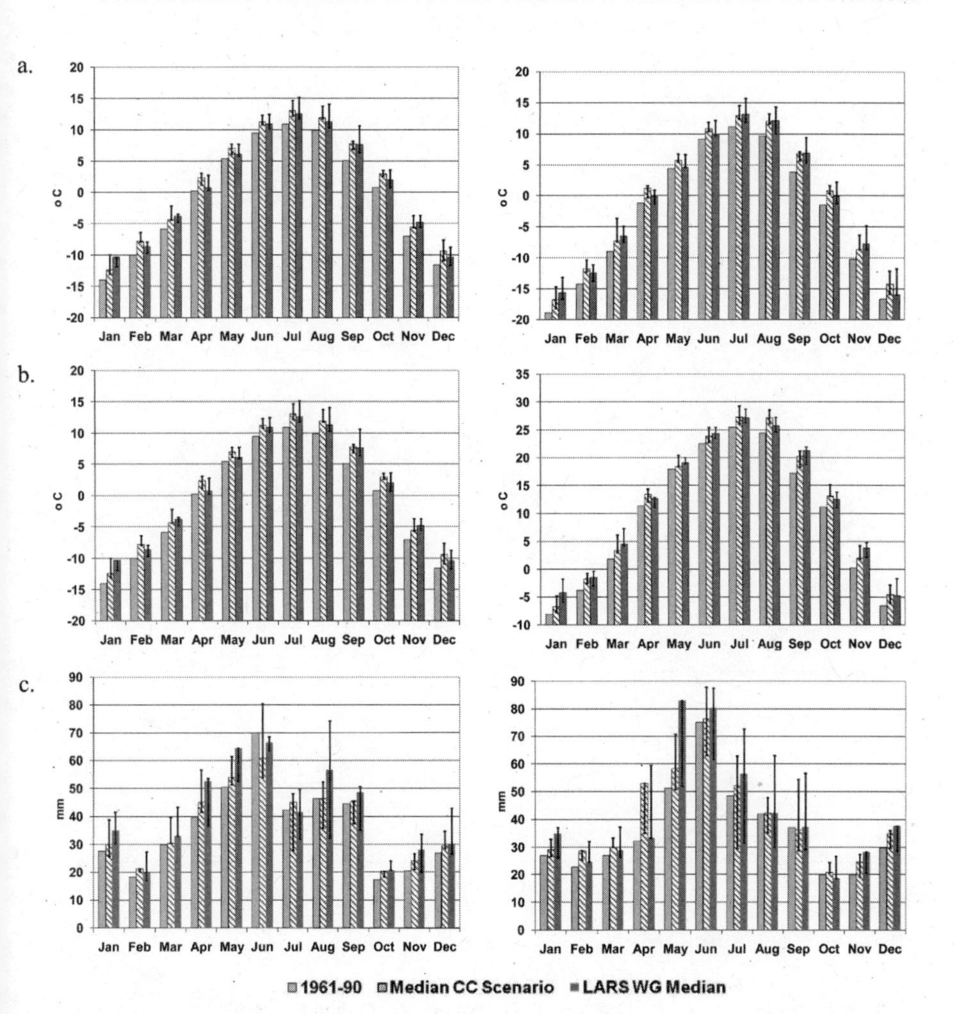

Figure 5. Lethbridge, AB (left side) and Swift Current, SK (right side) future climate scenarios: Solid grey bars represent the monthly averages for the baseline 1961–90 period and the hatched bars represent the median (CGCM3 B1(2)) scenario for the 2040–2069 period, derived directly from the GCM and downscaled using the LARS-WG. The error bars represent the full range of values from five GCMs. (a) minimum temperature (b) maximum temperature (c) precipitation.

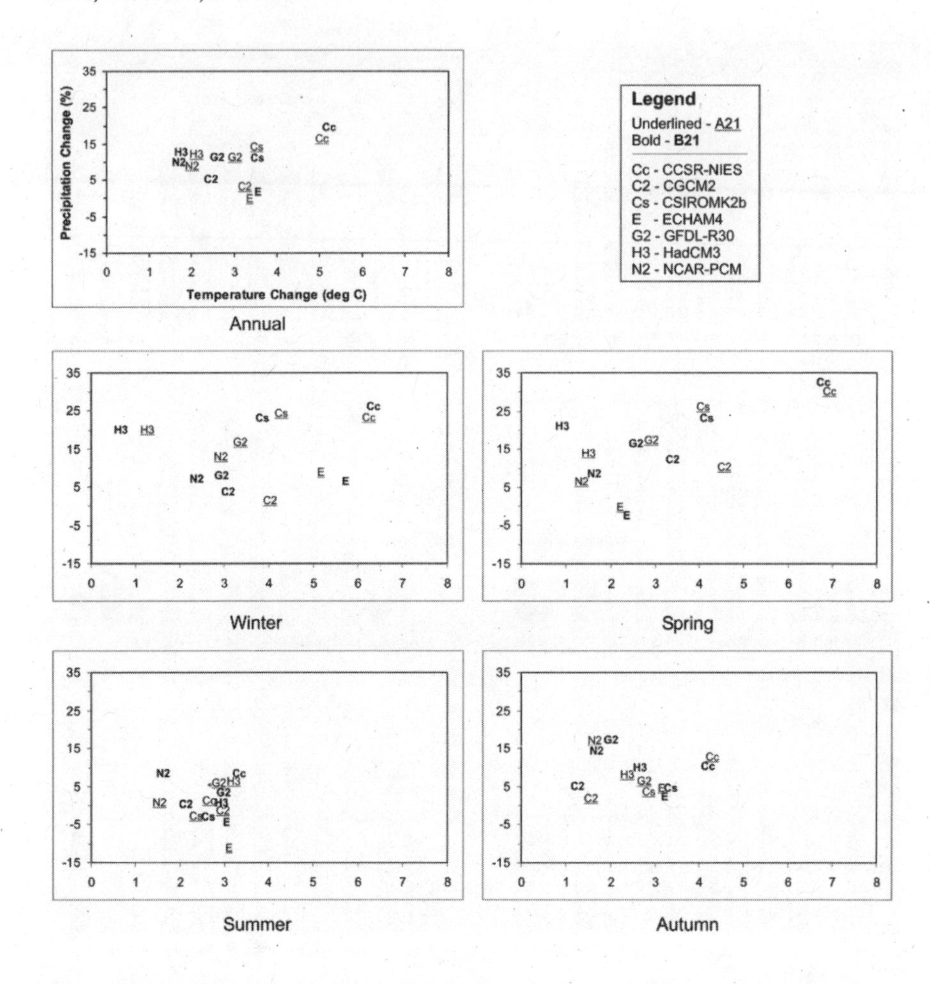

Figure 6. Plots showing the regional averages of projected annual and seasonal change in mean temperature and precipitation based on SRES A2 and B2 scenarios for the 2050 (2040–69) climate.

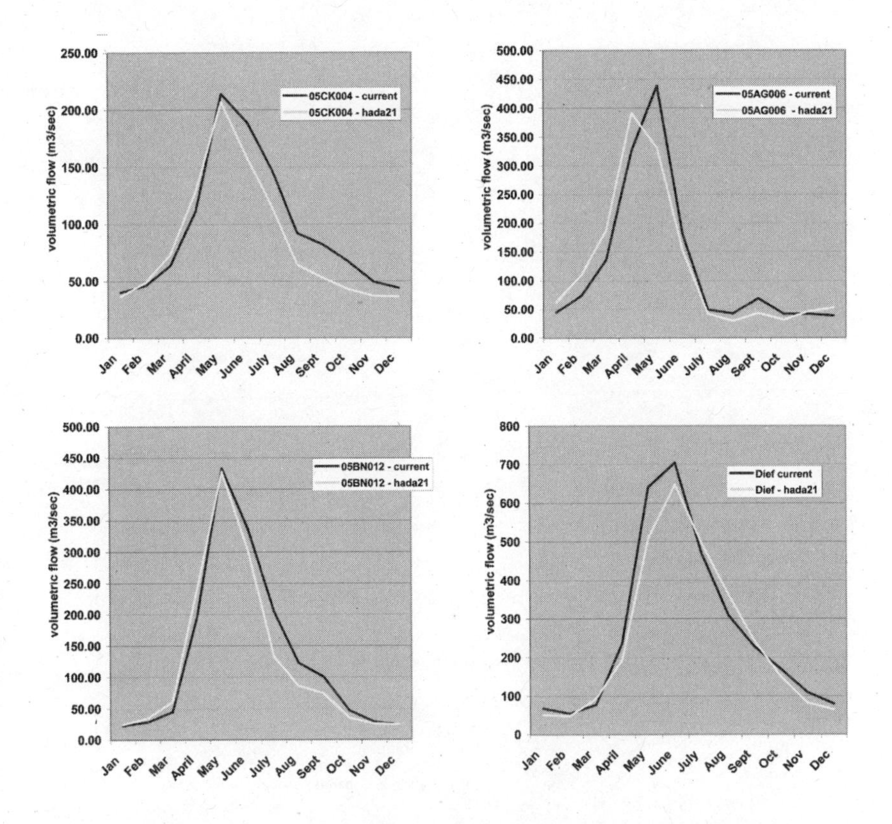

Figure 7. Response in annual future flows for the sub-basins: 05CK004 – Red Deer River at Bindloss; 05AG006 - Oldman River at Mouth; 05BN012 – Bow River at Mouth: Dief – South Saskatchewan River at Lake Diefenbaker.

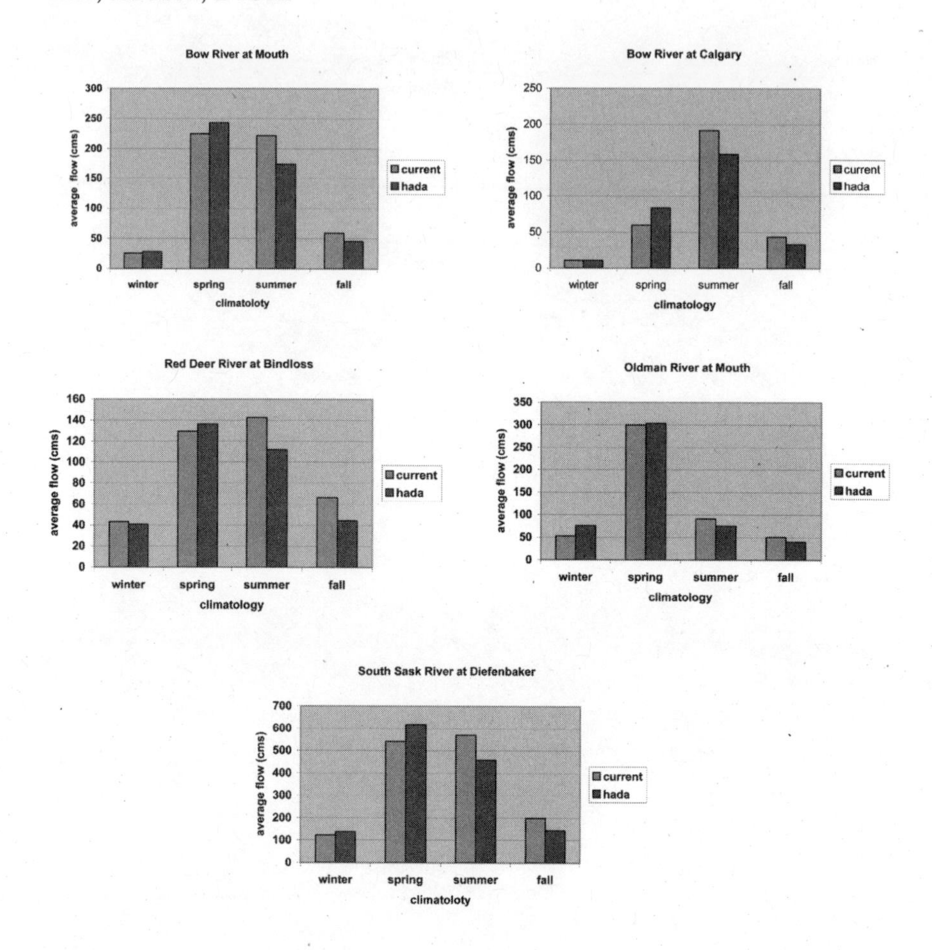

Figure 8. Seasonal response of modeled 1961- 90 stream flows compared to the modeled flows from 2040–2069.

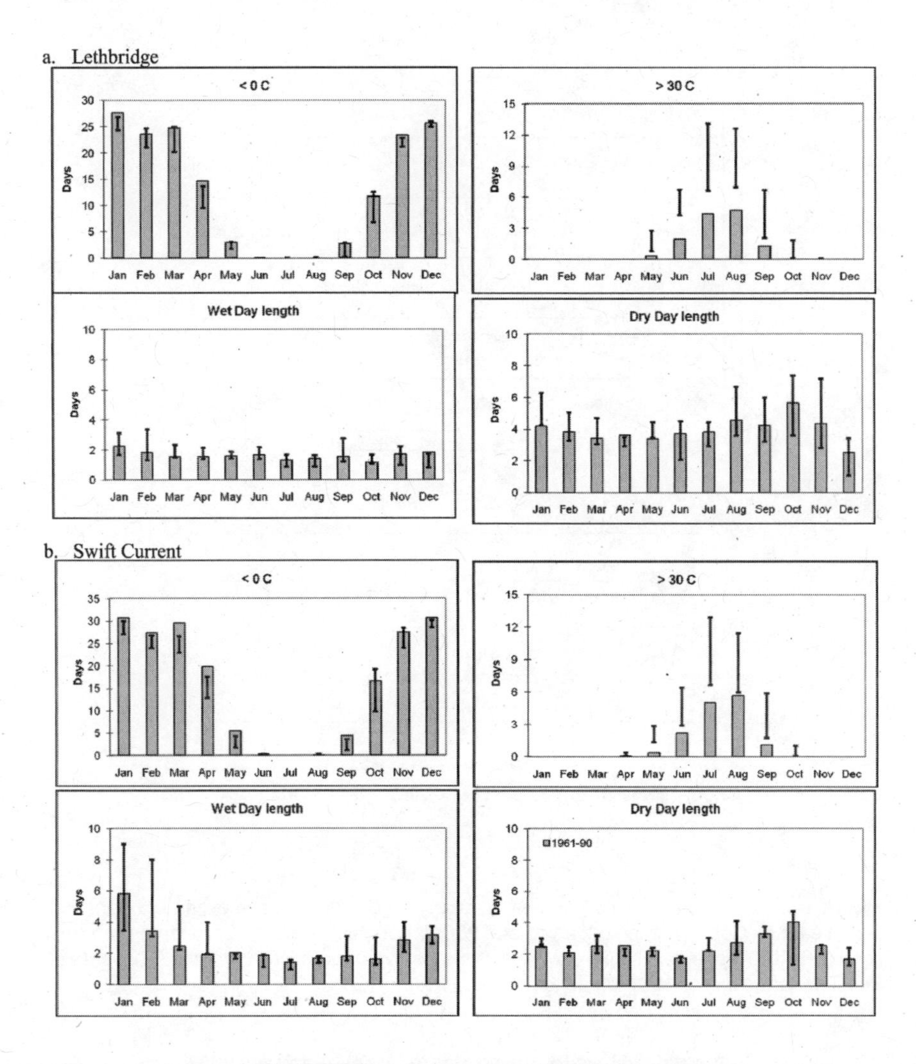

Figure 9. Results from LARS-WG at (a) Lethbridge and (b) Swift Current; grey bars represent the monthly values for the 1961–90 baseline period and the heavy vertical lines represent the range in 2050s projected climate for the five GCMs. Variables compared are the length of temperature spells below 0° C and above 30° C, and monthly dry-day and wet-day lengths.

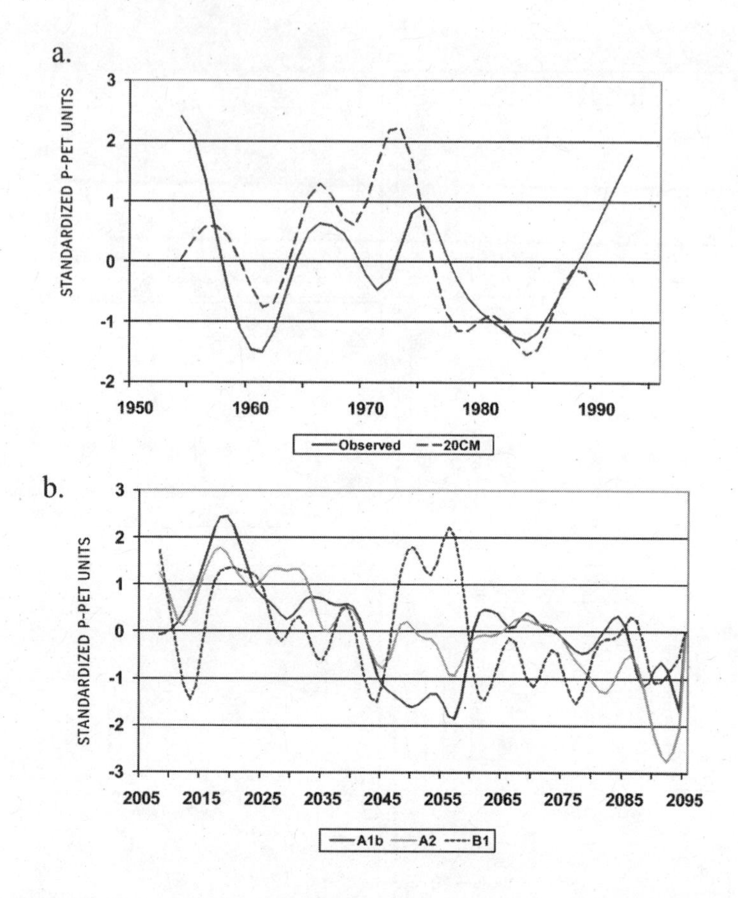

Figure 10. The first principal component of annual P-PET values for timescales greater than eight years. a. Compares the observed and 20CM modeled trends. b. Compares the future scenarios. The first eigenvalue explains 86% of observed and over 95% of the GCM annual P-PET variance for the 1961–90 period.

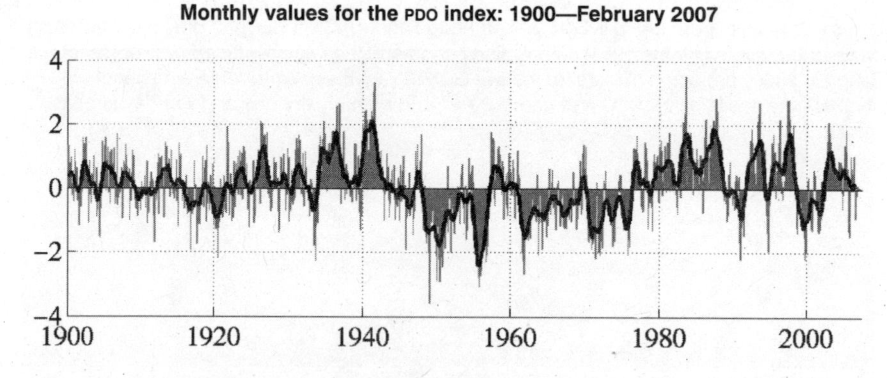

Figure 11. Monthly PDO index values: 1900–February 2007.[54]

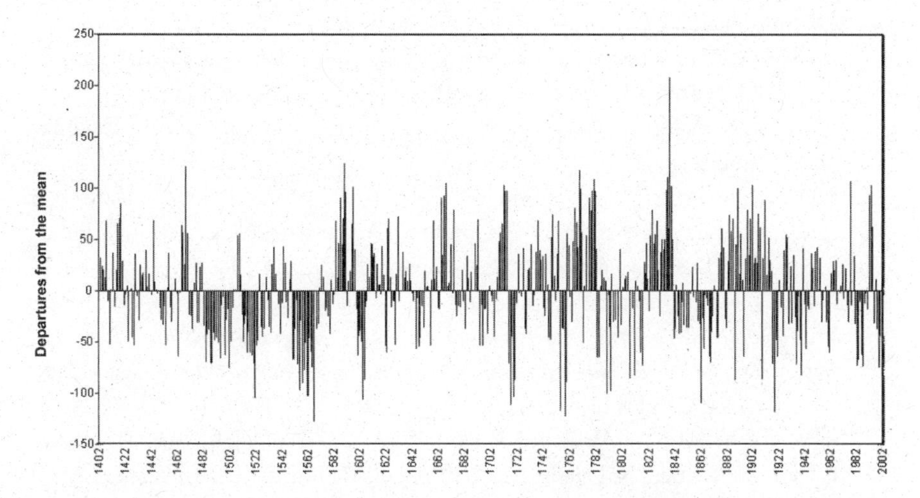

Figure 12. Reconstructed annual flow of the South Saskatchewan River for the period 1402–2002.[55] The vertical bars represent departures from the mean flow. Prolonged periods of low flows are evident prior to the instrumental period starting in the early 20th century.

Endnotes:

1. Hurlbert et al., this volume.
2. Intergovernmental Panel on Climate Change (IPCC), "Contribution of Working Group I to the Third Assessment Report of Intergovernmental Panel on Climate Change," in J.T. Houghton, Y. Ding, D.J. Griggs, M. Noguer, P.J. van der Linde, X. Dai, K. Maskell, and C.A. Johnson (eds.), *Climate Change 2001: The Scientific Basis* (Cambridge, U.K. and New York, N.Y., U.S.A.: Cambridge University Press, 2001.): 881.
3. IPCC, *Climate Change 2001: The Scientific Basis,* 881.
4. M.L. Parry and T.R. Carter, *Climate Impact and Adaptation Assessment: A Guide to the IPCC Approach* (London: Earthscan, 1998).
5. E.M. Barrow, B. Maxwell and P. Gachon, "Climate Variability and Change in Canada: Past, Present and Future," Climate Change Impacts Scenarios Project, National Report, Environment Canada, Meteorological Service of Canada, Adaptation Impacts Research Group, Atmospheric and Climate Sciences Directorate Publication (Canada: 2004): 114.
6. Intergovernmental Panel on Climate Change-Task Group on Scenarios for Climate and Impact Assessment (IPCC-TGCIA), "General Guidelines on the Use of Scenario Data for Climate Impact and Adaptation Assessment, Version 2." Prepared by T.R. Carter on Behalf of the Intergovernmental Panel on Climate Change Task Group on Data and Scenario Support for Impact and Climate Assessment" (2007): 66.

7. J.B. Smith and M. Hulme, "Climate Change Scenarios," in *Handbook on Methods of Climate Change Impacts Assessment and Adaptation Strategies. Version 2.0,* by J. Feenstra, I. Burton, J.B. Smith, and R.S.J. Tol (eds.), 3–1 to 3–40 (Vrije Universiteit, Amsterdam: United Nations Environment Programme and Institute for Environmental Studies, 1998).
8. D.A. Randall, R.A. Wood, S. Bony, R. Colman, T. Fichefet, J. Fyfe, V. Kattsov, A. Pitman, J. Shukla, J. Srinivasan, R.J. Stouffer, A. Sumi, and K.E. Taylor, "Climate Models and Their Evaluation," in S. Solomon, D. Qin, M. Manning, Z. Chen, M. Marquis, K.B. Averyt, M. Tignor, and H.L. Miller (eds.), *Climate Change 2007: The Physical Science Basis—Contribution of Working Group I to the Fourth Assessment Report of the Intergovernmental Panel on Climate Change* (Cambridge, UK and New York, NY, USA: Cambridge University Press, 2007).
9. IPCC, "Technical Summary," in J.T. Houghton, Y. Ding, D. J. Griggs, M. Noguer, P. J. van der Linden, X. Dai, K. Maskell, and C.A. Johnson (eds.), *Climate Change 2001: The Scientific Basis—Contribution of Working Group I to the Third Assessment Report of the Intergovernmental Panel on Climate Change* (Cambridge and New York: Cambridge University Press, 2001): 62.
10. N. Nakicenovic, J. Alcamo, G. Davis, B. de Vries, J. Fenhann, S. Gaffin, K. Gregory, A. Grübler, T.Y. Jung, T. Kram, E.L. La Rovere, L. Michaelis, S. Mori, T. Morita, W. Pepper, H. Pitcher, L. Price, K.Riahi, A. Roehrl, H.H. Rogner, A. Sankovski, M. Schlesinger, P.Shukla, S. Smith, R. Swart, S. van Rooijen, N. Victor, and Z. Dadi, *IPCC Special Report on Emissions Scenarios* (Cambridge, United Kingdom and New York, NY, USA Cambridge University Press, 2000): 599.
11. IPCC, "Contribution of Working Group I to the Fourth Assessment Report of the Intergovernmental Panel on Climate Change," in S. Solomon, D. Qin, M. Manning, Z. Chen, M. Marquis, K.B. Averyt, M. Tignor, and H.L. Miller, *Climate Change 2007: The Physical Science Basis* (Cambridge, United Kingdom and New York, NY, USA: Cambridge University Press, 2007): 996.
12. M. Hulme, E.M. Barrow, N. Arnell, P.A. Harrison, T.E. Downing, and T.C. Johns, "Relative Impacts of Human-Induced Climate Change and Natural Climate Variability," *Nature* 397 (1999): 688–91.
13. Barrow et al., 114.
14. IPCC-TGCIA, "General Guidelines."
15. IPCC, *Climate Change 2007,* 996.
16. Nakicenovic et al., *IPCC Special Report on Emissions Scenarios,* 599.
17. B. Bonsal and T.D. Prowse, "Regional Assessment of GCM-Simulated Current Climate over Northern Canada," *Arctic* 59, no. 2 (2006): 115–28.
18. B. Bonsal, T.D. Prowse and A. Pietroniro, "An Assessment of GCM-Simulated Climate for the Western Cordillera of Canada (1961–90)," *Hydrological Processes* 18 (2003): 3703–16.
19. J. Toyra, A. Pietroniro, and B. Bonsal, "Evaluation of GCM Simulated Climate over

the Canadian Prairie Provinces," *Canadian Water Resources Journal* 30, no. 3 (2005): 245–62.
20. Program for Climate Model Diagnosis and Intercomparison (2007). Available at: http://www-pcmdi.llnl.gov. (cited February 14, 2008).
21. IPCC–Data Distribution Centre (DDC). Available at: http://www.ipcc-data.org (cited December 12, 2007).
22. UK Meteorological Office, Hadley Centre. Climate Impacts LINK Project, British Atmospheric Data Centre, 2003. Available at: http://badc.nerc.ac.uk/data/link/. (cited November 9, 2008).
23. D.W. McKenney, J.H. Pedlar, P. Papadopol, and M.F. Hutchinson, "The Development of 1901–2000 Historical Monthly Climate Models for Canada and the United States," *Agricultural and Forest Meteorology* 138 (2006): 69–81.
24. J.H. Christensen, B. Hewitson, A. Busuioc, A. Chen, X. Gao, I. Held, R. Jones, R.K. Kolli, W.-T. Kwon, R. Laprise, V. Magaña Rueda, L. Mearns, C.G. Menéndez, J. Räisänen, A. Rinke, A. Sarr, and P. Whetton, "Regional Climate Projections," in *Climate Change 2007*.
25. C.J. Willmott and J.J. Feddema, "A More Rational Climatic Moisture Index," *Professional Geographer* 44 (1992): 84–87.
26. E.H. Hogg, "Temporal Scaling of Moisture and the Forest-Grassland Boundary in Western Canada," *Agricultural and Forest Meteorology* 84 (1997): 115–22.
27. E.M. Barrow and G. Yu, "Climate Scenarios for Alberta: a Report Prepared for the Prairie Adaptation Research Collaborative (Parc) in Co-operation with Alberta Environment" (University of Regina, Saskatchewan, 2005): 73.
28. N. Kouwen, E.D. Soulis, A. Pietroniro, J.R. Donald, and R.A. Harrington, "Grouped Response Units for Distributed Hydrologic Modelling," *Journal of Water Resources Planning and Management* 119, no.3 (1993): 289–305.
29. Kouwen, personal communication.
30. J. Toyra, A. Pietroniro, and B. Bonsal, "Evaluation of GCM Simulated Climate over the Canadian Prairie Provinces," *Canadian Water Resources Journal* 30, no. 3 (2005): 245–62.
31. IPCC-TGCIA, "General Guidelines," 66.
32. V. Venugopal, G.E. Foufoula, and V. Sapozhinikov, "A Space-Time Downscaling Model for Rainfall," *Journal of Geophysical Research: Atmospheres* 104, no. 16 (1999): 19–705.
33. IPCC, "Technical Summary," 62.
34. IPCC, *Climate Change 2001,* 881.
35. L. E. Hay, and M. P. Clark, "Use of Statistically and Dynamically Downscaled Atmospheric Model Output for Hydrologic Simulations in Three Mountainous Basins in the Western United States," *Journal of Hydrology* 282, no. 1–4 (2003): 56–75.
36. Canadian Centre for Climate Modelling and Analysis (2007). Available at: http://www.cccma.ec.gc.ca. (cited February 12, 2008).

37. M.A. Semenov and E.M. Barrow, "Use of a Stochastic Weather Generator in the Development of Climate Change Scenarios," *Climatic Change* 35 (1997): 397–414.
38. R.L. Wilby, C.W. Dawson, and E.M. Barrow, "SDSM—a Decision Support Tool for the Assessment of Regional Climate Change Impacts," *Environmental Modelling and Software* 17 (2002): 147–59.
39. Wilby, "SDSM," 147.
40. M.A. Semenov and E.M. Barrow, "Lars-WG a Stochastic Weather Generator for Use in Climate Impact Studies, Version 3.0., User Manual," 2002.
41. É. Mekis and W.D. Hogg, "Rehabilitation and Analysis of Canadian Daily Precipitation Time Series," *Atmosphere-Ocean* 37, no. 1 (1999): 53–85.
42. L.A. Vincent, X. Zhang, B.R. Bonsal, and W.D. Hogg, "Homogenization of Daily Temperatures over Canada," *Journal of Climate* 15 (2002): 1322–34.
43. IPCC-TGCIA, "General Guidelines," 2007.
44. IPCC-TGCIA, "General Guidelines," 2007.
45. B. Bonsal and R. Lawford, "Teleconnections between El-Nino and La-Nina Events and Summer Extended Dry Spells on the Canadian Prairies," *International Journal of Climatology* 19 (1999): 1445–58.
46. Nathan J. Mantua and Steven R. Hare, "The Pacific Decadal Oscillation," *Journal of Oceanography* 58, no. 1 (2002): 35–44.
47. Bonsal and Lawford, "Teleconnections between El-Nino and La-Nina," 1445.
48. B. Bonsal and E. Wheaton, "Atmospheric Circulation Comparisons between the 2001 and 2002 and the 1961 and 1988 Canadian Prairie Droughts," *Atmosphere-Ocean* 43, no. 2 (2005): 163–72.
49. Mantua and Hare, "The Pacific Decadal Oscillation," 35.
50. Jodi Axeslon, "Historical Streamflow Variability in the South Saskatchewan River Basin Inferred from a Network of Tree-Ring Chronologies" (University of Regina: M.Sc. thesis, 2007).
51. Hurlbert et al., this volume.
52. Magzul et al., this volume.
53. Wandel et al., this volume.
54. N.J. Mantua, S.R. Hare, Y. Zhang, J.M. Wallace, and R.C. Francis, "A Pacific Interdecadal Climate Oscillation with Impacts on Salmon Production," *Bulletin of the American Meteorological Society* 78 (1997): 1069–1079. Available from http://jisao.washington.edu/pdo/ (cited December 12, 2007).
55. Axeslon, "Historical Streamflow Variability."

GOVERNMENT AND CIVIL SOCIETY: ADAPTIVE WATER MANAGEMENT IN THE SOUTH SASKATCHEWAN RIVER BASIN

MARGOT HURLBERT, DARRELL R. CORKAL, AND HARRY DIAZ

Abstract

The existence of an effective institutional framework is essential for developing the capacities to reduce vulnerabilities to climate-induced water stress among rural communities. Based on data collected by different research activities in the Institutional Adaptation to Climate Change project, this paper discusses the current effectiveness of water governance as an institutional cluster, and how it fosters adaptive capacity of rural communities in the South Saskatchewan River Basin in Alberta and Saskatchewan. To set the theoretical framework of the paper, the first section discusses the roles of institutions as a determinant of adaptive capacity, the importance of water governance as an institutional cluster, and the challenges of water governance. After a brief description of the multi-level water governance systems in Alberta and Saskatchewan, the paper provides insights garnered from research activities and interviews of stakeholders and water institutions. Three issues are identified as being relevant to strengthening effective water governance in the present and in the future.

Developing a *long-term comprehensive climate change/adaptation policy framework* would provide a common baseline focus for all orders of government and water users dealing with climate-induced water stress. *Strengthening the roles of civil society in water governance* would help ensure local adaptations meet local needs. *Simplifying the complex nature of water governance* by the integration and coordination of institutions with water stakeholders would help build adaptive institutions and communities.

Sommaire

L'existence d'un cadre institutionnel efficace est nécessaire afin que les collectivités rurales puissent se doter des capacités requises pour réduire leur vulnérabilité au stress hydrique d'origine climatique. Cet article porte sur l'efficacité actuelle de la gouvernance de l'eau en tant qu'agglomération institutionnelle et de la façon dont celle-ci favorise la capacité d'adaptation des collectivités du bassin de la rivière Saskatchewan-Sud en Alberta et en Saskatchewan en s'appuyant sur des données recueillies par diverses activités de recherche du projet sur l'adaptation institutionnelle aux changements climatiques (Institutional Adaptation to Climate Change). Afin d'établir le cadre théorique de cet article, la première partie porte sur les rôles des institutions comme facteur déterminant de la capacité d'adaptation, l'importance de la gouvernance de l'eau en tant qu'agglomération institutionnelle, et les défis liés à la gouvernance de l'eau. La deuxième partie se consacre à une brève description des systèmes à multiples niveaux de gouvernance de l'eau en Alberta et en Saskatchewan. La dernière partie est consacrée à la lumière qui est jetée par les données recueillies de diverses activités de recherche et d'entrevues avec des intervenants et des institutions dans le secteur de l'eau. Trois enjeux sont pertinents au renforcement de la gouvernance de l'eau efficace maintenant et pour l'avenir.

L'établissement d'un cadre stratégique global à long terme au sujet de l'adaptation aux changements climatiques fournirait une direction de base commune pour tous les paliers de gouvernent et les utilisateurs d'eau qui sont touchés par un stress hydrique d'origine climatique. *Le renforcement des rôles de la société civile dans la gouvernance de l'eau* assurerait que les diverses formes d'adaptation locale répondent aux besoins locaux. *La simplification de la nature complexe de la gouvernance de l'eau* via l'intégration et la coordination des institutions avec les intervenants du milieu de l'eau permettrait de construire des institutions et des collectivités souples ayant une capacité d'adaptation.

Introduction

There is mounting evidence that global warming is already in progress and its impacts are expected to increase.[1] The impacts of global warming will vary among regions and social groups, depending on specific climate stimuli—including variability and extremes—and variations in adaptive capacity within social systems. Expected impacts in the South Saskatchewan River Basin (SSRB) involve increases in temperatures, reduction of glaciers, changes in rainfall and snow patterns, reduction in annual flow of the South Saskatchewan River, an increase in the intensity and severity of extreme climate events (drought, rainfall, hail storms), and more evapotranspiration, resulting in an increasing aridity of some areas. In this new climate context, water resources will be seriously impacted, both in terms of quantity and quality; a changing climate may also lead to new opportunities (e.g., higher value diversified crops).[2] Although climate change will affect everyone in the basin, rural communities are expected to be disproportionately affected given their dependency on natural resources. Many rural communities already face severe stress arising from political and economic processes (such as globalization and restricted fiscal policies) and the negative impacts of climate change may create a situation of "double exposure" resulting in exponentially increased vulnerabilities.[3]

An important determinant in the ability of a rural community to adapt to current climate variability and future climate change impacts is the institutional setting in which the community exists and functions, a setting that involves governance rules and organizations. Using information generated by different research activities in the Institutional Adaptation to Climate Change (IACC) project, this paper focuses on the water governance institutions in the SSRB and their capacity to reduce the vulnerability of rural communities. We commence with a brief description of the data sources used in this article, followed by a section focused on the concept of institution, and the roles institutions play in defining the level of vulnerability of social systems (i.e., rural communities), and the idea of water governance as an institutional cluster. The second section outlines the structure of water governance in the Alberta and Saskatchewan regions of the SSRB, and briefly describes the various organizations, and their relationships and mandates. Based on interviews with water users and representatives of all levels of government, the article concludes with insights of how water governance may be strengthened to increase adaptive capacity to climate-induced water stress.

Sources of Information

This paper is based on information gathered by the IACC project from three different research activities that took place in Alberta and Saskatchewan between 2005 and 2007, including several community vulnerability assessments, stakeholder meetings and structured workshops, and interviews with representatives from different governance institutions and civil society organizations participating in the management of water resources.[4]

The community vulnerability assessments, carried out in 2005 and 2006 in Saskatchewan and Alberta, were oriented to obtain information about the sensitivities and resiliency of several rural communities to a variety of stressors, including climate and water problems. Based on intensive ethnographic work, which included a large number of in-depth interviews, information was collected on the exposure and adaptive capacity of local residents, including the roles that governance organizations play in reducing local vulnerability.

These community vulnerability assessments were complemented with several stakeholder meetings which were attended by community residents, representatives from government agencies, water user associations, agricultural producers and organizations (e.g., irrigators), and non-government organizations (NGOs). The purpose of the stakeholder meetings was to disseminate information obtained during the community vulnerability assessments, and to discuss in more detail the existing institutional capacities in the areas of climate change and water management. Three of these stakeholders meetings inform the discussion in this paper: Lethbridge (December 2006), Outlook (January 2007), and Hanna (February 2007).

Lastly, in 2007, the IACC project started a governance assessment oriented to evaluate water governance organizations and processes in Alberta and Saskatchewan. Information for the assessment was obtained from multiple sources including public documents, focus group sessions, and in-depth interviews with approximately 60 representatives of water users, associations, watershed and environmental groups, community representatives, and all orders of government (local, provincial, and federal government agencies). Information about several organizational and procedural issues was gathered during the assessment, including:

- agency roles;
- institutional responses to situations of water stress;
- long-term planning in relation to future climate stress;
- types of climate and water information collected by agencies;
- organizational and legal resources;
- links to water stakeholders;
- accountability;
- the type and degree of participation (integration and coordination) of the organizations within the cluster of water governance; and,
- institutional mechanisms to support and satisfy water resource needs of rural communities and legal instruments to facilitate water governance roles.

From the governance assessment analysis, several initial insights have been integrated into the discussion in this paper.

Vulnerability, Institutions, and Adaptive Capacity

In the global environmental change literature considerable attention has been devoted to the idea of vulnerability.[5] Depending on disciplinary approaches and methodological orientations, vulnerability has been defined in many different ways. However, there is an increasing consensus about the need to move from a technocratic paradigm, with a hazard-centered interest in geophysical processes, into one that emphasizes the mutuality of hazard and social conditions.[6] Following the Intergovernmental Panel on Climate Change (IPCC), the IACC project defines vulnerability as the degree to which a system, such as a rural community, is susceptible to the adverse effects of climate change, including climate variability and extremes.[7] Vulnerability, in this definition, is a function of the exposure/sensitivity and the capacity of a rural community to adapt to climate stress.[8] Accordingly, the most vulnerable communities are those most likely to be exposed, more sensitive to perturbation, and who possess limited capacity for adaptation.[9]

Using this approach the vulnerability of a community is not a function of climate alone, but rather it is the result of multiple social conditions including environmental, social, economic, and political factors. It is extremely difficult to separate the impacts of climate events on a community from the impacts of other factors such as the price of grain, the existence or absence of grain subsidies, out-migration or other social and political factors. These social conditions affect the vulnerability of a community, as much as the biophysical aspects of climate. A corollary of this argument is that effective strategies for reducing vulnerability and adapting to future climate conditions need to incorporate the diversity of environmental, social, economic and political conditions which shape exposures, sensibilities, and adaptive capacities.[10]

Adaptive capacity is defined as the "ability to design and implement effective adaptation strategies, or to react to evolving hazards and stresses. . . ."[11] Thus, successful adaptation depends on the existence of resources available for adaptation and the ability or predisposition to use these resources in the pursuit of adaptation.[12] The Third Report of the IPCC has identified a set of determinants of adaptive capacity that "influence the occurrence and nature of adaptation and thereby circumscribe the vulnerability of systems and their residual impacts."[13] These determinants include the existence of economic resources, technology, information and skills, infrastructure, equitable social relations, and institutions (such as government bodies). In its discussion of institutions the report states that "countries with well developed institutions are considered to have greater adaptive capacity than those with less effective institutional arrangements," but it offers a less definitive statement about the capacities of established institutions to deal with the hazards of future climate change.[14] The governance assessment of the IACC project provides insights about these issues, especially in terms of the capacities of existing water governance systems in Canada and Chile to reduce the vulnerability of rural communities to both climate change and climate variability.

For the purposes of the IACC research, an institution is defined as a persistent, reasonably predictable arrangement, law, process, custom or organization structuring aspects of the political, social, cultural or economic transactions and relationships in a society.[15] Although inclusive of organizations, the term institutions is broader and ranges from highly formalized law systems to the more diffuse hegemonic discourses, informal rules, and spans from highly formalized settings to informal arrangements.[16]

Communities are places where formal and informal institutions coexist and interact. Regional and local governments, church organizations, and forms of social capital such as networks for mutual support, become structured into complex sets of relationships that contribute or disrupt the stability, viability, and in our case, the adaptive capacity of the communities.[17] As well, rural communities, like any other human setting, function within larger institutional political systems that link them with the larger society. These political systems, which are central to the development of a planned adaptive capacity given their purposeful mandate, permanency, social acceptance, and legal basis, pervade the lives of the community members by imposing a body of regulations, rules, processes, and resources on communities.[18]

An important political cluster is governance, which refers to "the patterns by which public power is exercised in a given context."[19] More specifically, water governance is an organizational form of public power that defines the organization and management of the interrelationships between society and water resources, determines how much water may be used, by whom and under which conditions, and assesses the impacts of society upon water resources in the natural environment. As expected, water governance assumes a diversity of forms, ranging from a highly centralized system to one that encompasses a diversity of public and non-public organizations. The dominant tendency, however, has been to define governance (and organize it) as a more inclusive system than solely government, defining it as form of political power that encompasses laws, regulations, public organizations, and includes those sectors of civil society that participate, interact with, or influence the management of water resources.[20] Inclusive water governance is a persistent, reasonably predictable arrangement, a process of providing structure to the transactions relating to water and its management in our society, and doing so with the participation of civil society and government networks. It is this form of water governance—subjected to different political interpretations and conditions—that has been predominant in Canada, and it is sometimes referred to as managing water by 'shared jurisdictions.' A graphic description of this definition of water governance is presented in Figure 1.

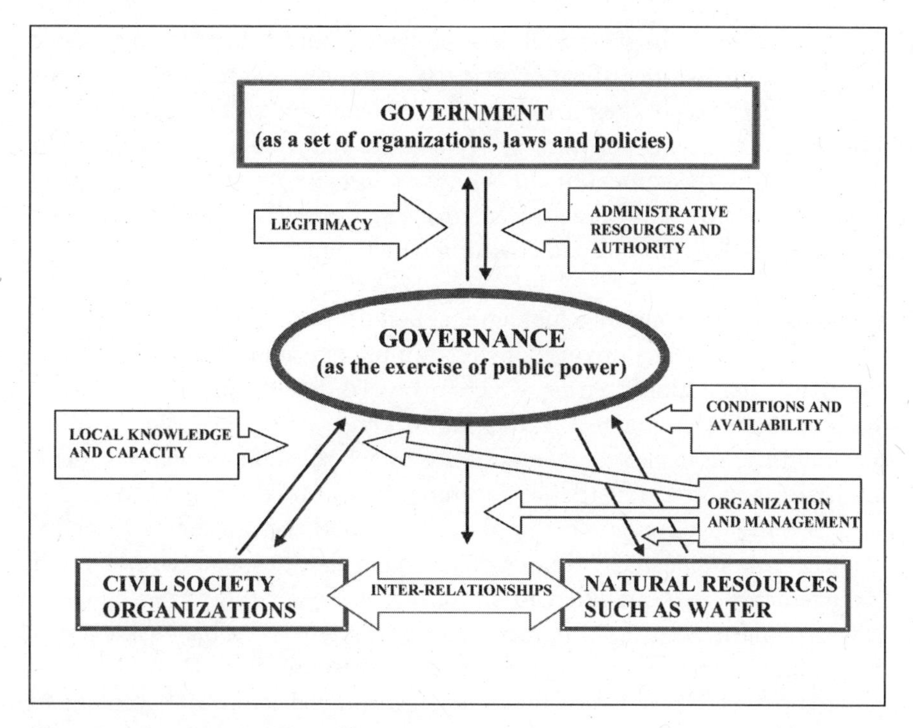

Figure 1. A Description of Water Governance.

Given that water governance involves a wide variety of organizations and stakeholders with an interest in water, it has the potential to become characterized by different degrees of power or authority and competing interests. In these terms, fundamental challenges in achieving effective water governance includes the clear articulation of the roles and mandates of the different organizations and orders of government, to avoid overlap, competing priorities, and unclear mandates, and, enables the coordination of policy development and implementation activities. Moreover, an effective integration of local stakeholders is also fundamental to ensure effective management of water resources at the local level. The reconciliation of these competing priorities is the central challenge of water governance.[21]

Climate change impacts on water resources bring additional challenges to governance. To foster and build adaptive capacity, water governance policies and programs will need to consider climate variability and long-term changing climatic conditions. In the context of adaptive capacity, water governance also requires flexibility to deal with the unanticipated conditions that may result from the impacts of climate change. The fundamental role of governance should be to provide an enabling environment that allows civil society to deal successfully with the challenges of climate variability and climate change. To be successful at improving adaptive capacity governance must allow for the identification and resolution of communities' problems and the satisfaction of their needs in a fair, efficient, and sustainable manner.[22] Thus, the effectiveness of water governance to reduce the vulnerabilities of rural people in the context of climate stress, rests on its ability to anticipate problems and to manage risks, challenges, and opportunities in a way that balances social, economic, and environmental interests.

The SSRB Water Governance Institutional Setting

The great geophysical expanse of the SSRB predetermines a complex water governance institutional setting, which involves different administrative jurisdictions and many actors including urban, rural, and First Nation communities, and in leading roles, various provincial and federal government ministries. Selected ministries and water agencies and their key responsibilities for water management within the SSRB are outlined in Table 1. The table is not a comprehensive listing of government institutions active in water management—it is provided to simplify and depict the rather complex arrangement of the various ministries and agencies responsible for the use of water and the protection of water as a natural resource.[23]

To understand the interplay of these institutional actors, it is germane to have a basic understanding of the constitutional setting and history of water governance in the provinces of Alberta and Saskatchewan. Water management was not treated as a single topic in the Canadian Constitution. In 1867, when federal and provincial powers were first negotiated, water management was not part of the negotiations.[24] The topic of water spans several heads of legislative power assigned to the federal and provincial governments. The provincial government has powers which relate to water because of its jurisdiction over natural resources (specifically the transfers of natural resources, which included water, to the Prairie Provinces after the depression of 1930).[25] This jurisdiction relates to protection of water resources in the natural environment, and regulating water quality and water quantity on provincial lands. Municipalities only have that jurisdiction delegated to them by provincial governments.[26]

The federal government has certain powers in relation to water, albeit historically more limited than the provinces. This includes water as it relates to other matters assigned to the federal government such as federal lands (i.e., national parks), trade and commerce, navigation and shipping, seacoast and inland fisheries, and lands reserved for First Nations.[27] The federal government is also responsible for ensuring the safety of drinking water within areas of federal jurisdiction, such as national parks and First Nation communities, and water quality in respect of inter-jurisdictional waters (water flowing

ORDERS OF GOVERNMENT	PRINCIPAL MINISTRIES/ AGENCIES	OTHER KEY MINISTRIES/AGENCIESWITH MAJOR WATER RESPONSIBILITIES					
Alberta	**Alberta Environment (AB Env.)**	Health	Agriculture	–	Transportation & Utilities	Alberta Research Council	Other Ministries; Watershed Groups
Saskatchewan	**Saskatchewan Watershed Authority (SWA)**	Health	Agriculture	Environ-ment	Sask Water	Saskatchewan Research Council	Other Ministries; Watershed Groups
Local Municipali-ties	**Utility Departments**	Environmental protection from development and land use; local and regional public health protection. Drinking water and wastewater treatment/distribution.					
Canada	**Environment**	Health	Agriculture	Natural Resources Canada	Fisheries & Oceans	National Water Research Institute	Other Ministries & Agencies
Simplified Summary of Main Water Respon-sibilities	**Alberta Env.; SWA**	**Authority for provincial water management**: water allocations, licens-ing, water use and apportionment, hydrology, planning, source water protection.					
	• *Environment:* environmental protection & research; weather & climate change science *Health:* public health protection (drinking water, wastewater); Regional Health District support • *Agriculture:* Agricultural programs & research; promotion of adoption of Agricultural Best Management Practices to protect the environment from potential agricultural contamination • *Provincial Utilities:* municipal water supply/distribution, hydroelectricity, energy • *Local Utilities:* potable drinking water & wastewater treatment/distribution • *Research:* water, environmental monitoring, contamination & protection, land use • *Fisheries and Oceans*: protection of freshwater and saltwater fisheries • *Natural Resources:* ground water mapping, forestry, energy, minerals, climate change adaptation • *Watershed groups/ river basin councils:* watershed planning and source water protection						
SELECTED KEY AGENCIES WITH INTER-JURISDICTIONAL WATER RESPONSIBILITIES IN THE SSRB							
Inter-govern-mental	International Joint Commission	The IJC represents the Governments of Canada and the United States. The IJC addresses water use and quality of boundary waters affecting both nations. With respect to the SSRB, the Boundary Water Treaty includes clauses for water flow in the St. Mary and Milk Rivers, and the inter-basin transfer of water from the St. Mary to the Milk. This agreement affects Montana (USA) and Alberta, Saskatchewan, and Manitoba (Canada).					
	Prairie Provinces Water Board	Federal-Provincial Board (Environment Canada, Agriculture and Agri-Food Canada-PFRA, Alberta Environment, Saskatchewan Watershed Authority, Manitoba Water Stewardship); administers, monitors and reports on inter-jurisdictional water issues in the Prairie Provinces of Alberta, Saskatchewan, and Manitoba (allocations, flows, water quantity and water quality).					
Unique Federal Pan-prairie Agency	Prairie Farm Rehabilitation Administration	The Federal government created PFRA in 1935 as an emergency response to catastrophic multi-year droughts. PFRA evolved from Agriculture and Agri-Food Canada's Dominion Experimental Farms, adapting scientific findings for crop production, enhancing land use and tree culture, devel-oping water supplies, etc. to help the prairies achieve greater economic security. PFRA continues to be responsible for soil and water conservation work, agricultural applied research, rural water (water supply/quality, irrigation, climate/drought adaptations), and is evolving into a national agri-environmental branch.					

Table 1. Key Water Agencies in the South Saskatchewan River Basin.

across provincial or international boundaries).[28] The federal government also has responsibilities for protection of water resources in the natural environment.

Although Alberta and Saskatchewan originally shared the same water law and governance, after the transfer of water to their jurisdiction in 1930 their laws have evolved differently. Two examples of these differences are priority respecting water allocations and transfer of water licenses.[29] The SSRB watershed and its water resources are defined by geographic boundaries but are separated by artificial federal, First Nations, provincial, and municipal boundaries representing different legal norms, rules and laws, or legal instruments.

It is evident from this simplified and very brief review of the constitutional and legal arrangements for water jurisdiction in Canada that the provincial, the federal, and the municipal governments (through provincial delegation), share roles and responsibilities for the protection and use of water resources and interrelated matters. Added to the complexity of government organizations involved in SSRB water governance illustrated in Table 1 are other organizations that participate in the management of water resources, such as water user organizations (irrigation districts, rural water utilities, local watershed groups), provincial parks, a variety of industries and economic sector associations, special interest groups, and non-government organizations concerned with the environment.

In Alberta, watershed groups and watershed councils exist, organized under the province's Water for Life Strategy, created in 2001 to achieve more sustainable management of water resources after Alberta recognized the province to be facing significant pressures on its water supplies (e.g., limited availability, water quality impacts, competing demands, etc.).[30] In Saskatchewan, the Watershed Advisory Committees, supported by the Saskatchewan Watershed Authority, participate in an advisory role in the preservation and management of water resources (e.g., developing source water protection plans).[31] In both provinces, these watershed groups have the potential to coordinate issues of local development and water planning decisions, and have been formed partly as a response to drinking water disease outbreaks in Canada (Walkerton in 2000, North Battleford in 2001, Kasheshewan in 2005).[32]

Successful institutional adaptations have occurred in the SSRB since the days of post-European settlement. The SSRB has a wide variability in climate and has historically been prone to drought, with at least 40 droughts recorded in the last 100 years.[33] Institutions such as the Special Areas Boards in Alberta and the Prairie Farm Rehabilitation Administration working across the prairies, were created to assist agriculture and Canadian society cope with climate stress by implementing adaptations to achieve greater economic security and environmental sustainability.[34] In 1948, the Prairie Provinces Water Board (with membership from Alberta, Saskatchewan, Manitoba and Canada) was created to jointly manage the basin's shared water resources flowing across the provincial boundaries.[35] The International Joint Commission represents Canada and the United States, and is designed to prevent and resolve water disputes between the two nations.[36] Even though many successful adaptations have occurred in the SSRB, water governance challenges continue to exist, and risk impeding water management and adaptation to impacts from climate change and climate variability.

Assessing Water Governance in the SSRB

This section discusses some of the initial insights about the strengths and weaknesses of water governance in the basin, emphasizing three issues. The first issue relates to the effectiveness of water governance in the context of the challenges presented by climate change impacts: *the need for a comprehensive long-term climate change policy* (one which addresses adaptive capacity to cope with the natural and changing wide-ranging climate variability of the region). The other two issues relate to the effectiveness of an inclusive water governance that ensures the sustainability of water resources in the context of regional development: *the need for integrating civil society into the water governance process* to ensure continued participation of local stakeholders, and, *the need to simplify the complex nature of water governance* resulting from the large number of government organizations involved.

The need for a comprehensive long-term climate change policy

Fostering adaptive capacity to climate change requires, as a first step, a more comprehensive consideration of climate in the governance agenda. Because of the ubiquitous nature of water, and the expected impacts of natural climate variability and climate change, overarching strategies to reduce the impacts of climate change on water resources are required. These strategies must also be long-term in order to more effectively address vulnerabilities (such as droughts and floods) and they must effectively combine both mitigation and adaptation activities. To date, most attention has been focused on climate change mitigation, while adaptation strategies and activities are only in their infancy.

The federal government's climate change plan, "Turning the Corner," is focused on mitigation: reducing greenhouse gas emissions, setting up a carbon emissions trading market, and establishing a market price for carbon.[37] Canada will not meet its Kyoto Protocol target of 6% below 1990 levels (a target of 563 Mt/yr for the years 2008–2012). However, Canada has committed to achieving a 20% reduction relative to 2006 levels a target of about 600 Mt/yr by year 2020.[38]

To date, a limited consideration to the development of stronger adaptive capacity is evident on the climate change agenda of Alberta and Saskatchewan. Alberta's 2008 Climate Change Strategy is focused on mitigation (energy efficiency, carbon capture/storage and greening of energy production) and makes a commitment to develop a provincial Climate Change Adaptation Strategy.[39] Alberta created Climate Change Central, a special agency whose key purpose is to expand Alberta's efforts on climate change and on reducing greenhouse gas emissions.[40] In 2007, Saskatchewan's previous government released its Energy and Climate Change Plan in response to climate change issues.[41] The plan focused mainly on mitigation of carbon dioxide emissions (conservation, carbon capture and storage, renewable energy, and carbon sinks) and contained a statement of intent to develop an adaptation strategy in the future. Although the plan was not retained after the 2007 election (as of fall, 2008, a new plan is under development), the new government has retained the mitigation targets from the previous plan.[42] Other studies confirm the need to more concretely integrate the variable climate of the region into the governance agenda, especially for decisions that will have long-term impact.[43]

Since the year 2000, there have been significant research activities that consider climate change impacts and adaptation. The federal government published reports on climate change impacts and adaptations in 2004 and 2008, both of which contribute knowledge and understanding.[44] There has also been a significant increase in the institutional research capacity in the area of water resources and climate change within the SSRB. Climate change research programs have been developed in regional institutions, such as the Prairie Adaptation Research Collaborative at the University of Regina, the Water Institute for Semi-Arid Ecosystems based at the University of Lethbridge, and the Alberta and Saskatchewan Research Councils. These programs have been jointly undertaken with the participation of various government agencies. This increasing production of knowledge, fundamental to assessing impacts and vulnerabilities and to developing adaptive capacities, still requires proper channels of integration into governance policies, management, programs, and practices. Towards this end, in the summer of 2008, the federal government announced a Regional Adaptation Collaboratives Program "to equip decision-makers with the information and advice needed to respond to regional opportunities and challenges presented by our changing climate."[45]

There have also been positive developments in the area of drought response in Alberta and Saskatchewan. Alberta Agriculture developed a

Drought Action Plan focused on drought preparedness, reporting, and drought response.[46] The 2002 draft *Drought Risk Management Plan for Saskatchewan* was designed to help government agencies develop a coordinated response to assist stakeholders (agriculture, forestry, municipalities) prepare for, mitigate and respond to drought.[47] The existence of these overarching policies are a significant step in developing a purposeful mandate for governance, and may define a common vision and set of goals for all organizations in relation to climate variability and change.

While these overarching policies represent an increasing level of awareness and concern about climate change, they have not yet been translated into a comprehensive approach that includes adaptation strategies to manage the risks and opportunities of new climate conditions. In light of both economics and climate change predictions, more emphasis on adaptation and on the interrelationship of water with development decisions is required to avoid political disagreements within the policy community and tensions within governance.[48]

The absence of a comprehensive governance approach to climate change impacts and adaptations translates into a lack of consensus within water governance agencies regarding climate change relevance and its proper integration into action agendas, a limitation that continuously emerges in our interviews with water users, stakeholders, and government agencies. For some water organizations climate change is an important challenge, and attention has been directed towards understanding the phenomenon, its impacts upon water resources, and identifying viable adaptation strategies. In many of these cases, however, a lack of targeted resources has been a significant obstacle in transforming issues of climate awareness into programs and organizational practices. Other agencies express a concern with climate change, but they are reluctant to integrate the issue into their agendas, claiming that scenarios of climate change are plagued with uncertainty and cannot yet be considered.

This desire for certainty in climate change predictions is problematic in strengthening adaptive capacity to a range of climate conditions where changes, surprises, and uncertainty are part of the nature of the problem.[49] Legal instruments, such as environmental legislation in both Alberta and Saskatchewan, have statements endorsing principles of precaution, or regulation and rule making prior to scientific certainty.[50] The endorsement of the precautionary principle is a positive sign, but other than the endorsement in principle found in non-binding 'guiding principles' it would appear that actual rules, regulations, policies, and strategies are constrained in advance of scientific proof and certainty. There is also a need to pay more attention to the principle of flexibility in policy and management approaches, to be prepared to experiment, and to find ways to prevent governance processes and approaches from becoming too rigid. Flexibility is essential to be able to respond and adjust to new knowledge in a timely manner. Without such flexibility, insufficient responses and adaptations may occur and may not address the natural wide range of climate variability in the SSRB, let alone address the potentially more severe future impacts from climate change.

The need for integrating civil society into the water governance process

In many Canadian provinces, water governance is essentially achieved in an inclusive manner, with provincial, and in some cases federal governments assuming leadership roles. The formation of local watershed groups indicates that the approach to include civil society is becoming more desirable and potentially more formal. However, as these watershed groups evolve, it is not clear what degree of authority or empowerment will be granted to them by legal or government entities. No less relevant for strengthening water governance is the need for full inclusiveness, where the complete range of stakeholders are involved in defining problems and solutions, identifying local needs and problems, balancing interests, and executing and implementing solutions.[51] Our initial assessment in this area is that there are strengths and weaknesses in terms of developing a clear and genuinely participatory structure and process for water governance within the SSRB. There is no doubt an important and exciting development is occurring in the engagement of civil society in water governance in both Alberta and Saskatchewan. Local organizations and community members are participating on watershed councils or committees, and are engaged in the development of water plans and source water protection plans. This integration of local people and communities into watershed decision-making processes is consistent with the best practices of water governance of the World Water Council[52] and is a central theme of integrated water resource management. The full implementation of this approach will contribute to better management of the risks and opportunities associated with climate change and also foster better adaptive capacities within the rural population of the SSRB.

The process of integrating local people into the management of watersheds is relatively recent, and significant work is still required in this area to ensure their full integration. There is a need to identify clear and definitive roles for these local organizations. In Alberta and Saskatchewan their roles have largely been limited to establishing a diagnostic of the state of the watershed and defining plans to deal with the most urgent watershed issues. There is no clear definition of the degree of management authority these civil society groups have in their participation in watershed governance, nor is there sufficient financial or technical support for their activities. Since the creation of these groups, their role so far is fundamentally an advisory one. In Saskatchewan there is also an issue of the legitimacy of these local organizations, in the sense that proper recognition of their importance is only still limited to a few government agencies, yet their watershed decisions could affect many government agencies.

In addition to the uncertainty surrounding the contribution and continuation of these local water governance initiatives, there are barriers to the ability of local groups and communities in participating on local watershed initiatives. Many stakeholders cited a lack of knowledge of water issues (they are not technical experts), a lack of knowledge of all of the water organizations (partly a result of the confusing maze of organizations as expressed below), and their inability to dedicate the necessary time to learn and acquire the necessary technical and managerial knowledge for informed participation in a watershed group. Furthermore, community members stated they could only contribute limited amounts of time to volunteer activities, even if these were important for watershed management decisions. Even if travel expenses were covered to attend important meetings, the cost associated with missing paid employment is a significant barrier for members, especially volunteers. Stakeholders and watershed groups realize that watershed management decisions undoubtedly require significant commitments of time and resources before effective decision-making can be expected.

A further problem exists in that the SSRB watershed groups generally have limited or no financial resources, and no taxation powers (they are not defined by administrative legally-formed boundaries). The lack of capacity to generate funds means that local watershed groups do not have financial resilience to sustain themselves as a volunteer organization, let alone generate funds to address water monitoring or take corrective actions needed within the watershed. Furthermore, most funding sources for watershed groups are from limited short-term programs, and hence are not likely to support long-term planning needs. There is no doubt that the long-term participation of civil society on integrated water resource management will require the development of increased technical support, and sustainable funding arrangements, to build the capacity of these watershed groups and citizen contributions.

The need to simplify the complex nature of water governance

Most of the rural water stakeholders interviewed felt confused by the multitude of organizations involved in water and water governance, and the various mandates and functions of these organizations. Their concerns could be summarized in the statement of one elder farmer who stated there was "too much water governance, yet not enough water governance," an expression that clearly reflects one of the shortcomings of existing governance.

The statement captures issues related to the confusion over the number of agencies involved with all orders of government, their level of coordination, and a perceived or real disconnection between governance institutions and local watershed issues.

Community members claimed organizations did not communicate with each other, or their functions and roles overlapped, making the links between communities and governance agencies even more challenging and difficult. Interviews with representatives from different governance organizations in Alberta and Saskatchewan also point in the same direction. Those interviewed often complain of a lack of proper communication channels, a lack of clarity on the distinctive roles of agencies, and the barriers that these issues create for policy and program coordination and information sharing.

The absence of clearly-established linkages among organizations certainly makes it difficult to recognize the interconnected nature of climate change and its potential impacts and development needs to address water issues. Stakeholders expressed concerns about water management decision-making, water policies and programs, agency responsibilities, levels of authority, and the need for strong provincial and federal institutions capable of functioning within competing demands. Without this, agencies are reluctant to act: "nobody wants to take the step . . . nobody moves, nobody listens."[53] One focus group noted that, in general, people and governments usually give the impression that 'the government has all the answers' when oftentimes it does not. Effective water governance needs to be able to identify challenges and obstacles, and ways to address these, without fear of political upheaval.

A related theme concerned the multitude of conflicting answers provided to a question depending on the governmental representative asked and providing the response. This situation makes it difficult for communities to plan as they have no concrete guidance, and answers are always changing depending on whom is being dealt with. Respondents accredit this to having too many organizations and departments. Having an abundance of departments makes it difficult for people to communicate with one another because there are so many different avenues of communication and it is very easy to get mixed messages.[54]

This sentiment is also expressed in other literature from numerous sources, ranging from academia to government. For over two decades, questions on water governance mechanisms in Canada have been reported, stemming back to the 1985 review of the federal role in relation to water and water strategy.[55] More recently, the 2005 Senate Report entitled "Water in the West: Under Pressure" identified 'unacceptable' information gaps, recommended that all orders of government need to work together, and stated that "decision-makers must pay urgent attention to water, especially in the semi-arid regions of western Canada where the impacts [of climate change] are already being felt."[56]

Respondents to the IACC research project likened climate change to a 'chronic issue,' and it is therefore less understood or perceived to be 'less critical.' While education and awareness are fundamental, it was noted that, perhaps a simpler form of water governance could be established, perhaps akin to a non-partisan water council (this is also noted in some of the literature) that would operate under a long-term mandate that would not be impacted by policy shifts resulting from elections. The goal would be to ensure "that long-term planning can be made in the interest of public safety because really that is what water is." [57]

Many of the complaints from rural people are the result of an increasing separation between rural and urban people, where rural people feel marginalized from the centers of power. A disconnection was expressed as existing between the local community and the provincial and federal levels of government. Politicians and government agencies are viewed to be both far away in physical presence and unavailable because of time pressures. Because of this perceived disconnection, people interviewed expressed that local concerns, challenges, and issues were not understood by these distant, non-local levels of government.[58] This seems to be the case also in relation to water quality standards. People in some communities expressed that provincial standards were resulting in higher costs to upgrade equipment and hire trained staff, costs that communities stated they could not afford. If changes were not made, the communities were under the risk of action by the government to take control of the community's water system, which again would raise the cost of water in the community and thereby increase the community's vulnerabilities.[59]

Community members expressed a preference for their local governments, and noted an interest in giving more emphasis to local authorities in the decision-making process of managing local water resources. For example, stakeholders cited the importance of local decisions during the drought of 2001–2002. Water allocations to Alberta irrigation districts were reduced from the average year amounts. [Irrigation allocations vary from year to year to suit the available amount of water, and are managed in accordance with inter-provincial agreements that ensure maintenance of minimum river flows to Saskatchewan.] Alberta's irrigation districts, the 'on-the-ground stakeholders,' then made participatory communal decisions on how to manage distribution of reduced irrigation water availability. Water trading and short-term changes to apportionment within the districts were determined to best suit the community of irrigators. In this fashion, stakeholders noted that they themselves need to drive the local solutions, and they should receive support from the government to meet these ends.[60]

Citizens are genuinely interested in participating on watershed management. However, they are interested in more than talk and wish action and implementation of improvements. Meetings to increase awareness are necessary, but some stakeholders stated they are tired of frequently meeting and not doing anything.[61] Whatever improvements can be developed in simplifying the maze of water governance, it will be imperative to sustain and support citizen engagement. Finding mechanisms that lead to implementing constructive changes, by undertaking actual activities that result from the contributions of citizen engagement, will be a necessary piece of the water governance puzzle.

Based on these research findings, the current governance institutional framework will need to adapt by finding new ways of working together. Traditional agency roles where respective agencies and departments are focused on their own mandates will almost certainly need to be broadened to engage participation with larger inter-agency partnerships that also include citizen and stakeholder contributions. Cross-departmental and inter-governmental committees will need to find new ways of working together and sorting out differences. Genuine citizen engagement will likely mean government agencies will need to take increasing risks with their own decision-making. Building trust will be essential for airing out differences, addressing potential conflicts, and achieving effective and efficient adaptive decisions. New mechanisms may be needed for cross-regional governance. Provinces may need to engage federal roles in new ways and to participate on broader initiatives looking beyond provincial boundaries. Federal organizations may need to find innovative ways of supporting provincial and local activities while facilitating national interests that balance economic, social, and environmental interests. Attention will need to be paid to efficient and flexible processes as potentially cumbersome arrangements may bog down decision-making. Broader frameworks with long-term planning for climate and water stress and a simplified water governance model may yet be possible with effective citizen engagement in the new watershed advisory committees and councils.

Conclusions

Inclusive governance involves all of the systems by which authority is exercised. The decision-making process in the management of water resources and the relationships between people and water resources are important aspects where this authority is exercised. Good governance is most often applied with a bottom-up approach that incorporates a proper degree of active participation of local people. The Canadian form of water governance has been essentially a top-down approach interacting with stakeholders. It is now changing by more fully integrating civil society in water management decisions by government agencies. As climate-induced water stress puts more pressure on the management of water resources, there is a need for a governance system that can identify problems of water management within civil society, seek solutions to the problems, and implement solutions that build adaptive capacities and reduce community vulnerabilities. Achieving this will require a water governance system that has a long-term vision, but one that is also flexible and capable of adapting to change itself.

Governance of water resources in the SSRB has already demonstrated its capability of fostering the development of adaptive capacity in response to a variety of stressors, including water scarcities and human-induced water problems. Examples of this existing capacity are numerous. The creation of special agencies, such as Alberta's Special Areas Boards and Canada's Prairie Farm Rehabilitation Administration, has helped the region and rural communities manage drought and adapt to climate. Responses to the crisis of the North Battleford drinking water disease outbreak have led to significant institutional changes to reduce vulnerabilities created by human errors (the Saskatchewan Watershed Authority was created from this event). Irrigated agriculture, initially established to increase the region's resiliency to drought, is continually adopting more efficient irrigation practices to conserve water, allowing for irrigation expansion and strengthening the region's adaptive capacity to climate and water stress. Rural and urban communities have adopted conservation practices to reduce water consumption and pollution. Sectors of civil society have been increasingly involved as participants in safeguarding watersheds and developing watershed plans. And, significant advances have been made in establishing strategies oriented to deal with climate change in both the federal and provincial governments. Many of these developments are for reasons other than solely concerns for climate change, but they each contribute to the accumulation of experiences that support regional adaptive capacity.

There is, however, a long, and perhaps, difficult road ahead, yet to be traversed by governance agencies to ensure sufficient capacity is developed to manage the risks and opportunities created by climate variability and climate change. Significant work has to be done in terms of integrating climate-induced water stress into governance policies and decision-making processes, as well as into managerial practices. The recent failure to establish a strong international climate change mitigation consensus in Bali in 2007, has demonstrated that the possibility to control the speed of global environmental change is less and less feasible, and adaptation strategies are therefore more and more necessary. Significant policy and organizational development within governance and within each of its organizations are critical aspects in achieving adaptive institutions and ultimately more resilient and adaptive rural communities.

A fundamental step in this direction is to incorporate the interest of all stakeholders in water management decisions. The creation of watershed stakeholder organizations, either in the form of Watershed Councils or Advisory Committees, is a step in the right direction. What is also needed is a greater degree of clarity of the roles of these organizations in the governance and management of water resources, and more solid financial and technical support to ensure their sustainability. There are also urgent measures to be taken to strengthen the decision-making process within governance, and to facilitate the communications between government agencies and rural people. There is a need for improved intra-governmental policy coordination (at the provincial and federal levels), as well as a need for better coordination between federal and provincial agencies. Clear roles and mandates are needed for federal institutions and their interaction with provincial agencies (and vice versa). All orders of government need to find ways to build flexibility into programs and policies, and to support local watershed management with appropriate flexibility and accountability. The ubiquitous nature of climate and water certainly demand the development of policies and practices across many agencies to ensure effectiveness. Moreover, given that social impacts of climate change are likely to be heterogeneous, different sectors have different degrees of vulnerability. There is a need for the development of comprehensive support mechanisms that improve the capacity of different sectors to adapt to climate change.

Solutions to climate change impacts and adaptation will undoubtedly require adaptive planning and adaptive institutions. Stakeholders in the South Saskatchewan River Basin have provided useful guidance for adaptation. The development of a comprehensive long-term climate change and adaptation strategy would be advantageous and would serve as a baseline focus; maintaining flexibility would increase effective and efficient responses and adjustments to the plan. The contributions of civil society, water users, and stakeholders have the potential to increase the likelihood of adoption of improved water management practices, streamline water governance, and, ultimately, strengthen the adaptive capacity of rural communities.

Endnotes

1. Intergovernmental Panel on Climate Change (IPCC), *Climate Change 2007: Impacts, Adaptation and Vulnerability* (Cambridge: Cambridge University Press, 2007).
2. See S. Lapp, D. Sauchyn, and B. Toth, "Constructing Scenarios of Future Climate and Water Supply for the SSRB: Use and Limitations for Vulnerability Assessment" in this volume of *Prairie Forum;* D. Sauchyn and S. Kulshreshtha, "Prairies," in D.S. Lemmen, F.J. Warren, J. Lacroix, and E. Bush (eds.), *From Impacts to Adaptation: Canada in a Changing Climate* 2007 (Ottawa: Natural Resources Canada, 2008), 275–328; N. Henderson and D. Sauchyn, *Climate Change Impacts on Canada's Prairie Provinces: A Summary of our State of Knowledge (Summary Document)* (Regina: Prairie Adaptation Research Collaborative (PARC), 2008); W. Schindler and W. Donahue, "An Impending Water Crisis in Canada's Western Prairies Provinces," *Proceedings of the National Academy of Sciences of the United States of America* 103, 19 (2006): 7210–7216.
3. N. Adger et al., *Fairness in Adaptation to Climate Change* (Cambridge: MIT Press, 2006); K. O'Brien and R. Leichenko, "Double Exposure: Assessing the Impacts of Climate Change Within the Context of Globalization," *Global Environmental Change* 10 (2000): 221–232; S. Olmos, *Vulnerability and Adaptation to Climate Change: Concepts, Issues, Assessment Methods* (2001), Climate Change Knowledge Network. Available at: www.cckn.net/va.asp (cited 16 December 2008).
4. The website of the Institutional Adaptations to Climate Change (IACC) project contains information about the approaches, procedures and results of all of these research activities. Available at: http://www.parc.ca/mcri/index.php (cited 13 November 2008).
5. See Jeanne Kasperson et al., "Vulnerability to Global Environmental Change," in J. Kasperson and R. Kasperson (eds.), *The Social Contours of Risk. Volume II: Risk Analysis, Corporations and the Globalization of Risk* (London: Earthscan, 2005), 245–287.
6. See John Birkmann, "Measuring Vulnerability to Promote Disaster-Resilient Societies: Conceptual Frameworks and Definitions," in J. Birkmann (ed.), *Measuring Vulnerability to Natural Hazards* (New York: United Nations University Press, 2006), 9–54; D. Hilhorst, "Complexity and Diversity: Unlocking Social Domains of Disaster Response," in G. Bankoff, G. Frerks, and D. Hilhorst (eds.), *Mapping Vulnerability. Disasters, Development and People,* (London: Earthscan, 2004), 52–66.
7. Olmos, *Vulnerability and Adaptation to Climate Change* ; IPCC, *Climate Change 2001: Impacts, Adaptation, and Vulnerability Technical Summary.* A Report of Working Group II of the Intergovernmental Panel on Climate Change (WMO and UNEP: New York, 2001).
8. D. Liverman, "Vulnerability to Global Environmental Change," in S. Cutter (ed.), *Environmental Risks and Hazards* (Englewood Cliffs: Prentice Hall, 1994); B. Wisner et al., *At Risk. Natural Hazards, People's Vulnerability and Disasters* (New York: Routledge, 2005); J.W. Handmer, S. Dovers, and T.E. Downing, "Societal

vulnerability to climate change and variability," *Mitigation and Adaptation Strategies for Global Change* 4 (1999): 267–281; B. Smit and J. Wandel, "Adaptation, Adaptive Capacity and Vulnerability," *Global Environmental Change* 16 (2006): 282–292.

9. N. Adger and p.m. Kelly, "Social Vulnerability to Climate Change and the Architecture of Entitlements, *Mitigation and Adaptation Strategies for Global Climate Change* 4 (1999): 253–266.
10. Olmos, *Vulnerability and Adaptation to Climate Change.*
11. I. Burton et al., *Adaptation Policy Frameworks for Climate Change: Developing Strategies, Policies and Measures* (Cambridge: UNDP/Cambridge University Press, 2005).
12. Burton, *Adaptation Policy Frameworks for Climate Change,* 168.
13. IPCC, *Climate Change 2001,* 893.
14. IPCC, *Climate Change 2001,* 896–897.
15. J. Henningham (ed.), *Institutions in Modern Societies* (Melbourne: Oxford University Press, 1995), as discussed in H. Diaz, A. Rojas, L. Richer, and S. Jeannes, *Institutions and Adaptive Capacity to Climate Change,* IACC Project Working Paper No. 9, Canadian Plains Research Center. (Regina: Prairie Adaptation Research Collaborative, 2005), 5. Available at: www.parc.ca/mcri/unit1E.php (cited 23 December 2007).
16. P. Diaz and A. Rojas, *Methodological Framework for the Assessment of Governance Institutions,* IACC Project Working Paper No. 33. (Regina: Prairie Adaptation Research Collaborative, 2006), 3. Available at: http://www.parc.ca/mcri/pdfs/papers/iacc033.pdf (cited 21 November, 2008).
17. N. Adger, "Social Aspects of Adaptive Capacity," in J. Smith, R. Klein, and S. Huq (eds.), *Climate Change, Adaptive Capacity and Development* (London: Imperial College Press, 2003); D. Halpern, *Social Capital* (Cambridge: Polity Press, 2005); Ann Dale and Jenny Onyx (eds.), *Social Capital and Sustainable Community Development: a Dynamic Balance* (Vancouver: UBC Press, 2005).
18. There is a large body of literature on this topic, ranging from theoretical to applied arguments. See U. Beck, *Risk Society. Towards a New Modernity* (London: Sage, 1992); J. Dryzek, *The Politics of the Earth: Environmental Discourse* (Oxford: Oxford University Press, 1997); M. Foucault, "Governmentality," in G. Burchell, C. Gordon, and P. Miller (eds.), *The Foucault Effect: Studies on Governmentality* (Chicago: University of Chicago Press: 1991); M. Howes, *Politics and the Environment. Risk and the Role of Government and Industry* (London: Earthscan, 2005); Elim Papadakis, *Environmental Politics and Institutional Change* (Cambridge: Cambridge University Press, 1996); J. Alcorn and V. Toledo, "Resilient Resource Management in Mexico's Forest Ecosystems," in F. Berkes and C. Folke (eds.), *Linking Social and Ecological Systems: Management Practices and Social Mechanisms for Building Resilience* (Cambridge: Cambridge University Press, 2000).
19. R. Jenkins "The Emergence of the Governance Agenda: Sovereignty, Neo-Liberal Bias, and the Politics of International Development," in V. Desai and R. Potter (eds.), *The Companion to Development Studies* (London: Hodder Arnold, 2002), 485.

20. Cited in Conference Board of Canada, *Navigating the Shoals, Assessing Water Governance and Management in Canada* (Toronto: Conference Board of Canada, 2007), 2–3. Available at: www.conferenceboard.ca/documents.asp?rnext=1993 (cited 16 December, 2008); see also A. Hall, "Water: Water Governance" in G. Ayre and R. Callway, *Governance for Sustainable Development* (London: Earthscan, 2005).
21. Conference Board of Canada, *Navigating the Shoals,* 2–4.
22. World Development Report 2003, *Sustainable Development in a Dynamic World: Transforming Institutions, Growth, and Quality of Life* (Washington: World Bank and Oxford University Press, 2003), 37–55.
23. A more systematic description of these organizations exists in D. Corkal, B. Inch, and P. Adkins, "The Case of Canada—Institutions and Water in the South Saskatchewan River Basin," IACC Working Paper (Regina: Prairie Adaptation Research Collaborative, 2006). Available at: http://www.parc.ca/mcri/pdfs/papers/iacc045.pdf (cited 21 November, 2008); see also, the IACC compendium of water institutions and their roles, links to other organizations, and legal instruments, presented in E. Orrego's "Water Governance Institutions in the South Saskatchewan River" (2007). Available at: http://www.parc.ca/mcri/pdfs/papers/iacc053.pdf.
24. *The Constitution Act, 1867* (U.K.), 30 & 31, Victoria, c. 3. being Schedule B to the *Canada Act 1982* (U.K.), 1982, c. 11.
25. When the provinces of Manitoba (1870), Alberta and Saskatchewan (1905) were created, the natural resources, including water rights, were retained by the Dominion Government. See Alastair R. Lucas, *Security of Title in Canadian Water Rights* (Calgary: Canadian Institute of Resources Law, 1990): 9. In 1930 these natural resources were transferred to the provinces in the Natural Resources Transfer Agreements, and water was confirmed as part of the transfer in Agreements of 1938 (ibid. at p. 11 referring to the *Natural Resources Transfer (Amendment) Act,* 1938, S.C. 1938, c. 36).
26. In traditional Canadian common law, water rights were transferred with the land with which it was associated. "Land" is defined as "every species of ground, soil or earth whatsoever, as meadows, pastures, woods, moors, waters, marshes, furs and heath." S. 91 of the *Constitution Act.* In addition the provincial heading of "local works and undertakings" helps support water structures located within a province on a particular body of water.
27. The term "First Nations" is used in the article. However, most of the Acts and legislation use the term "Indian" to refer to First Nations people/communities. See: *Indian Act,* R.S.C. 1985, c. I-5 and s. 91 of the *Constitution Act,* 1982.
28. The *Canada Water Act,* R.S.C. 1985, c. C-11.
29. For a more detailed description of these laws and similarities and differences see Margot A. Hurlbert, "Water Law in the South Saskatchewan River Basin (Alberta/Saskatchewan)," IACC Working Paper 27 (Regina: Prairie Adaptation

Research Collaborative, 2007). Available at: http://www.parc.ca/mcri/pdfs/papers/iacc027.pdf (cited 21 November 2008).

30. Government of Alberta, *Water for Life: Alberta's Strategy for Sustainability* (Edmonton: Alberta Environment, 2003), 31.
31. For a list of completed Source Water Protection Plans by Watershed Advisory Committees in Saskatchewan, see the Saskatchewan Watershed Authority's website for Watershed and Aquifer Planning. Available at: http://www.swa.ca/Stewardship/WatershedPlanning/Default.asp (cited 14 November 2008); see also Saskatchewan's *State of the Watershed Report March 2007*, which provides a basis for governments, industry, and communities to act in the interest of environmental sustainability (Regina: Saskatchewan Watershed Authority, 2007), 152.
32. Recent drinking water disease outbreaks in Canada have had a profound effect on water management in the country. Most governments are emphasising the management of water on a watershed basis, with increasing participation of stakeholders. For more information see D. Corkal et al., "The Case of Canada."
33. G.P. Marchildon, S. Kulshreshtha, E. Wheaton, and D. Sauchyn, "Drought and Institutional Adaptation in the Great Plains of Alberta and Saskatchewan, 1914–1939," *Journal of Natural Hazards* 45–391 (2008): 411.
34. The *Prairie Farm Rehabilitation Act,* R.S.C., 1985, c. P-17.
35. This arrangement is now considered a model for coordination and cooperation in managing inter-jurisdictional water flow. See Conference Board of Canada, *Navigating the Shoals.*
36. For the SSRB, the Commission makes rulings on the international water flows and water extractions in accordance with the *1909 Boundary Waters Treaty.* This is a part of the "Mission Statement" of the International Joint Commission, which is available at: http://www.ijc.org/en/home/main_accueil.htm (cited 4 May 2006).
37. Government of Canada, *Turning the Corner: Taking Action to Fight Climate Change* (Ottawa: Environment Canada, 2008). Available at: http://www.ec.gc.ca/doc-/virage-corner/2008–03/brochure_eng.html (cited 21 November 2008); see also Government of Canada, *Turning the Corner: Regulatory Framework for Industrial Greenhouse Gas Emission* (Ottawa: Environment Canada, 2008), 26. Available at: http://www.ec.gc.ca/doc/virage-corner/2008–03/pdf/COM-541_Framework.pdf (cited 17 December 2008).
38. GOvernment of Canada, *A Climate Change Plan for the Purposes of the Kyoto Protocol Act—2007* (Gatineau: Envirnoment Canada, 2007). Available at: http://www.ecgc.ca/doc/ed-es/p_123/CC_Plan_2007_e.pdf (cited 21 November 2008).
39. Government of Alberta, *Alberta's 2008 Climate Change Strategy: Responsibility/ Leadership/ Action* (Edmonton: Alberta Environment, 2008). Available at: http://environment.alberta.ca/2430.html (cited 16, December 2008).
40. Climate Change Central, *About Us.* Available at: http://www.climatechangecentral.com/about-us (cited 16 December 2008).

41. Government of Saskatchewan, "*Energy and Climate Change Plan 2007*" (Regina: Premier's Office, 2007).
42. Government of Saskatchewan, *Go Green Saskatchewan, Climate Change* (Regina: Saskatchewan Environment, 2008). Available at: http://www.environment.gov.sk.ca/Default.aspx?DN=9192fbe8–23fe-4077-ac7d-30b7b269bdbf (cited 17 December, 2008).
43. Social Dimension of Climate Change Working Group, Research Report, "*Social Dimensions of the Impact of Climate Change on Water Supply and Use in the City of Regina*" (Regina: University of Regina and Canadian Plains Research Center, 2005), 34–36; see also V. Wittrock, E.E. Wheaton, and C.R. Beaulieu, "*Adaptability of Prairie Cities: The Role of Climate: Current and Future Impacts and Adaptation Strategies,*" Limited Report, Environment Branch, Saskatchewan Research Council, Publication No. 11296–1E01 (Saskatoon: Saskatchewan Research Council, 2001), 9–7.
44. Government of Canada, "*Climate Change Impacts and Adaptation: A Canadian Perspective,*" (Ottawa: Climate Change Impacts and Adaptation Directorate, Natural Resources Canada, 2004), 174; see also D.S. Lemmen et al. (eds.), *From Impacts to Adaptation: Canada in a Changing Climate 2007* (Ottawa: Natural Resources Canada, 2008), 448. Some identified adaptations were: a need to conduct integrative studies of water resources planning to address the role and influence of water managers on adaptive capacity; building the understanding of current capacity of water management structures and institutions to deal with projected climate change impacts and the associated social, economic and environmental costs and benefits of future adaptations; and improving our understanding of environmental justice and equity consequences.
45. Natural Resources Canada, *Regional Adaptation Collaboratives—About the Program* (2008). Available at: http://adaptation.nrcan.gc.ca/collab/abosuj_e.php (cited 17 December 2008).
46. Alberta Agriculture, *Drought Action Plan.* Available at: http://www1.agric.gov.ab.ca/$Department/deptdocs.nsf/all/ppe9026 (cited 14 December 2007).
47. "Drought Risk Management Plan for Saskatchewan," in *Canada-Saskatchewan MOU (Memorandum of Understanding) on Water Committee Minutes* (Regina: Agriculture and Agri-Food Canada-PFRA, June 24, 2002).
48. The existence of these disagreements is clearly identified in A. Welstead, D. Davidson, and R. Stedman, "Assessing the Potential for Policy Responses to Climate Change," (PARC Project 011 2002). Available at: www.parc.ca (cited 21 November 2007).
49. International Institute for Sustainable Development, *Designing Policies in a World of Uncertainty, Change and Surprise. Adaptive Police-Making for Agriculture and Water Resources in the Face of Climate Change* (Winnipeg: International Institute for Sustainable Development, 2006).
50. *The Environmental Protection and Enhancement Act* of Alberta provides in s. 2, "The purpose of this Act is to support and promote the protection, enhancement

and wise use of the environment while recognizing the following: (c) the principle of sustainable development, which ensures that the use of resources and the environment today does not impair prospects for their use by future generations." Similarly, the Saskatchewan *Environmental Assessment Act,* S.S. 1979–80, c. E-10.1 provides in s.2 (k) "pollutant" means a substance, including a contaminant, which results, or is likely to result, in pollution of the environment.

51. World Development Report 2003, *Sustainable Development,* 37- 55.
52. A report prepared as a follow up to a World Water Forum organized by a World Water Council outlined these principles for assessing challenges and initiating change of water institutions. See World Water Council Water Action Unit, *World Water Actions: Making Water Flow for All* (London, England: Earthscan Publications Ltd., 2003), 22; David B. Brooks, *Water: Local-Level Management* (Ottawa: International Development Research Center, 2002).
53. Outlook Focus Group Stakeholder Workshop, Jan. 25, 2007. Prairie Adaptation Research Collaborative, Regina. A summary report is available at: www.parc.ca/mcri/outlook.php (cited Nov. 21, 2008).
54. Brett Matlock, *Report on the Community Vulnerability Assessment of Cabri and Stewart Valley, Saskatchewan* (IACC project paper, p. 21). Available at: http://www.parc.ca/mcri/cabstew.php (cited Nov. 21, 2008).
55. For a few examples: see P.H. Pearse, R. Bertrand, and J. W. MacLaren, *Currents of Change, Final Report, Inquiry on Federal Water Policy* (Ottawa: Environment Canada, 1985); the report of Oliver M. Brandes et al., *At a Watershed: Ecological Governance and Sustainable Water Management in Canada* (Victoria: Polis Project on Ecological Governance, University of Victoria, 2005); the Conference Board of Canada, *Navigating the Shoals;* the 2005 Senate report by Tommy Banks and E. Cochrane, *Water in the West: Under Pressure,* 3–5. Available at: http://www.parl.gc.ca/38/1/parlbus/commbus/senate/com-e/enrg-e/rep-e/rep13nov05-e.pdf (cited 16 December 2008); also E. Bakker *Eau Canada: The Future of Canada's Water* (Vancouver: UBC Press, 2007). The Government of Canada has numerous policy research initiative publications discussing water policy options, from Integrated Water Resources Management (IWRM) policies to market-based instruments to community water sustainability indices. One example is *Freshwater for the Future,* available at: http://www.policyresearch.gc.ca/doclib/v9n1_e.pdf (cited 17 December 2008). See also associated publications at: http://www.policyresearch.gc.ca/page.asp?pagenm=pub_index; the Gordon Water Group's *Changing the Flow* report and *Executive Summary* available at: http://www.gordonwatergroup.ca/ (cited 17 December 2008).
56. Banks and Cochrane, *Water in the West: Under Pressure.*
57. Outlook Focus Group Stakeholder Workshop, 11.
58. Monica Hadarits, IACC project, Summary Report, Outlook Stakeholder Workshop on Water and Climate (Regina: Prairie Adaptation Research Collaborative), 11.

Available at: http://www.parc.ca/mcri/outlook.php (cited 28 December 2007). The quotation was extracted from the focus group interviews supporting documentation.

59. Brett Matlock, *Report on the Community Vulnerability Assessment*, 21.
60. Lethbridge Focus Group Stakeholder Workshop, Dec 1, 2006 and Outlook Focus Group Stakeholder Workshop, Jan. 25, 2007, Prairie Adaptation Research Collaborative projects, Regina. Available at: http://www.parc.ca/mcri/lethbridge.php and www.parc.ca/mcri/outlook.php (cited Nov. 21, 2008).
61. Outlook Focus Group Stakeholder Workshop.

PART 3: CASE STUDIES OF VULNERABILITY

THE 2001–2002 DROUGHT: VULNERABILITY AND ADAPTATION IN ALBERTA'S SPECIAL AREAS

JOHANNA WANDEL, GWEN YOUNG, AND BARRY SMIT

Abstract

Insights into vulnerability and adaptation of particular cases can be gained through examining management of past stresses. This paper uses the example of the 2001–2002 drought in Alberta's Special Areas to examine the exposure-sensitivities and adaptive strategies of stakeholders within the region. Not all stakeholders were equally affected, as the existence of a secure municipal water supply mitigated the impact of the drought for urban users. Among the agricultural community, dryland farmers faced particularly great challenges, though all producers, including those with access to irrigation, were negatively affected. Locally-based institutions, who have a long history of drought management, facilitated some drought management strategies.

Sommaire

L'étude de la gestion des stress auxquels une collectivité a été soumise dans le passé fournit des renseignements utiles à l'égard de la vulnérabilité et de l'adaptation. Cet article s'appuie sur la sècheresse de 2001–02 dans les zones spéciales de l'Alberta afin d'étudier la sensibilité des intervenants de la région par rapport aux divers facteurs de stress auxquels ils sont exposés et les stratégies évolutives déployées par ceux-ci. Certains intervenants furent plus touchés que d'autres. La présence d'un réseau municipal sûr d'approvisionnement en eau a atténué de façon substantielle les impacts de la sècheresse pour les habitants des régions urbaines. Chez les collectivités agricoles, les producteurs en régime d'aridoculture ont dû faire face à de particulièrement grands défis, bien que tous les producteurs, incluant ceux ayant accès à l'irrigation, furent touchés de façon négative. Les institutions locales, qui depuis longtemps travaillent sur la gestion relative à la sècheresse, ont favorisé l'établissement de stratégies de gestion en cas de sècheresse.

1. Introduction

Recurring persistent dry periods have been documented for the southern Canadian Prairies for over 400 years,[1] including the well-known case of the "dust bowl" years of the 1920s and 1930s.[2] While persistent dry periods invariably have had economic and social impacts on Prairie residents, recent drought events have not had the same catastrophic implications, including loss of livelihood and displacement, as earlier events. This is particularly notable in the case study presented in this paper, the Special Areas of southeastern Alberta. As a result of the droughts of the 20s and 30s, the majority of the land within the present-day Special Areas reverted to the Crown as tax recovery land; however, the drought of 2001–2002 had no similarly dramatic and visible impacts. This difference in impact is attributed to changes in differential vulnerability of this area of the southern Prairies.

The concept of "vulnerability," commonly defined as the degree and manner to which a system of interest is susceptible to and unable to cope with the adverse effects of climate, incorporates both the characteristics of the system of interest and the nature of the stimulus that acts upon it.[3] In the context of this paper, the system of interest is a study area in southern Alberta and the stimulus is a drought event. The implications of drought on southern Alberta reflect the region's physical landscape, the existing human land-use system, and the human and financial capital of its occupants. A region's vulnerability to a persistent dry period is thus related not only to a deviation from hydrological norms but also to the human use of and reliance on water and the available coping mechanisms to manage periods of insufficient moisture.

Under climate change models, it can be expected that drought stress and surface water availability shortages in the southern Prairies will increase due to increased temperatures and decreased snow pack.[4] Consequently, there is considerable interest in identifying the ways in which drought is managed in southern Alberta. This paper investigates the vulnerability of the case study of Alberta's Special Areas during the 2001–2002 drought. Impacts of and responses to the extended dry period are documented at the local level, and conclusions are drawn with respect to adaptation, adaptive capacity and vulnerability in the study region.

2. Vulnerability and Adaptation to Climate Change in Agriculture

The southern Prairies have experienced several multi-year drought episodes since the mid-1980s.[5] The most recent drought of 2001–2002 is estimated to have had a range of impacts on the Canadian economy, including agricultural production losses of $3.6 billion and negative or zero net farm income in Alberta and Saskatchewan for 2002.[6] There is a recognized need for a better understanding of agricultural adaptation to climate variability and change to improve farmer ability to cope with or manage periods of stress.[7]

Considerable research effort has estimated production and economic implications of changing climate normals for the agricultural sector at the macro scale.[8] However, there has been less research on how farmers experience actual climatic conditions, the implications of particular conditions, the management strategies available to respond to these and associated constraints, and opportunities for enhancing adaptive capacity.[9] There is also comparatively little information on what particular conditions pose problems, what has been done in response in the past, and what lessons these provide for what could be done in the future. Some of the few examples of actual experience of climatic conditions and associated risks, opportunities and management strategies were framed using the concepts of vulnerability and adaptation.[10]

Vulnerability can be conceptualized as reflective of exposure-sensitivity and adaptive capacity.[11] Exposure relates to how a system experiences an external stress such as a persistent dry period, while the system's sensitivity to the stress reflects its use of affected resources and is thus a function of livelihoods and occupancy characteristics. Adaptive capacity includes available adaptation tools and strategies and the ability of individuals and institutions within the system to implement these given financial, managerial, time and other constraints. Research in recent years has highlighted the need to consider climate-related stimuli in the context of other stresses.[12] Considerations of exposure-sensitivity and adaptive capacity necessarily include opportunities and constraints introduced by other non-climatic processes including falling cattle prices due to BSE outbreaks and labour shortages given the upswing in Alberta's economy.

Key insights for understanding how a system may be affected by and respond to a climatic stress in the future can be drawn from the ways in which the stress is currently managed or has been managed in the recent past. Documenting the impacts of and responses to the 2001–2002 drought in the Special Areas provides a basis for identifying what magnitude of moisture stress is manageable, and what conditions fall beyond the available adaptation strategies. An analysis of what stresses occurred, how these impacted individuals and how these were managed, allows an identification of what management strategies can be further strengthened so as to increase the ability to manage problematic conditions in the future.

Documentations of vulnerability via examinations of exposure-sensitivities and adaptive strategies draw on a number of sources. Stakeholders themselves provide accounts of the ways in which they experience climatic (and other) stresses, how these have been managed in the past, possible unrealized opportunities, and speculations of impacts and adaptations in "what if" scenarios. The research for this paper is based in part on detailed narrative accounts of rural and urban households, water managers, and agricultural producers of climatic and non-climatic stresses during both a "normal" year and the 2001–2002 drought, and the implications of cumulative persistent dry periods. Further data inputs are derived from key institutional informants, secondary source material, and climatic records to fully understand exposure-sensitivities and adaptive strategies and capacities.

3. Case Study: Alberta's Special Areas

3.1. Methodology

Under climate change, it can be expected that moisture stress and surface water availability shortages will increase in the southern Prairies. This paper draws on case study research to document how vulnerability and adaptation are experienced at the local level during one period of stress, the 2001–2002 drought. Data sources include published secondary source material, census and instrumental records, and interviews with key informants and stakeholders.

Forty-three interviews were conducted by two female interviewers with residents of the western portions of the Special Areas, Special Area 2 and the Town of Hanna, during a five-week period. Initial contact with the community was made via the local Agriculture Canada/Prairie Farm Rehabilitation Administration field office, and the researchers familiarized themselves with the local context through attendance at water managers association and irrigation committee meetings, as well as through key informant representatives from the federal government, regional administration, local government, the municipal water commission, and an environmental non-governmental organization. Further study participants were selected by the researchers in collaboration with local key informants using a stratified sampling frame to ensure that the range of agricultural enterprises present in the study area were included in the research. Table 1 outlines the agricultural producers, non-farm residents, and key informants who were consulted.

Interviews followed a semi-structured protocol based on an interview guide. Interviews ranged from 30 to 150 minutes in length. All interviews were audio recorded and subsequently transcribed. Interview data were analyzed with the help of qualitative data management software.[13]

3.2. The Context of the Special Areas

Alberta's Special Areas comprise a tract of 2,024,292 hectares (ha) north of the Red Deer River and just west of the Saskatchewan border between 50.6° and 53.3° north and 110° and 112.5° west (see Figure 1). The majority of the municipality falls within Palliser's Triangle, with soils primarily in the Brown (GA) and Dark Brown (GB) Chernozem soil groups.[14] Most of the area

KEY INFORMANTS	#
Mayor, Town of Hanna	1
Special Areas Board Chairman	1
Special Areas Board Past Chairman	1
Special Areas Bullpound Pasture Manager	1
Special Areas Ditchrider/Water Plant Operator	1
Henry Kroeger Water Plant Personnel	2
Agriculture Canada Field Staff	2
Ducks Unlimited Field Staff	1
Oil Industry Representatives	2
Crop Insurance Agent	1
Sheerness Generating Station Engineer	1
PARTICIPANTS	**#**
Non-farm Residents	6
Dryland Farmers (no livestock)	1
Dryland Farmers, Mixed Operation (cash crops and livestock)	8
Ranchers (no cash crops, no irrigation)	6
Ranchers (no cash crops, irrigation for pasture/feed)	6
Irrigated Farmers (crops and livestock mixed)	2
TOTAL INTERVIEWS	**43**

Table 1. Sample Description.

receives less than 250 mm precipitation per annum[15] with a gradual increase toward the north and north-west, where long-term precipitation averages reach as high as 400 mm per annum (Figure 1). Evaporation generally exceeds precipitation throughout the study area, and consequently Solonetzic[16] soil groupings are common in parts of the study area, including 30% or more of all land in the Special Area 2.

This region of Alberta is part of the dry mixed-grass natural grasslands, characterized by treeless prairie with shrubs and trees only in relatively moist areas such as coulees and river valleys. The area was settled beginning in 1909 in accordance with the Dominion Lands Act, which granted settlers title to one quarter section of land[17] under the condition that a fee was paid, a house built, and one-fifth of the land area was cultivated within three years. Population of the area increased from 75 in 1901 to 24,164 in 1916 and an all-time high of 29,689 in 1921. Consequently, substantial areas of native grasslands were plowed and wheat was established as the dominant crop.[18]

A period of relatively abundant rainfall in southeastern Alberta ended in 1916, and the area experienced eight successive years of low precipitation and poor harvests. Poor harvest and wind erosion led to many settlers being unable to meet tax obligations and subsequent outmigration with claimed land reverting to the Crown The region was closed to settlement under the Dominion Lands Act in the 1920s. The Special Areas Act of 1938 replaced regional government in this region with a crown agency which manages the remaining never claimed crown and tax recovery lands and set up a governance structure which is distinct from the rest of thc southern Prairies.[19] Unlike the remainder of Alberta's counties and municipal districts, regional government falls directly under the authority of the Province. The Special Areas are under the direction of the Special Areas Board, which consists of the Chairman appointed by provincial cabinet and two board members which are nominated by the Special Areas Advisory Council and appointed by the provincial department of Municipal Affairs and Housing. The Special Areas Advisory Council itself has 13 elected members. The Special Areas Board has no set term of office, and thus administration of the Special Areas does not follow the three year election cycle of the remainder of the province.

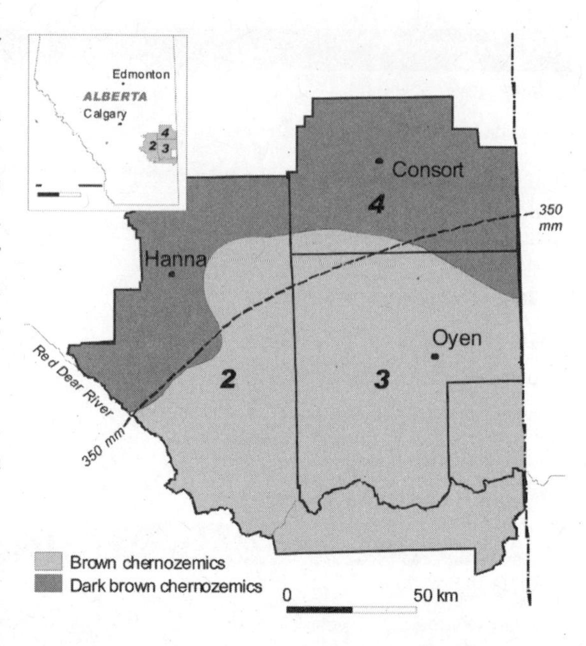

Figure 1. Alberta's Special Areas.

The Special Areas administration administers agricultural land with a view to drought management via cropping and grazing leases, parkland and community pastures. In addition to public land administration, the Special Areas Board is responsible for roads, water management, emergency and protective services, agricultural and rural economic development, and conservation.

The governance framework established by the Special Areas Act and the Agricultural Service Board Act (1945) allows the Special Areas Board to devote resources to prevent and mitigate some of the impacts of drought. For example, the Special Areas Agricultural Service Board has undertaken exten-

YEAR	RURAL	URBAN	TOTAL
1941	11794	3325	15119
1946	9542	3504	13046
1951	8430	4076	12506
1956	8723	4657	13380
1961	8799	5256	14055
1966	7974	5354	13328
1971	7050	5250	12300
1976	5824	5182	11006
1981	6042	5535	11577
1986	6261	5850	12111
1991	6010	5835	11845
1996	5765	5725	11490
2001	5756	5703	11459
2006	5314	5526	10840

(Urban areas include the Towns of Hanna and Oyen, Villages of Consort, Cereal, Veteran, Youngstown, and Empress.)

Sources: data up to 1976 from Gregory P. Marchildon, "Institutional Adaptation to Drought and the Special Areas of Alberta, 1909–1939," *Prairie Forum,* vol. 32, no. 2 (2007): 263; data after 1976 from Alberta Municipal Affairs Statistics.

Table 2. Rural/Urban Population in the Special Areas, 1941–2006.

sive research into, and the promotion of, zero tillage initiatives to decrease wind erosion during dry periods, and has recommended particular mixes of native grasses for reseeding grazing leases. To generate research with particular relevance to the drought-prone Special Areas, the Agricultural Service Board established the Dryland Applied Research Association now known as the Chinook Applied Research Association.

The Special Areas currently have a population of 10,840, of which just over half live in towns and villages. The rural population of the Special Areas steadily decreased from a record of 26,031 in 1921 to 5,314 in 2006, while the urban population (in the towns of Hanna and Oyen and the villages Cereal, Consort, and Empress) increased until 1961, but has been slowly declining since.

Primary industries remain as the dominant source of employment within the Special Areas, accounting for 39% of all jobs. Table 3 illustrates the urban/rural difference in employment by sector, with agriculture as the dominant sector in the rural areas, and sales and service related occupations leading in towns and villages. Towns and villages are classic service centres for the rural population with a diversified wage base. The rural areas are still primarily agricultural areas by employment, though the oil and gas industry has gained some prominence in recent years. Currently 7% of all jobs within the Special Areas (for both rural and urban residents) are oil and gas related (extraction, service, manufacture of related lubricants, absorbents etc.) In addition, Sheerness Generating Station and the adjacent Luscar coal mine provide employment for over 100 people.

	TOTAL	RURAL	URBAN
Total employed	6,240	3,340	2,900
By sector (%)			
Management	6	1	12
Business/administration	10	8	13
Science related	2	1	2
Health	4	3	6
Education/government	5	4	6
Arts/recreation/sport	1	1	2
Sales/service	17	8	27
Trades	15	10	19
Primary (total)	39	62	13
Agriculture	36	59	10
Manufacturing/processing	3	3	2

Source: Census of Population, 2001.

Table 3. Employment by Sector.

3.3. Current Land Use in the Special Areas

Currently, 60% of the Special Areas is made up of public land (30% never-claimed crown lands and 30% tax recovery lands). Public land is primarily used as community pasture (6%) and leased (91%) under various terms, with the majority of the remainder designated as parkland. The most common arrangements are grazing leases, which are for 20 years, and 10-year cultivation leases, although there are also a number of small oil and gas development leases.

The 2001 Census of Agriculture data indicate that 96% of the total land area of the Special Areas was used for agriculture, with 73% of the public land area leased by individual producers for agriculture. Since the Dust Bowl years, the area tilled has decreased dramatically—in 2001, only 24% of the total farm area was under crops. Wheat remains the most common crop (52% of farmers reporting / 10% of the total farmland area), followed by alfalfa and alfalfa mixtures (43% reporting / 4% of land area), oats (36% reporting / 2% of farmland area), barley (25% reporting / 3% of land area), and tame hay (15% reporting / 1% of land area). Other crops reported by more than 100 of the 1,437 farmers but comprising less than 1% of the total farmland area are mixed grains, rye and canola.

Of the five leading field crops, four can be and are commonly used as livestock feed. Given that that the majority of the public land leased to farmers is classified as grazing leases, it is not surprising that 74% of all agricultural producers in the Special Areas report cattle. While there are a handful of dairy farms (16 total), the majority of the cattle producing enterprises are cow-calf operations, stocker operations, and feedlots. In many cases, cow-calf operations also function as stocker operations, and feedlots also include stocker operations.

3.4. The 2001–2002 Drought in the Special Areas

The extent of the early 21st century droughts in the Canadian Prairies is well documented.[20] Based on the Palmer Drought Severity Index, the Special Areas suffered a drought in both 2001 and 2002.[21]

This case study is focused on exposure-sensitivities and adaptations at the producer scale in the Special Areas. Consequently, moisture availability is considered in light of farm operations, that is, consideration of moisture is needed and what other sources are available at critical times. Both crop production and ranching are particularly sensitive to early growing season (April-June) moisture shortages, since crops and tame pastures require rain to germinate and establish themselves.[22] Furthermore, if there is insufficient precipitation during the late growing season (July-September), crops wither and lose their value. Native prairie grass retains its nutritive value during droughts so long as it is well established. Impacts on native species are particularly severe if there is little soil moisture at the beginning of the growing season, and this is related to the previous year's conditions and, most importantly, to snowfall in the early part of the calendar year. Figure 2 illustrates the particularly low early and late season growing

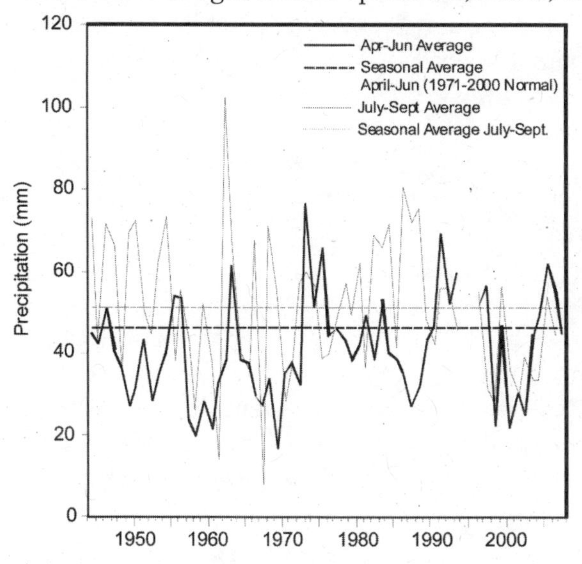

Figure 2. Early and Late Growing Season Precipitation at Coronation.

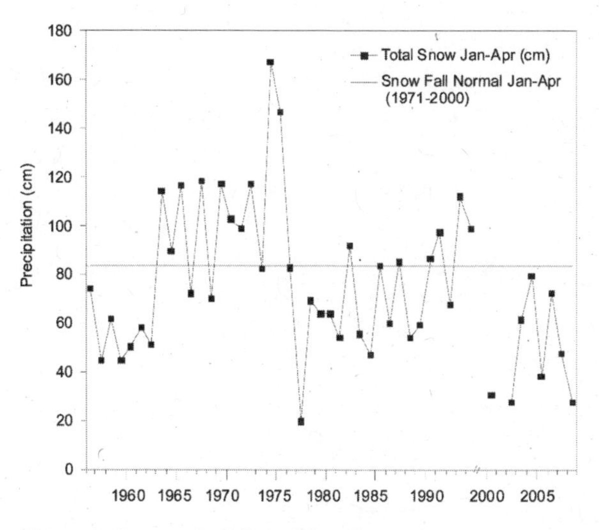

Figure 3. January-April Snowfall at Coronation.

seasons for 2000–2002 inclusive, and Figure 3 shows that snowfall was less than one third of the long-term seasonal average in the spring of 2001 and considerably below normal in 2002.

Moisture stress in the Special Areas was compounded by grasshopper outbreaks in 2002. Grasshopper populations are particularly high after two to three years of hot, dry summers and snow-free falls.[23] Conversely, wet and cool springs provide conditions which curtail grasshopper hatching, and serve to stop an outbreak.[24] Conditions during 2001–2002 provided the basis for record numbers of grasshoppers in 2002, and set the stage for another year of outbreaks if the spring of 2003 was dry.[25] Grasshoppers denude areas of all vegetation, including crops and tame and native grasses. Given the low precipitation values throughout the growing season that year, little regrowth was possible and thus both crop farmers and livestock producers relying on grazing and feed production were affected.

3.5. Water Use, Exposures, and Adaptation in the Special Areas

Water for household, industrial, and farm use in the Special Areas comes from three sources: groundwater, precipitation and associated spring run-off, and surface water via the Red Deer River and small-scale diversion projects. Water uses include household and drinking water, stock water for livestock, irrigation, recreation, and industrial uses for thermal power generation and in oil and gas drilling.

3.5.1. Industrial Water Use

The largest single industrial water user in the Special Areas is the Sheerness Generating Station. ATCO Power built Sheerness Thermal Generating Plant in 1986. The plant, which was originally designed to produce 380 megawatts (MW) of electricity, was expanded in 1990 to boost the coal-fired plant to 760 MW. Thermal generating stations produce electricity by heating water in large boilers to create steam, the steam spins turbines, and the turbines create electricity. The Sheerness plant was built next to its coal supply, the adjacent Luscar mine. Water is required both for thermal generation, and for cooling. To meet this need, ATCO built a raw water pipeline from the Red Deer River to the plant at the time of construction (see Figure 4).

Intake pumps on the Red Deer River can only divert water during the ice-free period (April-October). Water is pumped to the Sheerness Cooling Pond. The plant then withdraws water from the pond for its own fire water system, for treatment which in turn produces potable water and demineralized water for equipment cooling and the boiler, for the condenser and for general service. In 2005, ATCO diverted a total of 19,324 megaliters (ML) from the Red Deer River. Of that, 1,933 ML were withdrawn from various pipeline tap-offs before reaching the Cooling Pond. 17,391 ML entered the cooling pond. The largest consumptive water use at this stage was evaporation (9,213 ML), which was only partially offset by precipitation (1,239 ML). 221 ML were consumed by the generating plant.[26] The largest portion of the water, once it had served its cooling functions, was released via the Blowdown Canal (Figure 4).

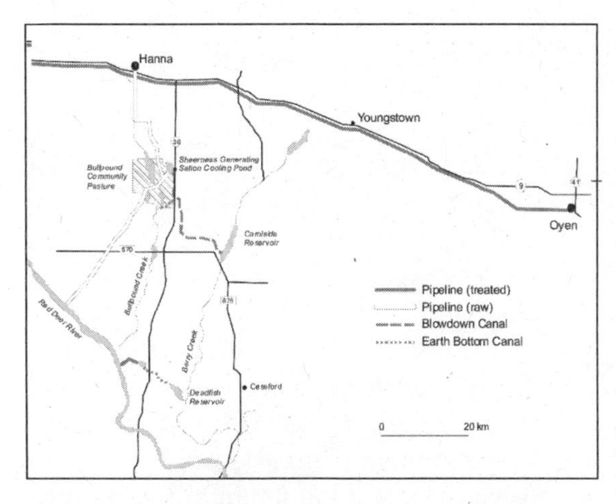

Figure 4. Surface Water Delivery in the Special Areas.

Since ATCO can only pump during ice-free months, the cooling pond serves as long-term storage. However, it is constructed larger than needed for the plant alone and serves as a recreational facility for the Special Areas population. Similarly, water releases via the Blowdown Canal are higher in volume than the plant needs for operation as part of ATCO's agreement with Special Areas to contribute to irrigation (see below). The determining factor for pumping from the Red Deer River is turbidity, and turbidity increases with precipitation. Thus, so long as there is sufficient water in the Red Deer River to meet ATCO's water license requirements without compromising inter-provincial allocation, the generally temporary disruption in pumping caused by turbidity is the only limiting factor. To date, the Red Deer River is not over-allocated, and there has never been too little flow for ATCO to pump, and low precipitation actually facilitates pumping through low turbidity. Consequently, ATCO Power was not greatly affected by the drought beyond less rainfall input into the cooling pond and thus slightly higher pumping volumes.

The oil and gas industry represents the other large industrial water use. Water is used in both drilling and "fracing" (a process which fractures layers of formations to gain access to further deposits). While the oil and gas sector is allocated just over 7% of Alberta's total water licensing (and 1.1% of all groundwater allocations), this allocation is not currently used in full.[27] Actual water use in the oil industry at the municipal scale is not well documented. Key informants note that drilling often uses untreated, highly saline groundwater. A local oil industry support company supplies treated water for fracing using municipal water as an input supply. An industry informant noted that treated water is expensive, and companies commonly source slough and dugout waters as close to the drilling site as possible. In some cases, the oil companies will pump Red Deer River water.

3.5.2. Municipal Water

Since the 1980s, the Towns of Hanna and Oyen and the Highway 9 corridor have been supplied with treated water via the Henry Kroeger Regional Water Services Commission. The Commission operates a treatment plant in Hanna which is supplied via raw water pipeline from the Town of Hanna reservoir near the Sheerness Generating Station (see Figure 4). This reservoir is filled by a tap-off from the ATCO raw water pipeline. The Commission can only fill the town reservoir if ATCO is pumping, which in turn depends on the generating station's needs and turbidity at the intake on the river. Pumping is limited to the summer months (April to October), meaning that the Commission needs to store a minimum of five months' worth of water in the reservoir.

The Hanna Reservoir was recently expanded and currently has a capacity of 913,000 cubic meters. In 2005, the Henry Kroeger water treatment plant diverted 1,006,000 cubic meters with an output of 833,023 cubic meters of treated water.[28] In 2005, 55% of the treated water was consumed during the spring and summer (April to September), though this figure is higher during particularly dry years due to lawn and tree watering among municipal water users. Given that current annual demand for water is less than the total storage capacity in the reservoir, the supply is considered secure. Similarly, the plant is rated at a maximum capacity of 7,500 cubic meters per day. The plant manager estimates that the Henry Kroeger treatment plant, including its supply, is operating at only 50% of its total capacity.[29]

In the event of a power failure, the plant's generators are insufficient to continue treating water. A plant technician estimates that the plant's storage of treated water, the "clear well," can hold approximately one day's worth of winter demand—the same volume would be exhausted over approximately eight daytime hours in the summer. Thus, the municipal water supply is exposed to anything which disrupts the continuous operation of the plant such as interruption of electrical power supply, major maintenance operations, or a failure in the raw water pipe from the Hanna Reservoir. The Henry Kroeger Regional Water Services Commission is currently planning to expand the clear well to address this. In addition, the three main clients of the plant (the Town of Hanna, and the pipelines running east and west) maintain their own storage facilities which can serve as a buffer in these situations. Smaller clients use coin operated treated water tank filling stations which access the pipelines in various locations. Since the plant has been in operation, end users have always been able to access water since the plant has been in operation, though there have been multiple instances where water treatment was halted for more than 24 hours and backup storage reservoirs had to be used.

In addition to the Henry Kroeger plant, there are smaller municipal water plants in the Special Areas. For example, the Cessford treatment plant serves a school and under 30 households. It stores a year's worth of water in two dugouts adjacent to the plant. The dugouts are recharged in the spring from the Carolside Reservoir (see Figure 4 for the location of canals and reservoirs), which in turn is recharged via a canal originating in the Sheerness cooling pond. While the dugouts have never been empty in the spring, these small plants have very limited capacity and could not accommodate large users.

The exposure-sensitivities in municipal water supplies are currently not related to precipitation given that municipal water primarily originates in the Red Deer River, which in turn originates in the Rocky Mountains and is heavily influenced by accumulated snowpack. Currently, the Red Deer River has greater capacity than is needed for existing water licenses, even during dry years. Consequently, the only additional stresses placed on the municipal water system during the 2001–2002 drought related to increased consumption. Given the high capacity of the system relative to use, higher consumption was easily met using existing resources.

The six non-farm residents consulted for this research were all connected to Henry Kroeger water. Their recollection of the 2001–2002 drought focused on the unpleasantness of the grasshopper infestation. One non-farm resident noted that her water bill went up substantially with increased consumption as she chose to water her yard trees. None of the residents decreased water consumption during this time.

The stability of the municipal water supply and the unused capacity for further supply are considered one of the key selling points for attracting business to Hanna by its mayor. He noted that a secure municipal water supply is "obviously very important for rural communities to survive. If you don't have quality water, it's difficult to bring in industry or any kind of residency of that nature"[30] and the Henry Kroeger plant put the Town of Hanna "in a position because of this to build for more lifestyle which of course encompasses quality of water, encompasses industry that needs volumes of water . . . (it puts) us in a position as a community to continue to look forward to economic growth."[31]

Due to the combination of the ATCO pipeline and the Henry Kroeger system, droughts are completely mitigated for urban residents and Hanna businesses. For urban water users, water is provided by the tap, and the tap has never run dry and is not expected to do so considering the reserve capacity built into the system.

3.5.3. Groundwater

There are an estimated 8,000 water wells accessing the groundwater resources under the Special Areas.[32] Close to 90% of these wells were completed for household and stock watering; the remaining wells serve uses such as irrigation, municipal water supply (for towns and villages not on the Henry Kroeger pipeline), industrial use and observation.

Well depth ranges from 0.6 to 260 meters, and just under half of the wells draw water from surficial deposits with the majority under 30 meters.[33] The balance of the wells consists of bedrock water wells, and most of these are located in the north and west of the Special Areas. Surficial wells are generally high in dissolved iron content, while bedrock water wells carry a very high average sodium load.[34] The Guidelines for Canadian Drinking Water set maximum recommended dissolved solids limits for sodium, sulfate, chloride, and fluoride. Water quality, on average, exceeds federal guidelines for sulfate and sodium.[35]

Twenty-three rural farm residents participated in the research. Twenty-one of these used one or more wells on the property. Of the 19 agricultural producers who had wells, five used well water only for stock watering and hauled their drinking water, either bottled water or water obtained with a water truck at a tank filling station. An additional two users were close enough to the Highway 9 treated water pipeline corridor to access this water for household use, though they used dugout and well water for stock.

The 2001–2002 drought had no observable impact on either flow rates or water quality in the wells of the producers consulted for this study. However, groundwater is always an issue of concern in the Special Areas as a result of poor quality and insufficient supply. This past stress meant that reliance on groundwater for drinking and stock water is already low, and residents already have numerous coping mechanisms in place to mitigate further quality and quantity decreases. In some cases, coping strategies such as hauling water presented short-term adaptations to surface water shortages as well.

3.5.4. Precipitation and Spring Runoff

Both cash crop and livestock producers are highly dependent on precipitation. In the words of one producer, "our biggest limiting factor in this country is water."[36] Furthermore, the timing of precipitation events matters. Key times of year are the early growing season (April-June, germination and initial growth) and the late growing season (July-September, continued growth and maturation). In addition, the snowfall in the early part of the calendar year (January-March) matters as this is the source of spring runoff, which is important for soil moisture and dugout recharge. Figures 2 and 3 illustrate that all three of these seasonal parameters were well below the long-term averages during the drought.

Good spring runoff is important for cattle producers for grass growth and stock watering. Local spring snowmelt is important for both soil moisture and dugout recharge. With little spring runoff, dugouts will not fully recharge, and ranchers can run into stock water shortages on grazing parcels later in the year. Adequate moisture is required during the spring for grass to establish itself and get some good growth. During the drought, there was inadequate grass growth for both feed production and pastures. Ranchers and mixed farmers reported difficulties meeting their feed supplies.

Most cattle producers routinely carry feed supplies from year to year, and "reserve" native grass pastures for contingency use through rotational grazing. However, prior to the drought, most ranchers only carried one year's worth of stockpiled hay. These supplies were insufficient. Six producers reported decreasing herd size to cope with less available feed and three others increased their pasture acreages. Since native grasses are more drought tolerant and retain their nutritive value even during extreme dry events once they have established themselves, three livestock producers have actively begun to reseed native grasses since the drought. Nine out of 12 ranchers reported carrying pasture insurance to offset the costs of purchasing more feed supplies. Finally, four of the eight mixed farmers realized that their cash crops would not bring them any profit, and turned cattle onto those fields to at least use the crops for feed.

Crop farmers felt similar impacts as livestock producers. The spring of 2001 was considered adequate for germination and crop establishment. However, cash crops require continued moisture availability to mature. One producer noted that "we start praying through the end of June."[37] Without early growing season moisture or sustained moisture throughout the growing season, yields and crop quality decrease. For crop farmers, this meant either taking an income hit or drawing on crop insurance.[38] Most mixed farmers and dryland farmers reported that they had decreased their areas under crops and some had increased the ranching aspects of their operation. Dryland farmers commonly reported increasing the area under zero tillage to control wind erosion.

In 2002 feed supplies and reserves were already low, and cash crops were severely impacted by the compounding stress of the grasshopper epidemic. One mixed farmer explained, "drought you can deal with. You know you can maybe save grass here and there, and you know we saved some here and two days later the grasshoppers went through and there's nothing. What you have kinda stockpiled you may as well say is gone."[39] For farmers who were hardest hit, this "double whammy" of drought and grasshoppers meant having to import feed, in many cases from out of province. Ontario farmers sent hay to the southern Prairies during this time, though Special Areas farmers felt that these shipments were of low quality.

The low snow cover during the drought years had implications beyond soil moisture for ranchers. Although many have access to wells on the home farm quarter, cattle spend the majority of the year out in pasture. Grazing is generally rotational, meaning that stock water supplies must be available throughout the pasture areas. This is met via dugout construction. Dugouts are generally recharged during the spring snowmelt, and supply enough water for an entire snow-free season. However, during the drought, low snowfall meant that dugouts were only partially recharged. With successive dry years, dugouts become further drawn down. Agriculture Canada data indicate that dugouts in the Special Areas were generally dry by September 2002. [40]

Livestock producers responded to low stock water levels in a number of ways. Reductions in herd size by culling older cows without replacing these with younger animals reduced some of the demand. Some ranchers hauled stock water, an expensive and labour-intensive undertaking, while others kept their herds close to more secure water supplies and hauled feed. Dugout improvements, which include fencing off the dugout and pumping water on demand via nose pumps, help preserve water quality in the dugout and allow a larger proportion of the volume to be used for stock watering. The local Agriculture Canada (PFRA) field office had actively promoted dugout improvements over the previous decade, and several ranchers reported that these improvements helped them manage their dugout levels during the drought. In addition, some ranchers had introduced non-precipitation dugout recharge. Seven farmers had either already or have since installed shallow pipelines for dugout recharge or stock water troughs. This is an option only if there is a secure water supply in reasonable proximity, and this was provided in some cases by the Blowdown Canal, in others by good wells on the home quarter.

Managing stock water needs during the drought placed an additional management and labour burden onto livestock farmers. Producers needed to consider their possible stock water supplies, and adjust rotations and herd sizes accordingly or, in the worst case scenario, resort to cost-prohibitive water hauling.

3.5.5. Irrigated Farming

The Special Areas do not contain irrigation projects on the scale of the southern Alberta Irrigation Districts. However, the Berry and Deadfish Creek watersheds in the southern portion of Special Area 2 provide some irrigation opportunities. The Special Areas Board maintains reservoirs on both creeks (the Carolside Reservoir on Berry Creek and the Deadfish Reservoir on Deadfish Creek). There are currently 26 irrigators in the Special Areas irrigation project. Of these, 11 are along the Blowdown Canal or on Berry Creek above its confluence with Deadfish Creek, five are on Deadfish Creek above the confluence, and ten are between the confluence and the Red Deer River. The Blowdown Canal is important as natural runoff into Carolside Reservoir in most years equals evaporation from the reservoir, which is also used for recreation and thus is kept at a minimum level, and thus little if any water would have been available for irrigation during the drought without the ATCO contribution of almost 10,000 ML. The Deadfish Reservoir is filled via intake pumps on the Red Deer River.

Farmers who have irrigation agreements were affected by the drought in two ways: their own irrigation requirements increased due to poor natural precipitation. In the words of one producer, "the rain helps with irrigation, cause I don't care what anybody says, rain is way better than any irrigation, you can pump and pump it seems like irrigation won't go as good . . . even about three inches of rain just makes a load of difference with production."[41] Without precipitation, irrigators had their pumps running continuously and irrigation water supplies were insufficient. All of the irrigators included in this research were either mixed farmers or ranchers, and used irrigation for only a small portion of their farm enterprise. In many cases, irrigation was used to produce secure feed supplies. However, when water supplies ran low during the drought, all farmers were asked to reduce their water use and were threatened that they would be cut off completely if they did not comply. This meant that those who had relied on irrigation as a drought management strategy could not entirely meet their needs in this way. Furthermore, irrigators noted that the fuel costs of pumping water to pivots were too high for this to be feasible in successive drought years. A logical adaptation during the drought would have been to use more irrigation water, but this was limited by a lack of availability.

Since the drought, Alberta Environment has upgraded the pumping capacity for the Deadfish Reservoir to provide more secure water supplies. The 15 irrigators who are located on Deadfish Creek or below the confluence of Deadfish and Berry Creeks have access to more secure irrigation supplies, but this capacity upgrade does not improve the situation for the 11 producers above the confluence.

4. Discussion and Conclusions

The Special Areas include urban non-farm residents and businesses, industry and agricultural producers. These groups had differential exposure-sensitivity to drought given the ways in which they rely on water and the means in which they secure water supplies. The drought was primarily an inconvenience for non-farm residents and businesses. Agricultural production, on the other hand, was severely impacted and required adaptation.

Agricultural producers in the Special Areas are accustomed to managing dry periods. The 2001–2002 drought was problematic because successive dry years resulted in cumulative effects: soil moisture continued to decrease, dugouts were not substantially recharged during the entire drought, and feed reserves were exhausted. Furthermore, the compounding stress of the grasshopper epidemic meant that even those who had planned for shortages found their reserves ruined. However, this drought episode, while it created management challenges and resulted in financial impacts for many producers, did not have nearly the same impacts as the catastrophic results of the Dust Bowl years which led to the establishment of the Special Areas. Relative to the earlier periods, farmers were far less vulnerable during the most recent drought episode.

Reasons for this decreased vulnerability can be noted by examining the exposure-sensitivities and adaptive capacities of agricultural producers in the Special Areas. While all of the producers were exposed and sensitive to drought in that it had substantial impacts for their operations, this exposure-sensitivity was much lower than during the 1930s when most farmers were cash cropping much smaller parcels and dependent on crop growth. Cash cropping requires breaking the native prairie sod, leaving it far more exposed to drying and wind erosion, and eliminates the relatively more drought-resilient native prairie grass. Furthermore, the small parcel size during the 1930s meant that a much larger population had to be sustained on the same land base, and the carrying capacity of the land may have simply been exceeded during this time. Furthermore, crop farmers have changed their farming practices beyond increasing parcel size. A local PFRA employee notes that "farming practices have changed a lot since the 30s and moisture, rainfall is probably the same—it's probably the farming practices that are a lot better."[42] This theme is echoed by another respondent, who explains the role of minimum or zero tillage in cropping: " . . . in the 30s . . . you know, you'd see these clouds of dust. We didn't get those clouds of dust this time around. And largely in part because they were taking the straw off."[43]

Furthermore, farmers currently have a much higher adaptive capacity. In large part this is due to the fact that almost all of the farmers currently in the Special Areas are the descendants of those who successfully weathered earlier drought epidemics. These producers are the product of a long line of successful drought managers, and this expertise is highly developed in the Special Areas. Earlier, less severe, drought episodes in the 1980s and 1990s meant that farmers with relatively low adaptive capacity had already left the industry and remaining farmers already had contingency plans. In the words of the past Special Areas Board Chairman:

> The experience from the 30s was carried over in a lot of the old timers and the people of the area they have adapted to drought. we get periodic droughts. Not like it was in the 30s, but we still get them. I mean, we're only one day's rain away from drought at any time in this country.
>
> I mean, we could end up coming into a drought so easy in a year. But the people are more apt to be able to handle it. PFRA was a big pusher on dugouts and this type of thing and so people have water available in most cases. There's now a lot of water being piped around the country. We've got shallow pipes for filling cobs and dugouts and this type of thing.[44]

Adaptive capacity includes access to resources and capital. Water is one of the most important resources in the Special Areas. However, the importance of non-agricultural income sources including off-farm work and oil and gas leases cannot be discounted. In several instances, producers chose to farm as a way of life, but earned greater income from the oil and gas wells located on their owned or leased lands. Many of the common adaptation strategies required absorbing an income loss or using capital to meet feed needs, and this would not have been feasible for many without additional income.

Locally-based institutions play a significant role in managing drought in the Special Areas. The ATCO pipelines and Henry Kroeger Regional Water Services Commission mean that municipalities and those who have access to tap-offs from pipelines have guaranteed water supplies (though at a cost). One local institution which in the past played a large role in increasing adaptive capacity was the local Agriculture and Agri-Food Canada Prairie Farm Rehabilitation Administration (AAFC-PFRA) field office, which was instrumental in dugout design and improvements. Unfortunately, AAFC-PFRA has drastically reduced its field presence—during the time of this research the Hanna field office only had three staff members for all programs combined.

The Special Areas Board maintains community pastures to alleviate some of the stress on owned and leased parcels and manages the Carolside-Deadfish irrigation project. The Board maintains rental equipment for minimum tillage and constructing shallow water pipelines. While other municipalities provide some of these services, the governance structure of the Special Areas in particular facilitates long-term planning for increased adaptive capacity. Despite the fact that it is a creature of the provincial government, the Special Areas Board is not subject to regular provincial election cycles, nor does it need to engage in programs which are designed to appeal to voters. The current Special Areas Board Chairman notes:

> I think we tend to think in longer term things. We don't, we're not like, the government's all on a year to year cycle . . . we have a long-term plan for a lot of things . . . we just tend to think in terms of longer term. And we're not trying to go back and get money from the government every year, and if we don't spend it, we're not making stupid expenditures. We can carry (a fiscal surplus) over and that's allowed us to be really flexible.
>
> You get in a crisis situation, you do crisis-management. But after you're done crisis, like the time right now, we don't need the water right now. The time to build it is now. The time to build water storage is now because we have run-off. Get the hole in the ground, so you can catch the water because in the drought it's not coming. After a drought, you come out of it and you say, what can we do different?[45]

Finally, it should be noted that the drought had one net positive benefit for cattle producers in the Special Areas. In the fall of 2001, the first case of Bovine Spongiform Encephalopathy (BSE, colloquially known as mad cow disease) was found in Alberta. Since cattle producers in the Special Areas had already responded to the drought by reducing their stocking rates (and in most cases, this meant selling older cattle), they had sold cattle at a time before cattle prices plunged. Furthermore, the average age of remaining herds was substantially younger than is normally the case. When the United States cattle market re-opened to Canadian beef, only cattle less than 12 months in age were acceptable. As a direct result of adaptations undertaken in response to the 2001–2002 drought, Special Areas cattle producers were in a better position to survive this compounding stress than their counterparts in less drought-stricken areas.

The case of the 2001–2002 drought in the Special Areas of Alberta highlights a number of key themes in vulnerability research. First, vulnerability is not experienced equally by all members of a community or residents of a region, even if they are subject to the same climatic stress. Dryland farmers were more exposed to this drought event than other producers, and town residents were merely inconvenienced. Second, a climatic stress such as a drought cannot be seen in isolation of other stresses and opportunities. There were those compounding impacts which were not precipitation related but nevertheless related to the drought (e.g., grasshoppers) but also stresses which were unrelated to climate (e.g., cattle price changes as a result of BSE and opportunities for non-farm income due to the oil and gas sector). Third, adaptation occurs at individual levels but is frequently facilitated by institutions. For example, many producers chose to improve their dugouts or implement shallow pipelines, but federal and provincial organizations such as the PFRA and the Special Areas Board contributed expertise and equipment for these projects. Fourth, previous climatic stresses build adaptive capacity for future stresses. This region was in a very good position to manage drought given its management of recurring persistent dry periods since it was settled, and this most recent drought event prompted even higher adaptive capacity through strategic adaptations such as the construction of shallow pipelines or carrying greater feed reserves. Finally, climatic stresses such as a dry season need to be considered in a cumulative sense: one dry year, while unfortunate, is not a disaster in this region. Two to three dry years begin to approach the limits to regular management strategies as feed reserves are exhausted and dugouts run dry, and it is questionable whether even an area as highly adapted to persistent dry periods as the Special Areas would be able to maintain agricultural productivity during a longer event.

Endnotes

1. D. Sauchyn, "Climate change impacts on agriculture in the Prairies," in E. Wall, B. Smit, and J. Wandel (eds.), *Farming in a Changing Climate: Agricultural Adaptation in Canada* (Vancouver: UBC Press, 2007), 67–80.
2. M.L. Khandekar, *Canadian Prairie Drought: A Climatological Assessment* (Edmonton: Alberta Environment, Publication No. T/787, 2004).
3. B. Smit and J. Wandel, "Adaptation, Adaptive Capacity and Vulnerability," *Global Environmental Change* 16/3 (2006): 282–292.
4. J.H. Christensen, B. Hewitson, A. Busuioc, A. Chen, X. Gao, I. Held, R. Jones, R.K. Kolli, W.-T. Kwon, R. Laprise, V. Magaña Rueda, L. Mearns, C.G. Menéndez, J. Räisänen, A. Rinke, A. Sarr, and P. Whetton. "Regional Climate Projections," in *Climate Change 2007: The Physical Science Basis. Contribution of Working Group I to the Fourth Assessment Report of the Intergovernmental Panel on Climate Change* (Cambridge: Cambridge University Press, 2007), 849–940.
5. Khandekar, *Canadian Prairie Drought.*
6. E. Wheaton, V. Wittrock, S, Kulshrshtha, G. Koshida, C, Grant, A. Chipanshi, and B. Bonsal, *Lessons Learned from the Canadian Drought Years of 2001 and 2002: Synthesis Report* (Saskatoon: Saskatchewan Research Council, Publication No. 11602–46E03, 2005).
7. M. Kandlikar and J. Risbey, "Agricultural Impacts of Climate Change: If Adaptation is the Answer, What is the Question?" *Climatic Change* 45/3–4 (2000): 529–539; and E. Wall, B. Smit, and J. Wandel (eds.), *Farming in a Changing Climate.*

8. For example, R.M. Adams, R.A. Fleming, C.-C. Chang, B. McCari, and C. Rosenzweig, "A Reassessment of the Economic Effects of Global Climate Change on U.S. Agriculture," *Climatic Change,* 30/2 (1995): 147–167; R. Mendelsohn, W.D. Nordhaus, and D. Shaw, "The Impact of Global Warming on Agriculture: a Ricardian Analysis," *American Economics Review,* 84/4 (1994): 753–771; M.L. Parry, C Rosenzweig, A. Iglesias, M. Livermore, and G. Fischer, "Effects of Climate Change on Global Food Production under SRES Emissions and Socio-Economic Scenarios," *Global Environmental Change,* 14/1(2004): 53–67.
9. S. Reid, S. Belliveau, B Smit, and W. Caldwell, "Vulnerability and Adaptation to Climate Risks in Southwestern Ontario Farming Systems," in E. Wall, B. Smit, and J. Wandel (eds.), *Farming in a Changing Climate,* 173–186; S. Reid, B. Smit, W. Caldwell, and S. Belliveau, "Vulnerability and Adaptation to Climate Risks in Ontario agriculture," *Mitigation and Adaptation Strategies for Global Change,* 12/5 (2007): 609–637; B. Smit and O. Pilifosova, "From Adaptation to Adaptive Capacity and Vulnerability Reduction," in *Climate Change, Adaptive Capacity and Development* (London: Imperial College Press, 2003), 9–27.
10. For example, S. Belliveau, B. Bradshaw, and B. Smit, "Comparing Apples and Grapes: Farm-Level Vulnerability to Climatic Variability and Change," in E. Wall, B. Smit, and J. Wandel (eds.), *Farming in a Changing Climate,* 157–172; S. Belliveau, B. Smit, and B. Bradshaw, "Multiple Exposures and Dynamic Vulnerability: Evidence from the Grape Industry in the Okanagan Valley, Canada," *Global Environmental Change,* 16/4 (2007): 364–378; S. Reid et al., *Farming in a Changing Climate;* S. Reid et al., *Mitigation and Adaptation Strategies for Global Change;* C. Bryant, B. Singh, and P. André, "The Perception of Risk to Agriculture and Climatic Variability in Québec: Implications for Farmer Adaptation to Climatic Variability and Change," in E. Wall, B. Smit, and J. Wandel (eds.), *Farming in a Changing Climate,* 142–154.
11. B. Smit and J. Wandel, *Global Environmental Change;* B.L. Turner, R.E. Kasperson, P.A. Matson, J.J. McCarthy, R.W. Corell, L. Christenson, N. Eckley, J.X. Kasperson, A. Luers, M.L. Martello, C. Polsky, A. Pulsipher, and A. Schiller, "A Framework for Vulnerability Analysis in Sustainability Science," *Proceedings of the National Academy of Sciences,* 100/14 (2003): 8074–8079.
12. B. Bradshaw, "Climate Change Adaptation in a Wider Context: Conceptualizing Multiple Risks in Primary Agriculture," in E. Wall, B. Smit, and J. Wandel (eds.), *Farming in a Changing Climate,* 103–114.
13. Interviews were transcribed and the information organized with the use of NVivo software, which allows hierarchical coding of particular themes.
14. Soil Classification Working Group, *The Canadian System of Soil Classification,* 3rd ed., Agriculture and Agri-Food Canada Publication 1646, Revised (Ottawa: National Research Council Research Press, 1998).
15. Climate normals are based on the 1971–2000 average.

16. Solonetzic soils are characterized by a B horizon that is very hard when dry and swollen and compact if wet, making cultivation difficult.
17. One quarter section is 160 acres/65 ha of land.
18. J. Gorman, *A Land Reclaimed: The Story of Alberta's Special Areas* (Hanna: Gorman and Gorman for the Special Areas Board, 1988).
19. While the Special Areas originally included substantial lands south of the Red Deer River, the current boundaries are as noted in Figure 1 and include the former Sullivan Lake, Berry Creek, Sounding Creek, and Neutral Hills Special Areas, as well as portions of the former Tilley East.
20. Wheaton et al., *Lessons Learned;* V. Wittrock, D. Dery, S. Klushreshtha, and E. Wheaton, *Vulnerability of Prairie Communities' Water Supply During the 2001 & 2002 Droughts: A Case Study of Cabri and Stewart Valley, Saskatchewan.* (Saskatoon: Saskatchewan Research Council, Publication No. 11899–2E06, 2006).
21. See Figure 4 in Wittrock et al., *Vulnerability of Prairie Communities' Water Supply.*
22. Tame pastures include all non-native pasture mixes including grasses and legumes grown for feed: such as timothy, brome and alfalfa.
23. Alberta Agriculture, Food and Rural Development (AFRD), "Grasshoppers: Life Cycle, Damage Assessment and Management Strategy," *Agri-Facts: Practical Information for Alberta's Agriculture Industry* (Edmonton: Agdex, 2002), 622–24.
24. L. Powell, A.A. Berg, D.L. Johnson, and J.S. Warland, "Relationships of Pest Grasshopper Populations in Alberta, Canada to Soil Moisture and Climate Variables," *Agricultural and Forest Meteorology,* 144/2 (2007): 73–84.
25. Agriculture and Agri-food Canada (AAFC), *$10.5 Million Federal/Provincial Grasshopper Program Targets Areas of Greatest Need* (Ottawa: AAFC News Release, June 12, 2003).
26. Consumptive uses include steam losses, water losses during treatment, and general service and drinking water.
27. Alberta Energy, "Water Use by Alberta's Oil and Gas Industry," in *Talk About Oil and Gas* (October 2006).
28. The reservoir has evaporative losses, and the treatment process itself requires water.
29. Personal communication, Garth Turner, Manager, Henry Kroeger Plant, November 2006.
30. Personal communication, Pat Burns, Mayor, Town of Hanna, November 2006.
31. Personal communication, Pat Burns.
32. Hydrogeological Consultants Ltd., *Special Areas 2, 3 and 4, and M.D. of Acadia. Part of the Red Deer, and the South and North Saskatchewan River Basins. Regional Groundwater Assessment.* (Edmonton: Hydrogeological Consultants Report 99–101, 2006.
33. Hydrogeological Consultants, *Special Areas.*
34. Hydrogeological Consultants, *Special Areas.*
35. Average loads of 544 and 606 mg/L versus maximum recommended levels of 200 and 500 mg/L for sodium and sulfate respectively.

36. Personal communication with a pasture manager.
37. Personal communication, rancher, November 2006.
38. Six out of nine farmers who grew cash crops without access to irrigation carried crop insurance.
39. Personal communication, mixed farmer, November 2006.
40. Agriculture and Agri-food Canada (AAFC) Drought Watch, *The 2002 Prairie Drought Summary—December 2002* (Ottawa: AAFC, 2002). Available at: http://www.agr.gc.ca/pfra/drought/drought02sum_e.htm (cited October 2007).
41. Personal communication, irrigated mixed farmer, November 2006.
42. Personal communication, Ken Forsyth, Prairie Farm Rehabilitation Administration staff, Hanna office, November 2006.
43. Personal communication, Jay Slemp, Chairman of the Special Areas Board, November 2006.
44. Personal communication, Ab Grover, Past Chairman of the Special Areas Board, November 2006.
45. Personal communication, Jay Slemp.

THE OLDMAN RIVER DAM CONFLICT: ADAPTATION AND INSTITUTIONAL LEARNING

ALEJANDRO ROJAS, LORENZO MAGZUL,
GREGORY P. MARCHILDON, AND BERNARDO REYES

> *"The link between environmental degradation, water scarcity and violent conflict is a serious threat. Water is becoming a commodity that even peaceful neighbours are willing to battle over."*[1]

> *"Global climatic change will affect water availability in many ways, although the precise nature of such changes is still obscure. Our challenge is to identify those cases in which conflicts are likely to be exacerbated and to work to reduce the probability and consequences of those conflicts."*[2]

Abstract

Water conflicts can increase the vulnerability to climate change impacts for affected communities, but they can also positively influence the values, behaviors, and institutional arrangements in such a way as to enhance the adaptive capacity of social systems to respond to present and future impacts of climate change. We suggest a process through which such lessons can be integrated into decision-making and policies for an improved management and governance of water arrangements. This paper reports on the findings of a study on the Oldman River Dam conflict. The study suggests that the vulnerability of the Piikani nation, one of the main actors in the conflict, to the impact of future climate change has increased. The process of conflict negotiation and compensation and the environmental impacts of the construction of the Oldman Dam have negatively affected the adaptive capacity of the community to respond to climate related stresses, demonstrating the negative impacts of poorly designed, multi-stakeholder consultations which are insensitive about power differentials among parties in conflict.

However, some positive consequences include increased water use efficiency by irrigation districts, improved consultation by government with stakeholder groups in decision-making water governance, and improved coordination of provincial and federal efforts for environmental protection and monitoring.

Sommaire

Les conflits dans le secteur de l'eau augmentent la vulnérabilité des collectivités touchées par les impacts des changements climatiques. Cependant, ils peuvent aussi exercer une influence positive sur les valeurs, les comportements et les mesures institutionnelles de façon à améliorer la capacité d'adaptation des systèmes sociaux à faire face aux impacts présents et futurs des changements climatiques. Les auteurs suggèrent un processus par lequel les leçons tirées peuvent être intégrées dans la prise de décision et les politiques afin d'améliorer la gestion et la gouvernance des ententes sur l'eau. Cet article présente les résultats d'une étude au sujet du conflit du barrage de la rivière Oldman. L'étude indique que la vulnérabilité de la Première nation Piikani, un des principaux acteurs dans ce conflit, aux impacts des changements climatiques a augmenté. Le processus de résolution de conflit et de compensation et les impacts sur l'environnement de la construction du barrage Oldman ont eu un effet négatif sur la capacité d'adaptation de la collectivité à faire face aux divers stress liés aux changements climatiques. Cela démontre les impacts négatifs des consultations mal conçues et menées auprès de plusieurs intervenants qui ne sont pas sensibilisés aux déséquilibres de pouvoir qui existent chez les parties impliquées dans un conflit. Cependant, parmi les conséquences positives qui découlent de ces consultations, on retrouve des mesures d'économie d'eau dans les districts d'irrigation, des consultations améliorées de la part du gouvernement avec les divers groupes d'intervenants quant à la prise de décision au sujet de la gouvernance de l'eau, et une coordination améliorée des efforts des gouvernements fédéral et provinciaux quant à la protection et la surveillance de l'environnement.

Introduction

This paper describes and explains the importance of the study of water conflicts and how the lessons gained from these conflicts can positively influence the values, behaviors, and institutional arrangements in such a way as to enhance the adaptive capacity of social systems to respond to present and future impacts of climate change. Studying past water conflicts offers the opportunity to avoid, mitigate or better resolve future conflicts due to water related scarcities. We suggest a process through which such historical lessons are integrated into decision-making and policies for an improved management and governance of water. In particular, we use the case of the Oldman

River Dam conflict to point out the factors that can result in increased vulnerability to climate change, or enhance the adaptive capacity of water governance institutions and water users in the South Saskatchewan River Basin (SSRB) to adapt to climate change induced water scarcity in the future.[3]

Climate Change and Conflict in the South Saskatchewan River Basin

Droughts in southern Alberta and Saskatchewan may become more frequent as a result of higher rates of water evaporation caused by an increase in temperature. Historic trends and current climatic change impacts in the South Saskatchewan River Basin reveal that from 1900–98 there has been an average increase in the annual air temperature of 0.9°C, which has contributed to the increased evaporation during the summer and the melting of snowpack during the winter.[4] While precipitation has increased over most of southern Canada, this has not been the case in southern Alberta and Saskatchewan.[5]

The Prairie Provinces produce 80% of Canada's agricultural output, and water is an essential input to this production. In Alberta alone, 2.2 billion cubic metres of water are used annually for irrigation. This figure accounts for most of the 75% of all agricultural water withdrawals, and is equivalent to 28% of the total annual river flow in the basin.[6] Increased water scarcity will likely generate conflict among the various water users in the SSRB, as well as tension between water users and governmental organizations. This potential for discord is great because of the power differentials among stakeholders based on the disproportionately high water withdrawals by the irrigation districts relative to other water users who are forced to compete for the remaining water flow.[7] In addition to the high water withdrawals for agriculture, in the Prairie Provinces different legal rules exist in each province in respect of water priorities. In Alberta, a very definite scheme of priorities is legislated which generally grants priority to licences obtained on a first-in-time, first-in-right basis.[8] This rigid water licence priority scheme grants historical licences precedence over licences for what might be regarded as a more socially important use (e.g., urban drinking water use). Rigid priority regimes together with the apportionment agreement between provinces constitute a recipe for conflicts in times of water scarcity. Since senior rights are mostly held by irrigation districts, in times of water scarcity these irrigation districts could exercise their senior rights and use up most of the water, while urban users and the ecosystem's in-stream flow needs will be left with little or no water at all. Given this regime, strife is predictable in southern Alberta, where water flow from the rivers such as the Waterton, Belly, and St. Mary are already over allocated, water shortages are common, and competition over water use is increasing.[9]

The apportionment agreement between Alberta and Saskatchewan requires Alberta to let 50% of water flow move eastward into Saskatchewan.

This is based on the total "natural water flow"; however, the calculation of the "natural flow" in times of water scarcity could become problematic because of the impounding of water in man-made reservoirs. Therefore, in order to reduce tensions over what constitutes "natural flow" between upstream and downstream users, new monitoring and data models must be developed.[10] Even more problematic is the fact that Aboriginal title to water, especially for non-consumptive riparian use, has not been recognized in the provincial Water Act, or in the inter-provincial apportionment agreement. As a consequence, some First Nations are contesting the principle of first-in-time, first-in-right allocation.[11]

Some of the problems associated with the prior appropriation of water allocations have been addressed through the water allocation transfers provision in the Water Act that permits water allocations to be transferred, either temporarily or permanently, from one water allocation licence holder to another licence holder. This is a change from the past, when water licences were, for the most part, permanently attached to the piece of property for which the water allocation was originally issued. However, Cheryl Bradley, a member of the Oldman River Basin Advisory Committee whose task is to set water management and water related land use planning priorities in the Oldman watershed, argues that irrigation districts monopolize river allocations. According to Bradley, individuals or groups should be allowed to purchase water allocations for environmental reasons, but irrigation districts refuse to transfer water rights to non-members.[12]

Methodology and Conceptual Framework

In the summers of 2005 and 2006, fieldwork was conducted in the Oldman River Basin to identify the manner in which the initiation and management of the Oldman River Dam project may have affected the adaptive capacity of water governance institutions in the Basin. Institutions involved in water governance are those which most directly influence decision-making, including government organizations responsible for water management and allocation, and the ample range of water users, beneficiaries, and other civil society organizations influencing water use. In-depth interviews were conducted following a highly structured but flexible interview guide based on the vulnerability-adaptive capacity model created by our wider collaborative study on institutional adaptations to climate change.[13] These interviews were conducted with individuals representing water governance institutions and stakeholders (e.g., Alberta Environment and Alberta Agriculture and Food) and water users (e.g., individual farmers, the Northern Lethbridge Irrigation District, the Alberta Irrigation Projects Association, the city of Lethbridge, the town of Fort McLeod, the members the Piikani First Nation, and the Oldman

River Watershed Council). Some of the information or opinions provided by the individuals interviewed are of a sensitive nature, therefore in this article the anonymity of the respondents is maintained by citing them by an assigned code (e.g., OMRBST1 for Oldman River Basin stakeholder number 1, OMRBST2 for Oldman River Basin stakeholder number 2, etc.) rather than by their name.

Water conflicts and the manner in which they are resolved affect the vulnerability and adaptive capacity of communities and water governance institutions. For instance, power differentials among stakeholders may result in those endowed with less power becoming more exposed (lose access to water), or having less capacity (lose economic opportunities), and, therefore, are increasingly vulnerable to impacts of climate change. Conversely, successful water conflict resolution requires a genuine and conscious effort to understand and include a broad spectrum of interests and values. In such cases, the outcomes could lead to more equitable access to water and a resulting decrease in overall vulnerability. Thus, it is critical for the agency to take concrete steps by facilitating dialogue and negotiation, to allow parties in conflict to achieve some degree of power symmetry in the dialogue, or else the concerns of some stakeholders may not be heard or merely considered cosmetic, ultimately making it less likely that they will be acted upon.[14]

In order to avoid disproportionate power symmetry among stakeholders, conditions must be established to ensure they perceive each others' concerns and interests as legitimate, regardless of differences in values and interests.[15] There is a range of ways that multi-stakeholder approaches can be conducted, facilitating anything from relationships built upon stakeholder empowerment, to more manipulative applications.[16] Problems can arise when multi-stakeholder assessments are not informed by participatory methodologies, such as in the design of the analysis or identification of relevant stakeholders, and the assessments only reflect the agendas of the researchers or decision-makers doing the analyses.[17] Stakeholder assessment is often used inconsistently throughout the duration of a project. Too often multi-stakeholder assessment is only used at the beginning of a project to identify stakeholders in a problem or conflict.[18] In such cases, this limited assessment is used to gather "rapid snapshots of social reality" generating bits of information that are then reviewed out of context by researchers, decision-makers, and project administrators.[19] Indeed, some argue that "the term 'participation' has been hijacked by everyone who finds it convenient to do so and its meaning can vary from contracted forms of interaction to a true co-learning mode of operation," whereby stakeholders are not involved at all research stages, such as designing, implementing, planning, and evaluating research.[20]

Even if a given organization conducts a complete multi-stakeholder consultation to gather input on a decision, the process of input gathering may not be equally participatory among all stakeholders. Participant views are easily understated or overstated. In the end, the final decision may benefit only a few stakeholders even though the organization claims to have conducted a comprehensive multi-stakeholder consultation. Power differentials between stakeholders in multi-stakeholder consultation, even those informed by participatory methods, is often overlooked, particularly in cases where the assumption underlying the participatory methodology presumes that "knowledge is equally distributed and freely accessible" to all stakeholders.[21] Such an assumption can mask significant power differentials among the stakeholders.

Although water conflicts can trigger violent confrontations between resource users, or even civil wars within nations, they do not always mean military or armed confrontations.[22] For some, conflict means competition, disputes, and tensions over access to and use of water.[23] For others, conflicts involve "any disagreements or dispute over or about water, where social, economic, legal, political or military intervention has been needed, or will be required, to resolve the problem."[24] There is a deep-rooted assumption that water scarcity, precipitated by climate change and combined with an increasing demand for water as a result of population growth, will lead to conflicts among water users.[25] Beyond distributional disputes, conflicts can also arise as a result of new water management initiatives, including the construction of reservoirs for irrigation or flood control purposes. These water management schemes "often lead to the displacement of local populations, adverse impacts on downstream water users and ecosystems, changes in control of local resources, and economic dislocations. These impacts may, in turn, lead to disputes among ethnic or economic groups, between urban and rural populations, and across borders."[26]

The factors which determine the intensity of a conflict include: 1) economic decline, which creates tensions because of a decrease in the quantity and quality of the natural resource; 2) population migration, which can lead to fighting between natives and immigrants, as the latter move to other regions in search for resources; 3) weakening of political institutions, resulting from civil unrest and tensions as people lose confidence in their institutions and governments; and, 4) political instability, where the scarcity of a resource can aggravate political tensions and divisions.[27]

Effects of Conflicts on Adaptive Capacity

Just as the impacts of climate change on a social system depend largely on the existing socio-economic and environmental stresses to which the social system is exposed as well as on the capacity of its water governance institutions to conduct effective managerial and policy decisions, so too do the

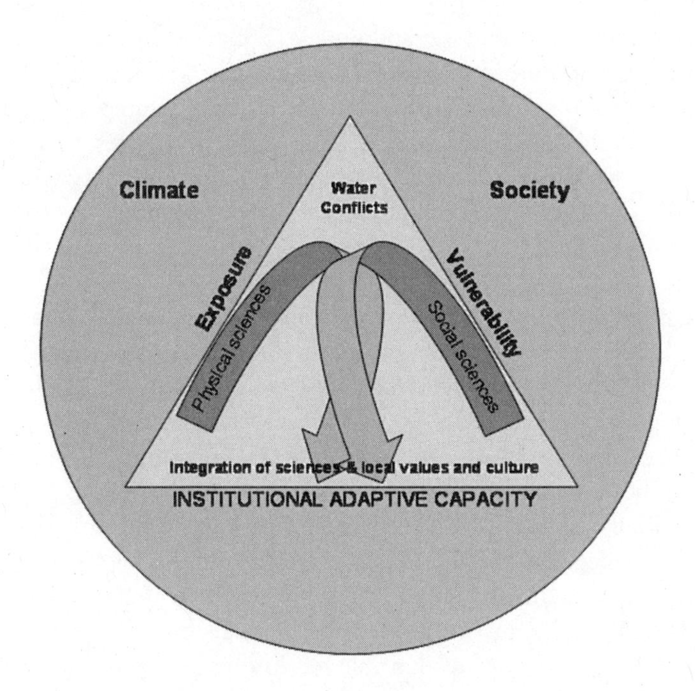

Figure 1. Relation: Exposure—Vulnerability and Adaptive Capacity in Water Conflicts.

impacts of water conflicts. Conflicts have greater impacts over a social system experiencing a wide range of other environmental stresses, resulting in a debilitated social fabric that increases the vulnerability of the weakest stakeholders in conflict. In developing countries, for example, conflicts divert labour away from normal economic activities into (sometimes) armed conflicts and even death.[28] In Kenya, for example, conflicts over environmental resources in dryland areas affect livelihoods and result in migration, landlessness, impoverishment, and lawlessness.[29] While conflict does not produce such extreme results in developed countries, at a minimum, livelihoods and sense of well-being can be seriously affected by such disputes. To mitigate, negotiate and resolve such conflicts requires a significant expenditure of fiscal resources and social capital.

Figure 1 provides a visual representation of the vulnerability of rural communities as a function of their exposure to climate change and the adaptive capacity of their water governance institutions. In addition, water conflicts affect the exposure and adaptive capacity because they mediate and influence institutional adaptive capacity through the manner in which the conflicts are managed and resolved, which in turn can dictate completely different outcomes for a given conflict.[30] Institutions that integrate lessons learned from these conflicts into decision-making and policies enhance their adaptive capacity, whereas those that do not may lose adaptive capacity. Given that water is an essential element for economic development, ecosystem functions, and the fulfillment of local values and cultures, the figure illustrates that successful resolution of conflicts can enhance institutional adaptive capacity, which in turn reduces the vulnerability of communities to conditions of water scarcity.

Experiences and Lessons Learned from Conflicts and Integrated in Water Governance Decision-Making and Policies

Figure 2 illustrates how a water governance system is enhanced when experiences and lessons learned from past water conflicts, including the assumptions, behaviours, attitudes, and processes underlying past conflict resolution, are integrated into future decision-making and policies, including future conflict resolutions. The ability to learn from the lessons generated by conflict will also shape the capacity of governments to improve on existing water governance systems. Of course, since the efficacy of any water governance regime depends in part on the extent to which water users and stakeholders trust the institutions and individuals involved, the interaction between the organizations responsible for enforcing any given water governance regime with water users and stakeholders in a period of conflict becomes the litmus test for future relations. Trust and legitimacy are gained when institutions are open to genuine participation by communities in water management decision-making.

The historical experience of dam construction in times of water scarcity offers a generally negative illustration. According to the World Commission on Dams, approximately 300 dams on average were constructed each year between 1986 and 1993, resulting in the displacement of 40–80 million people.[31] Although they have a major impact on communities, these dams have generally been initiated with minimal opportunity for community participation in planning and decision-making. Indeed, the decision to construct dams in order to impound scarce water has often been imposed on communities to the detriment of their livelihood and security.[32] These communities often lack the information concerning the potential negative consequences of the development, much less the resources to fight the proposed development. Some of the negative consequences for those more directly affected often include physical and social displacement, as well as loss of existing livelihood. In addition, unilateral decisions to construct dams have often failed to appreciate the importance of water to all water users and have foregone insightful perspectives for managing the resource.

As illustrated in the upper portion of Figure 2, unilateral decision-making, contested or uncontested, results in a poorer water governance system and in

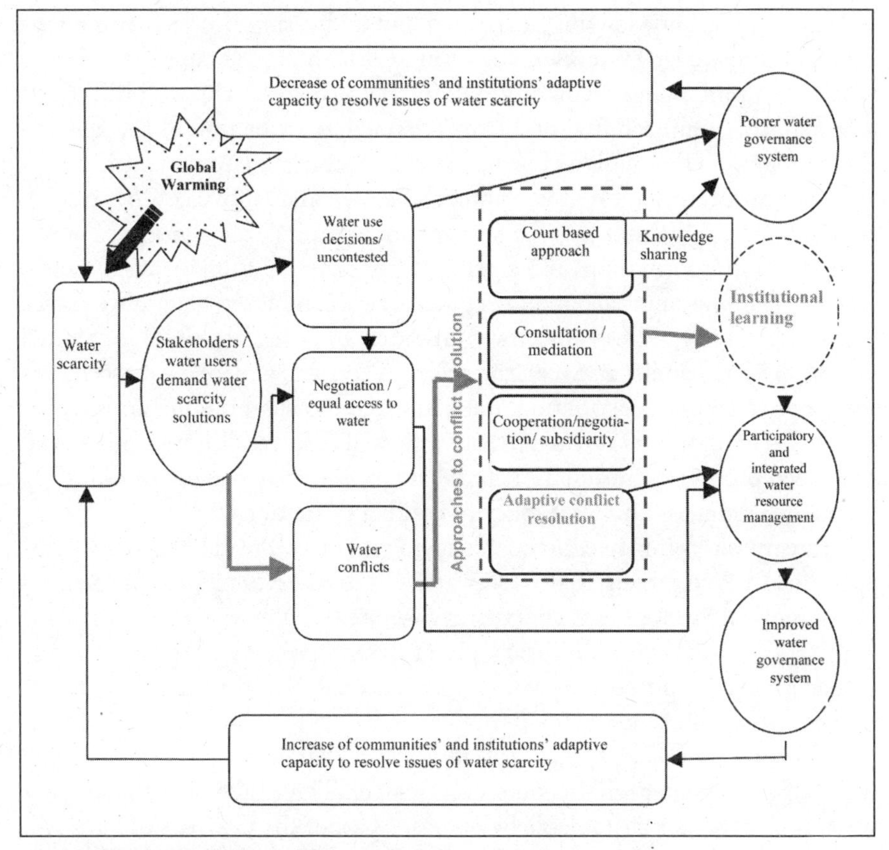

Figure 2. Path to an Enhanced Water Governance System.

a decrease in the capacity of affected communities of users and stakeholders to adapt to future water scarcity. However, when water governance institutions, and the governments that are ultimately responsible for them, enter into a genuine dialogue with the communities to reach a mutually acceptable outcome, the water governance system is improved, and both communities and institutions gain adaptive capacity for future scenarios of water scarcity. The second, more "virtuous" cycle, is shown in the middle portion of Figure 2. The lower portion of Figure 2 illustrates that even in the case of conflict, there are approaches to conflict resolution that allow knowledge sharing and institutional learning from lessons and experiences. If integrated effectively into future decision-making and policies, then the lessons obtained from such conflicts can actually lead to an improved water governance system and a more robust adaptive capacity on the part of water users, water stakeholders, and water governance decision-makers. The model shows a range of conflict resolution approaches, from a resolution imposed by the courts, to one where a resolution is reached by acknowledging, accepting and addressing power asymmetries, different knowledge systems, values and ideas. The former approach is one where negotiation and mediation fail, forcing resolution through the courts. Through this approach, the problem is defined as a violation of the law and the ultimate solution designed principally to meet legal requirements.

In contrast, this alternative approach, known as adaptive conflict resolution, requires significant investment in human capital, the sharing of power, resources and information, and a genuine political will to make it possible.[33] In particular, this approach requires governments and their water governance institutions to: a) redress power asymmetries; b) create elements for shared visions of sustainability between the actors in conflict; and, c) create scenarios for adaptive conflict negotiation, mediation, consensus-building and resolution.[34] Participation is not a panacea that will resolve all issues but a well organized process of consultation and dialogue that can increase significantly the appropriateness and legitimacy of conflict resolutions.

Integrated Water Resource Management (IWRM) principles reflect this approach by recognizing that fresh water is a finite resource that is essential for economic development and the environment, and that its management requires the participation of users, planners, and policy makers. Since 2002, Canada has committed to developing a national plan for managing water resources through IWRM principles.[35]

With the trends in climate variability and change observed in the SSRB, it is likely that water will become even scarcer. The challenge of managing water resources on a sustainable basis in this part of the Canadian Plains will become greater as a result of the climatic impact of global warming. The best scientific evidence available indicates that it will be warmer and drier in the SSRB in the 21st century than was the case in the 20th century. In particular, there will be lower summer stream flows, and summers will be especially dry, thus affecting farmers and ranchers throughout this extensive river basin.[36] With the combination of even lower soil moisture and already high water usage levels from available rivers in the basin, increased conflict among users and stakeholders is almost inevitable. The real question concerns how that conflict will be conducted and mediated, and to what extent lessons from previous conflicts will inform future water governance decision-makers to meet the needs of the various water users in the basin, while maintaining the natural integrity of the prairie ecosystem.

The Oldman River Dam Conflict: Impacts and Lessons Learned

The Oldman River Dam conflict is a prime example of a conflict in a dryland region that emerged out of the scarcity of water supply and availability impacting water users' sense of security (North Lethbridge Irrigation District farmers), concern for the ecosystem (Friends of the Oldman River, or FOR) and issues regarding equality and equity (displaced farmers and the Piikani). It is a conflict that, at least for the Piikani First Nation and some former members of FOR, remains unresolved. For the Piikani, the full impacts of the dam on the physical environment have not been assessed, and most of the members of the Piikani feel that their claims to water rights, based on the principle of equitable access and their status as the original human occupants of the region, have been consistently ignored, though they negotiated a settlement with the Alberta government. For FOR, an environmental organization that emerged from a coalition of members of local environmental organizations opposed to the construction of the Oldman Dam, there are impacts downstream that need to be evaluated and continuously monitored. For both FOR and the Piikani First Nation, the process followed by the government to address the conflict lacked legitimacy and these stakeholders felt that their voices were marginalized from the beginning. The shared opinion is that the government minimizes or ignores evidence that the dam construction negatively affected some aspects of the river's ecosystem to the detriment of displaced farmers and the Piikani. In sharp contrast, in the opinion of the water governance institutions, including the Alberta Irrigation Project Association and the Northern Lethbridge Irrigation District, the conflict has been resolved. From the perspective of these organizations as well as the government of Alberta, water security has been restored.

Impacts and Stakeholders

The Oldman River rises primarily from melting snow on the eastern slopes of the Rocky Mountains and flows eastward across the foothills and plains towards Saskatchewan. On its eastward journey, the Oldman River gathers additional waters from tributaries, including the Crowsnest, Castle, Waterton, Belly, and St. Mary rivers until it reaches the town of Brooks where it joins with the Bow River to form the South Saskatchewan River. The flow of the Oldman River varies throughout the year, correlating with snow accumulation and melting. In July, the flow peaks as a result of the melting snowpack in the mountains and precipitation in the spring and early summer. By mid-November, the flow is at its lowest, with the drop in air temperature as the rains turn to snow and the ground freezes. In addition to annual variability, the river flow also varies from year to year depending on the amount of snowfall in the mountains. Finally, compared to the mean annual volume of flow of the Ottawa and the Fraser rivers, the Oldman River flow is very small: its mean annual flow is ten and five percent of the Ottawa and Fraser respectively.[37]

The variability and uncertainty associated with the Oldman River flow has historically affected the sense of security of the agricultural community in the region. In fact, since the 1920s, the Northern Lethbridge Irrigation District has sought to enhance security by increasing water withdrawals from the Oldman River. Although discussions around the construction of the Oldman River Dam did not occur until the 1970s, other dams were constructed in order to secure water for irrigation districts in the region. Established in the mid-1930s, the Prairie Farm Rehabilitation Administration led the way in the construction of St. Mary River Dam in the 1940s, and the Bow River Dam in the 1950s. Discussions regarding the construction of the Oldman River Dam were triggered by the challenges facing contiguous communities during drought years in the 1970s and 1980s. In particular, irrigation farmers from the Northern Lethbridge Irrigation District (NLID) lobbied the Alberta government in the 1970s to build an on-stream reservoir on the Oldman River.[38] In response to this lobby, the newly elected Progressive Conservative government of Peter Lougheed increased investment in irrigation development and facilitated the development of a diversified agricultural sector. Then, in 1976, the Alberta Environment Department announced the need to construct a dam on the Oldman River for irrigation purposes, and listed nine potential sites for the dam, including the Three Rivers site where the dam is presently located.[39]

Following the announcement of the need to construct a dam on the Oldman River, studies were conducted by various government entities to assess the best site in terms of economic feasibility and environmental impact. A study by the Government of Alberta ranked the Three Rivers site sixth in terms of environmental impacts, but first in terms of economic feasibility and operation. However, the Oldman River Study Management Committee (ORSMC), a ten-person non-governmental committee that led further studies on water supply alternatives after strong opposition to the provincial government's 1976 recommendation of constructing a dam at the Three Rivers site, found that the Brocket (on the Piikani Reserve) and Three Rivers sites were comparable in terms of technological, economic, and environmental considerations, and thus decided to delegate the final choice to the politicians.[40]

Facing increasing discontent from various groups, the Alberta government asked the Environmental Council of Alberta (ECA), a four-person advisory panel, to convene public hearings on the construction of the Oldman Dam and to report their findings. In 1979, the ECA released a report stating that an onstream reservoir on the Oldman River was not needed to support the expansion of irrigation in the NLID; instead, it stated that off-stream storage would be sufficient. However, the ECA concluded that if a dam were to be constructed, the site at Brocket was preferred over Three Rivers.[41] Despite the recommendations of the ECA, the Alberta government announced in 1980 that the dam would be constructed at Three Rivers. After years of opposition from

Figure 3. Oldman River Basin: The Dam and Its Tributaries.

the Piikani and farmers at Three Rivers whose land would be flooded by the project, dam construction began in 1986. The dam was eventually located at Three Rivers, about 10 kilometres northeast of the town of Pincher Creek (Figure 3), and officially opened in 1992. The reservoir that captures the river flow flooded 43 kilometres of the river valley. The reservoir has a storage capacity of 740,000 cubic decametres of water, which increased the potential for irrigated lands in the basin by an additional 121,000 hectares, for a total of 344,000 hectares. According to Alberta Environment, the function of the Oldman River Dam is to impound water for present and future water demands of municipalities, industries and agriculture, and provide good quality water and river flow downstream of the dam for ecological purposes including aquatic and riparian environments, as well as for recreational opportunities and hydro-electric power generation.[42] In addition, a key function of the dam is to provide the province of Alberta with the flexibility to meet its inter-provincial apportionment agreement with Saskatchewan.

Although some members of the Piikani community initially supported the construction of the dam, the decision to locate the dam at Three Rivers represented a rejection of their own aspirations for potential economic development and benefits from the dam, and resurrected past injustices for which they now sought compensation. One of these alleged injustices was the 1922 agreement to locate the head-works of irrigation canals to divert water for the benefit of the LNID, and for which they were not consulted, and now demanded compensation.[43]

Perhaps the most lasting consequence of the conflict was the exacerbation of divisions in the Piikani community, which stemmed from differences in values and priorities about the assessment of both the process of conflict resolution and its outcome. Those in favour of the dam sought to address economic development challenges in the community, and accepted the potential negative environmental impacts. Those opposed to the dam emphasized the spiritual and cultural values associated with the Oldman River and its ecosystem, claiming that a change of the water flow regime would jeopardize cultural and spiritual values attained by their connection with the natural environment.[44] This division continues to drive a wedge in the community and for some the consequences of the construction of the dam has made them worse off than before:

> I think we are worse off . . . It has divided the community. Cut the community in half and created conflicts . . . A lot of the issues we are seeing are the direct effect of the way the issues were dealt with by the federal government and the province. And I would also say by our own people too. Our own people had been part of that orchestration as well, and ultimately I can see that our membership in a vote ratified the agreement.[45]

Although the Piikani Nation has negotiated a $64 million settlement agreement with the Alberta government as compensation for the impacts of the Oldman Dam over their territory, some members of the community feel that there was never equality at the negotiating table. One of the respondents stated:

> I think—and I am not trying to be too cynical, but I think that . . . being equal in negotiating with the federal and provincial governments is only a theory; I don't think that the province and the federal government had ever intended to have equal negotiations in anyway . . . I think that was just rhetoric, there was nothing substantive in the discussions that would suggest that we were ever in a position of equality and developing a mutual agreement that would equally benefit the province and the band.[46]

The sense of injustice and inequity experienced by the Piikani was reaffirmed by the findings of the Environmental Assessment Review Panel which, by order of the Supreme Court of Canada, conducted an environmental impact assessment of the dam. The Panel found that the "Oldman Dam project has many social and cultural consequences for the Piikani. Most significant among these are the following: the Peigan [Piikani] were not sufficiently involved in key decisions about the project: many resources important to their culture, such as cottonwoods, fish, game, and willows will be affected by the project."[47]

Farmers who were displaced because their land was flooded by the reservoir organized the Committee for the Preservation of the Three Rivers (CPTR) to also oppose the construction of dam. Despite their efforts in lobbying public support and their MLAs to stop the construction of the dam, the 16 affected families only learned about their displacement when the Alberta government publicly announced the construction of the dam at Three Rivers and were not given any personal notification.[48] Similar to the Piikani, the displaced farmers felt a profound sense of injustice and inequity in dealing with the government. One of the farmers echoed the views of the Piikani concerning the Alberta government's lack of good faith in its stakeholder consultations:

> Yes, we were heavily involved in that [negotiations], and we went to all the public hearings. The sad thing is that the decision was already made before they went to the public hearings, as far as we were concerned. They had made the commitment to build the dam, because there was more political pressure from the east [farmers from NLID] that needed the water, than from the west that was trying to preserve the watershed. So we went through the process as though we were part of the decision making, but we weren't, we were just being exercised. And we put a lot of energy into it. And it was hard to leave our place.[49]

Winners and losers were created during the course of this conflict. One displaced farmer concluded that the sense of water security gained by the NLID farmers was at the expense of water security lost by other farmers:

> If there was a drought, you know the only guy they can make money off? The guy that it's in the drought [area], another farmer; they don't make it off the government, they don't make it off the chemical company, they don't make it off the fuel company. The only time that they can sell their hay for $120 a ton is when I don't have any. I am

> the only guy that they can make a killing on, the only guy, nobody else . . . we pay the price. We pay the price to them, that's the benefit [for them]; it's not a fair thing.[50]

In addition to the opposition from the displaced farmers and the Piikani, the Alberta government also faced a formidable opposition from the Friends of the Oldman River. FOR was established soon after the construction of the dam began, and their concerns over the environmental impacts of the dam, as well as concerns over the disregard of the Piikani and displaced farmers, compelled them to go to court and question the legal basis of the licence for the construction of the dam issued by the Alberta government. FOR's legal challenges ranged from a failure by the Alberta government to follow the appropriate procedures for issuing a licence for the construction of the dam, to the destruction of fish habitat and the need for an environmental impact assessment of the dam.[51] Just like the Piikani and displaced farmers, FOR maintained that the government had decided to construct the dam regardless of the impact, and regardless of the ECA's conclusion that there was no need to build an on-stream reservoir:

> For the construction of the Oldman dam, the government spent millions of dollars on a consulting team to study the situation, the consulting team told them "you don't need to build a dam, you need to improve the holding capacities of existing reservoirs. You don't need a dam, and the worst possible place to build the dam is at Three Rivers." Which is exactly where they built the dam. And what is infuriating about that is the absolute scorn for science and the naked political agenda.[52]

According to a member of FOR, the lobbying from the irrigation farmers was very powerful and their interests superseded the interests of all other water users and the environment:

> I think what was there before [flooded portion of the river valley] was a gem, and it was very productive. We had successful ranchers and farmers along the river that were displaced by the dam so that we can provide water to grow sugar beets around Lethbridge. It's strictly political . . . You didn't see the sugar beet farmers around the Lethbridge area come out and say, why are we clear cutting in the Oldman river watershed? Why are we causing the water to flow off sooner than it would if we left tree and bush cover there? Why are we doing that? They didn't say that . . ."Build us a dam to look after

> our interests," was their claim. And there was no sense, there is no sense, in a lot of these people of any ecological principle, there is no sense of being connected to a wider environment.[53]

Lessons on Enhancing Adaptive Capacity

As this description of the Oldman River Dam conflict illustrates, there was a degree of consultation, although a number of users and stakeholders would argue the provincial government's openness did not extend to genuine participation in decision-making. Even in such an ambiguous instance, however, the Oldman River case can be used to test the proposition that the lessons learned from past conflict can contribute to improved water governance and enhance adaptive capacity in the future. Thus, as indicated by one of the decision-makers the conflict can be seen as a turning point in terms of the attitude of government:

> I have made presentations all over and I used that as a milestone, the dam [conflict]; that there was a change in attitude and action from that point. I think that we are at another crossroads right now, it comes around the whole issue of climate change, and I won't call it a water shortage, but a recognition that we are not in a situation where water is free. We are now at a situation, at a crossroads of saying "ok, there is an industry, but as basin stakeholders in general we need a new strategy to further improve conservation and effectiveness measures, but also deals with the issue of water quality and the whole of environmental sustainability issue in this basin.[54]

Of course, there are differences of opinion among the various stakeholders concerning the impact of the conflict on the knowledge of the general public in the region about water issues and the need for sustainable use of resources. Some claim that the conflict has made people more aware of the issue of water scarcity, resulting in private measures to minimize water use (e.g., the use of more efficient irrigation technology). However, others believe that this awareness is a product of growing global and national environmental concern in general, and not specifically a result of this conflict.

According to the Alberta Irrigation Project Association (AIPA), which represents most water districts in Alberta, farmers in southern Alberta have increased water use efficiency by 30% over the last few years. This efficiency is, to a large extent, due to the use of sophisticated data that allows the monitoring of soil moisture conditions on every individual farm, as well as using data on maximum and minimum temperature, precipitation, reflective radiation, and wind speed. This information can then be used to determine water use according to the irrigation method employed and the type of crop grown. While improved information has facilitated adaptive capacity, one representative of AIPA also pointed to the lessons learned from the Oldman River Dam conflict:

> We recognized that economic growth and population growth were really moving at a fast pace in Alberta, and we . . . had . . . verbal conflicts about who gets the water and who should use it. And what we wanted to do was to really find out what the status of our industry was? How good a job were we doing? How much better could we do? Because if we are not able to show that we are wise users of water, then it's pretty easy to take it [water] from us and give it to somebody else.[55]

This perspective was also reflected in the comments by a representative of Alberta Agriculture and Food:

> The one thing that dam did was galvanize the irrigation industry . . . because the challenge was put to them by the public saying, "we think the industry can do better in managing the water." The industry, including ourselves, have taken that very seriously, and since then I think the industry has more than met the challenge, to the point that it has become very efficient and could become more efficient in the next 20–30 years . . . So the building of the dam and the discussions around that, I think were a key trigger for industry to say "we can do better." Everybody became very focused so that efficiency and effectiveness of water use and management become a high priority, and it is to this day, and we continue to talk about, what can we do? How can we improve? Most of our irrigation farmers now are utilizing low pressure pivots systems. In terms of sprinkling systems, it is probably the most efficient system that there is to apply water to large areas of crops.[56]

It is not necessary to accept the view that the Alberta government had made up its mind in advance of the consultations, or that it simply failed in its consultative efforts on the Oldman River Dam. Rather, it may well be the case that the Alberta government had neither the experience (nor perhaps the capacity) to conduct and hold meaningful consultations with stakeholders and users. However, according to a former Alberta Agriculture minister, the government's approach to engaging stakeholders in water governance has changed as a result of the conflict:

> I think we've made a lot of improvements over the last number of years, in the industry and government, in being able to handle consultation

> process because the consultation process was not even really talked about much until about 1980; before that, when projects were done there was a bit of consultation and you built the thing and that was it; there wasn't a lot of kafuffle then. [Then] you had environmental groups . . . and they cranked up the atmosphere and so public consultations and the communication strategies increased dramatically.[57]

Since the conflict, water governance decisions of the Oldman River Basin have involved more inclusive stakeholder consultations through the Oldman Watershed Council, in which government agencies, businesses, NGOs, and environmental organizations participate. First Nations communities have been invited to participate, but they have declined, arguing they are not just another stakeholder or user in the region, but a sovereign First Nation with constitutional rights based upon their original occupation of the region and the subsequent treaties.

It remains to be seen whether the involvement of stakeholders in the decision-making process in the Oldman Watershed Council is meaningful enough to have the potential for improved water governance. As one representative of the Alberta Riparian Habitat Management Society noted:

> The watershed basin councils that are being formed, the goal is that they are made up of local stakeholders; I think it is too early to say that their voices will be equitably heard around the table, some of them are very much government agency focused, some councils are much more with various community stakeholders mixed with government agencies, so each one is different, but I think it's too early to say that they are definitely heard, and I am not saying that they aren't going to be heard because it's too early, the basin council down here is brand new; I think that there is definitely the potential for at least good input.[58]

An advisory structure that includes the Alberta Water Council, watershed advisory councils, and watershed stewardship groups oversees the council. The active participation of stakeholders in this consultative structure is, in part, the legacy of the Oldman River Dam conflict. This new stakeholder process is described as follows by a provincial government representative:

> One of the things we have [become] very good at is engaging the public . . . I think that's critical, and having gone through the South Saskatchewan River Basin review process all of the basin's Advisory

> Committee members that we had—and they had environmentalists to irrigators to academics [representatives] . . . there were certain things they disagreed on, there is no question about that, but when finally they got down to, what can we recommend? Given the constraints and everything, they were able to come to a consensus on what they could recommend and when that is done . . . it removes the public adversarial component; the environmentalists can't say "ok, here is our recommendations" and then go out of the room and say "you can't do this because . . ." Well, you just signed it off.[59]

> One example that I can use is the South Saskatchewan River Basin review, a water management plan that has come out to define what the minimum flow should be in the rivers, a lot of those decisions were made by advisory committees that are a cross-section of stakeholders . . . and they said, "here is the issue that we know of, here is the best information that we have, as objective as we can make it. You make recommendations in terms of the balance that you would like to see between economic development and environment." If you were to ask a group something like that 25 years ago, I don't think that you could ever come to a consensus because they would have had this conceptual polarization amongst them, they would not have given an inch; they would have argued that all the water should be used for development or for environmental protection.[60]

This government official also claims that lessons from the Oldman River Dam conflict have influenced federal-provincial collaboration and environmental policy in Alberta:

> I think also since the Oldman Dam there has been a lot more conscious efforts at the provincial and federal levels to integrate environmental policies into practices. I think that's an outcome that can be to some degree traced back to the Oldman Dam. I think there is a lot more cooperation now than there ever was in the past. There is a lot more harmonization than there ever was. There are still points of conflict; they are usually constitutional in nature, but in the day to day operating I think there is a lot more integration than prior to the Oldman, which is a positive thing.[61]

According to the Environmental Law Centre (ELC)[62], the Alberta government's recent policy framework on water, *Water for Life,* is a positive step towards a more effective and inclusive water management process. Despite some continuing reservations about the implementation of this policy, the ELC emphasizes that the new framework has facilitated stakeholder engagement with decision-makers, and has built relationships among stakeholders. The framework has also increased the opportunity for the involvement of water users, as well as the general public in water management planning.

The experiences and lessons gained in monitoring and mitigating environmental impacts of the Oldman Dam will be very useful in potential future similar scenarios. This 'learning' was aptly summarized by the same provincial official:

> Today, thanks to the Oldman Dam in some respects, if we were to build another dam and somebody asks: do we have any experience with mitigating environmental problems related to the building of a dam? We now have information because it's been a priority for us. Back then I think that it was very much conjecture; it was not that anybody was withholding information per se, nobody had it.[63]

> What we really learned from the environmental side is that by doing the continual monitoring of the Oldman Dam and our other projects, the Pine Cooley and Twin Valley, is that we now know a lot more than we did before and so when people come forward and say "this will cost the trees to die and fall," we can say "well actually no, or yes, it will," depending on what the case is, so we have learned a lot, and I think we have got to the point now where we know what we don't know and that's important too, and to recognize it and admit it.[64]

Conclusion

The study of water related conflicts in general, and of the Oldman River Dam conflict in particular, provide a particular opportunity to identify and understand how the impacts of conflicts can increase the vulnerability of affected groups vis-à-vis access to water in the context of climatic change in the South Saskatchewan River Basin. When these lessons are acted upon in effective and meaningful ways, the adaptive capacity of communities and water governance institutions can be enhanced. These responses, including a more transparent and accountable decision-making process, also enhance the legitimacy of state action in the eyes of stakeholders which, in turn, makes stakeholders and communities more open to accepting strategies of water management in times of water scarcity.

The history of this conflict, like so many others from around the world, is an example of how governments can yield too quickly and easily to powerful stakeholders. The substantial electoral weight and economic influence of some stakeholders (e.g., irrigation farmers and their organizations) led the Canadian and Alberta governments to proceed with the construction of the Oldman River Dam, while ignoring decisions of the courts that were based on sound scientific and technical environmental and social considerations. Weaker parties who felt either discriminated against or ignored, like the Piikani nation and the Friends of the Oldman River, were forced to resist the Alberta government's decision through the courts and through political protest.

Although FOR remains unhappy with the outcome, the environmental group's principled opposition has helped to open the door for a national policy that now makes conducting environmental impact assessments prior to the implementation of any major development project imperative. As for the Piikani First Nation, many of the resources important to their culture, such as cottonwoods, fish, and game were negatively affected by the dam. The conflict has left the Piikani divided and bitter, with the authority of their political institutions in jeopardy. In the end, the compensation paid for this dislocation ended up further fracturing the Piikani's solidarity. Even with the Piikani, however, there remains the opportunity to shape a more positive future using lessons learned from the conflict, something recognized by the Alberta government minister originally responsible for the dam:

> You can't all of a sudden state that the Oldman Dam is there and the mitigation is working and it is fine and let's go to sleep. There are a lot of things that need to be done to the Oldman River Dam as yet and there are opportunities there, things that did not work out quite right that need to be fixed, and that is an ongoing thing.[65]

Endnotes

1. B. Charrier, S. Dinar, and M. Hiniker, "Water, Conflict Resolution and Environmental Sustainability in the Middle East," *Arid Lands Newsletter* 44 (1988). Available at: http://ag.arizona.edu/OALS/ALN/aln44/charrier.html (Accessed Jan. 18, 2009).
2. P. H. Gleick, "Water and Conflict: Fresh Water Resources and International Security," *International Security* 18:1 (1993): 96.
3. Although the focus of this paper is the Oldman River Dam conflict, it relates to the broader goal of the Institutional Adaptation to Climate Change (IACC) project that seeks to understand the role of institutions in dry-land regions in Canada (South Saskatchewan River Basin) and Chile (the Elqui Valley, Fourth Region) in the resolution of water conflicts and their capacity to adapt to climate change related water insecurities. See http://www.parc.ca/mcri/

4. B. Bonsal, G. Koshida, E.G. O'Brien, and E. Wheaton, "Droughts," in Environment Canada, *Threats to Water Availability in Canada,* National Water Research Institute, Burlington, Ontario, WRI Scientific Assessment Report Series No. 3 and ACSD Science Assessment Series No. 1 (2004): 21–22.
5. Ibid.
6. B. Harker, J. Lebedin, M. J. Gross, C. Madramootoo, D. Neilsen, B. Paterson, and T. van der Gulik, "Land Practices and Changes—Agriculture," in Environment Canada, *Threats to Water Availability in Canada,* National Water Research Institute, Burlington, Ontario, WRI Scientific Assessment Report Series No. 3 and ACSD Science Assessment Series No. 1 (2004), 50.
7. Natural Resources Canada, Climate Change Impacts and Adaptation Directorate, *Climate Change Impacts and Adaptation: A Canada Perspective* (2004). Available at: http://adaptation.nrcan.gc.ca/perspective_e.asp (cited October 2006).
8. F. Quinn, J.C. Day, M. Healey, R. Kellow, D. Rosenberg, and J.O. Saunders, "Water Allocation, Diversion and Export," in *Threats to Water Availability in Canada* (Burlington, Ontario: National Water Research Institute, 2004); also A. Marshall, 2004, "Cold Shower for New Water Strategy," [online], *Wild Lands Advocate* 12:1 (February 2004). Available at: http://AlbertaWilderness.ca (cited 10 September, 2007).
9. Ibid.
10. Quinn et al., "Water Allocation," 7.
11. Quinn et al., "Water Allocation," 2–3.
12. Marshall, "Cold Shower,"4.
13. A. Rojas, B. Reyes, L. Magzul, H.L. Morales, R. Borquez, and E. Swartz, *Analysis of Role of Institutions in the Resolution of Water Conflicts: Final Report.* Appendices: Interview Guide, 74–75. Available at: http://www.parc.ca/mcri/ iacc056.php (Accessed September 2008).
14. A. Rojas, "The Mountain's Scar, the People's Wound: The GasAndes Conflict in Cajon del Maipo River in Central Chile," in A. Rojas and B. Reyes (eds.), *From Fires of Destruction to the Flames of Creativity: Towards an Adaptive Environmental Conflict Resolution. Lessons from Chile and Canada* (Vancouver: University of British Columbia, Liu Centre, 2003). CD ROM, Institute of Political Ecology, Santiago, 2003.
15. Ibid.
16. J. Chevalier, "Stakeholder Analysis and Natural Resource Management" (Ottawa: Carleton University, June 2001). Available at: http://www.carleton.ca/~jchevali/ STAKEH.html (Accessed June 21, 2006).
17. Ibid.
18. R. Grimble, M.K. Chan, J. Aglionby, and J. Quan, "Trees and Trade-offs: A Stakeholder Approach to Natural Resource Management." Gatekeeper Series 52. (London, UK: International Institute of Environment and Development, 1995).
19. Chevalier, "Stakeholder Analysis," 5.

20. D. Gibbon, "Systems Thinking, Interdisciplinarity and Farmer Participation: Essential Ingredients in Working for More Sustainable Organic Farming Systems" (2002). Available at: http://www.organic.aber.ac.uk/library/Systems%20thinking,%20interdisciplinarity%20and%20farmer%20participation.pdf (Accessed August 17, 2005).
21. See note 21.
22. Gleick, "Water and Conflict," 93.
23. Gleick, "Water and Conflict," 81; and Charrier et al., "Water, conflict resolution," 2.
24. Peter Ashton, "Southern African Water Conflicts: Are They Inevitable or Preventable?" in Hussein Solomon and Anthony Turton (eds.), *Water Wars: Enduring Myth or Impending Reality, 2000* (Umhlanga Rocks: African Centre for the Constructive Resolution of Disputes), 65–102.
25. Gleick, "Water and Conflict," 79; S. Postel and A.T. Wolf, "Dehydrating Conflict," 79. Available at: http://www.globalpolicy.org/security/natres/water/2001/1001fpol.htm; John W. Maxwell and Rafael Reuveny, "Resource Scarcity and Conflict in Developing Countries," *Journal of Peace Research,* Vol. 37, No. 3 (May, 2000), 303; N. LeRoy Poff, J. David Allan, Margaret A. Palmer, David D. Hart, Brian D. Richter, Angela H. Arthington, Kevin H. Rogers, Judy L. Meyer, Jack A. Stanford, "River Flows and Water Wars: Emerging Science for Environmental Decision Making," *Frontiers in Ecology and the Environment,* Vol. 1, No. 6 (August, 2003): 298.
26. Gleick, "Water and Conflict," 93.
27. Maxwell and Reuveny, "Resource Scarcity," 303; John Barnett and Neil Adger, "Security and Climate Change: Towards an Improved Understanding," Human Security and Climate Change International Workshop, Tyndall Centre for Climate Change Research, University of East Anglia, June 2005, 11.
28. Maxwell and Reuveny, "Resource Scarcity," 303.
29. Siri Eriksen, Kirsten Ulsrud, Jeremy Lind, and Bernard Muok, "The Urgent Need to Increase Adaptive Capacities: Evidence from Kenyan Drylands," African Centre for Technology Studies, Conflicts and Adaptation Policy Brief 2, November 2006, 1–2.
30. A. Rojas, and B. Reyes (eds.), *From fires of destruction to the flames of creativity: Towards an Adaptive Environmental Conflict Resolution, Lessons from Chile and Canada, 2003,* University of British Columbia, Liu Institute, Vancouver, CD ROM, Institute of Political Ecology, Santiago.
31. World Commission on Dams, "Dams and Development. A New Framework for Decisión-Making," The Report of the World Commission on Dams, (London and Virginia: Earthscan, 2000), 17.
32. World Commission on Dams, "Dams and Development," 98.
33. In *Water Blues in Climate Change: The Role of Institutions in Water Conflicts and the Challenges Presented by Climate Change,* IACC Project, working paper, 2006, A. Rojas, B. Reyes, L. Magzul, and H.L. Morales provide a more in-depth discussion of the various conflict resolution approaches shown in the model. Available at:

www.parc.ca/mcri; A. Rojas, J. Grandy, and J. Jamieson, "Toward and Adaptive Resolution of Environmental Conflicts: Lessons from Clayoquot Sound," in A. Rojas and B. Reyes (eds.), *From fires of destruction.*.

34. Rojas et al., *Water Blues in Climate Change,* 16.
35. Darrell R. Corkal, Bruce Inch, and Philip E. Adkins, "The Case of Canada? Institutions and Water in the South Saskatchewan River Basin," Institutional Adaptations to Climate Change project, 2007, 9–10. Published online at: http://www.parc.ca/mcri/iacc045.php (Accessed Jan 18, 2009).
36. Prairie Adaptation Research Collaborative (PARC), *Climate Change Impacts on Canada's Prairie Provinces: A Summary of our State of Knowledge* (Regina: Prairie Adaptation Research Collaborative, 2008).
37. J. Glenn, *Once Upon and Oldman: Special Interest Politics and the Oldman River Dam,* (Vancouver: UBC Press, 1999), 13–14.
38. Ibid., 22–24.
39. Ibid., 27.
40. Ibid., 28–29.
41. Ibid., 33.
42. Alberta Environment, [online]. Available at: http://www.environment.gov.ab.ca (accessed November 2006).
43. Ibid., 30.
44. Ibid., 195–204.
45. Piikani community member (OMRBST1), personal communication, 2005.
46. Piikani community member (OMRBST1), 2005.
47. Environment Canada, "Oldman River Dam: Report of the Environmental Assessment Panel," Federal Environmental assessment and review process No. 42, May 1992.
48. J. Glenn, "Once Upon an Oldman," 47–48.
49. Displaced farmer (OMRBST2), personal communication, 2006.
50. Displaced farmer (OMRBST2), 2006.
51. J. Glenn, "Once Upon an Oldman," 94–106.
52. Friends of the Oldman River member (OMRBST3), personal communication, 2006.
53. Friends of the Oldman River member (OMRBST3), 2006.
54. Alberta Agriculture and Food representative (OMRBST4), personal communication, 2006.
55. Alberta Irrigation Project Association representative (OMRBST5), personal communication, 2005.
56. Alberta Agriculture and Food representative (OMRBST4), 2006.
57. Farmer and former minister of Alberta Agriculture (OMRBST6), personal communication, 2006.
58. Alberta Riparian Habitat Management Society representative (OMRBST7), personal communication, 2005.

59. Alberta Environment representative (OMRBST8), personal communication, 2006.
60. Ibid.
61. Alberta Agriculture and Food representative (OMRBST4), 2006.
62. Environmental Law Centre, "RE: Water for Life Strategy Renewal," File: 33, August 24, 2007 (accessed October 2007), 2.
63. Alberta Environment representative (OMRBST8), 2006.
64. Ibid.
65. Farmer and former minister of Alberta Agriculture (OMRBST6), 2006.
66. Alberta Environment [online]. Available at: http://www.environment.gov.ab.ca (accessed November 2006).

COMMUNITY CASE STUDIES OF VULNERABILITY TO CLIMATE CHANGE: CABRI AND STEWART VALLEY, SASKATCHEWAN[1]

HARRY DIAZ, SUREN KULSHRESHTHA , BRETT MATLOCK,
ELAINE WHEATON, AND VIRGINIA WITTROCK

Abstract

This paper discusses the vulnerability of two dryland agricultural communities—Cabri and Stewart Valley—located in the southwestern area of Saskatchewan. The paper synthesizes two interrelated studies. The first is a vulnerability assessment of both communities based on interviews and local focus group meetings. The second study focused on the biophysical and economic impacts of the 2001–2002 drought. The integration of results provides an understanding of the vulnerability of rural communities to climate variability and extreme climate events. Results suggest that these rural communities are affected by a number of stressors, including climate; furthermore, climate-related stressors are important since they are reflected in economic and social stressors. Since both communities are in agricultural regions, changes in production and markets have serious impacts on them. Also, social changes related to community depopulation, and centralization of services are no less significant. These stressors are instrumental in affecting the long-term sustainability of these communities. The contrasting exposures and responses of the communities provided a valuable examination of adaptation and vulnerability.

Sommaire

Le travail suivant traite de la vulnérabilité de deux collectivités agricoles située sur des terres arides, Cabri et Stewart Valley, du sud-ouest de la Saskatchewan. Le travail fait une synthèse de deux études interdépendantes. La première est une évaluation de la vulnérabilité des deux collectivités s'appuyant sur des entrevues et des réunions de groupes de discussion organisées avec divers groupes locaux. La deuxième portait sur les impacts biophysiques et économiques liés à la sècheresse de 2001–02. L'intégration des résultats permet de comprendre la vulnérabilité des collectivités rurales à la variabilité climatique et aux évènements climatiques extrêmes. Les résultats semblent indiquer que ces collectivités rurales sont affectées par un nombre d'agents stressants, dont le climat. De plus, les agents stressants liés au climat sont importants puisqu'ils sont reflétés dans les facteurs de stress de nature sociale et économique. Puisque les deux collectivités sont situées dans des régions agricoles, les fluctuations de leur production et du marché ont de sérieux impacts sur elles. Aussi, les changements sociaux liés au dépeuplement et à la centralisation des services sont tout aussi importants. Ces facteurs de stress jouent un rôle quant à la viabilité à long terme de ces collectivités. Le contraste entre les collectivités par rapport aux facteurs auxquels elles sont exposées et leurs réponses ont fourni les données permettant d'étudier les capacités d'adaptation et la vulnérabilité.

Introduction

Global warming is expected to have a significant impact on the water resources of the South Saskatchewan River Basin. Although this impact will affect everyone in the region, it is expected that it will severely affect rural communities and rural households given their dependency on natural resources. Specially affected are communities exposed to the worst impacts of climate conditions, either because of their locations in areas of dryland agriculture or their higher exposure to other stressors, such as globalization, restricted fiscal policies, or depopulation.

A community's vulnerability is a function of numerous characteristics including environmental, social, economic, and political, each linked with the other. Thus, a proper assessment of a community's vulnerability requires it to be grounded in empirical contexts and circumstances that define the daily life of the people in the community. This paper discusses the vulnerability of two dryland agricultural communities—Cabri and Stewart Valley—located in the southwestern area of the province of Saskatchewan, their increased vulnerabilities to the drought of 2001–2002, and the adaptation strategies implemented in both under "status quo" conditions and drought conditions.

The paper is organized into four main sections. The first section describes the methodology utilized to undertake a vulnerability assessment of the communities. The second describes some characteristics of the communities that are pertinent to the study. These are the climatological conditions of the localities, their water resources, economic profile, and employment pattern. The third section focuses on the results of the community vulnerability assessment and discusses the main exposures and adaptive strategies of both communities to climate variability and other stressors. Finally, the last section discusses the severity of the 2001–2002 drought, its impact upon the communities, and the adaptation strategies utilized by the communities to deal with it.

Methods

The assessment was based on the notion that a crucial aspect of vulnerability assessment is to understand the perspective of local people on their own exposure-sensitivities and adaptive capacity. As such, the community assessment used a participatory approach requiring researchers to live in the communities and experience and observe the daily life activities. The community vulnerability assessment of the localities of Cabri and Stewart Valley was carried out between June and November of 2005[2] and it involved three main data collecting activities. First, archival research was carried out in both the Saskatchewan Archives and the local library to compile information about the history of the communities and historical information about water and climate related issues. Second, a number of semi-structured interviews with 30 people in Cabri and 11 in Stewart Valley were conducted with residents in and around the two communities. Respondents included business people, local authorities, senior citizens, and farmers and ranchers living in the vicinity of the two communities. Finally, as a follow-up to the interviews, two focus groups with participants from both communities were conducted in Cabri in November 2005 to discuss issues related to the drought of 2001–2002, and the relationships between the communities and governance organizations. The assessment followed the protocol established by Institutional Adaptation to Climate Change (IACC) project researchers, which included meeting with town council members, engaging with the community, talking with key informants, and ensuring that all interview responses remained confidential and anonymous. All of the interviews and focus group dialogues were recorded and transcribed verbatim, and then coded and analyzed using a qualitative data analysis software package. Qualitative data analysis was then conducted and the findings were interpreted.

The vulnerability assessment of both communities was complemented with a study of the impacts of the 2001–2002 drought on the rural municipalities (RMs) surrounding the two communities, that is, Riverside and Saskatchewan Landing. The method combined a review of previous drought studies,[3] other secondary literature relevant to this study, primary and secondary data-based research (which included an assessment of the historical patterns of the droughts in the region), features of 2001–2002 drought, the associated bio-physical and economic impacts, and data from the communities collected by a combination of focus group meetings and interviews conducted during the vulnerability assessment of the communities.

The Communities

Stewart Valley and Cabri are two rural communities located in southwestern Saskatchewan adjacent to the South Saskatchewan River. Cabri is located in the RM of Riverside, while Stewart Valley is situated in the RM of Saskatchewan Landing. The only other community in the RM of Riverside is Pennant, a town with smaller population than Cabri (which had a population 439 in 2006), while Stewart Valley (population 100 in 2006) is the only community in the RM of Saskatchewan Landing. The two communities are close to each other, approximately 46 km apart, and are near the city of Swift Current, a small urban centre (Figure 1).

Both communities are similar in several respects, such as their low populations and having the agriculture industry as their main economic driver. Most businesses in the region are affected either directly or indirectly by fluctuations in the physical and economic conditions facing agriculture.

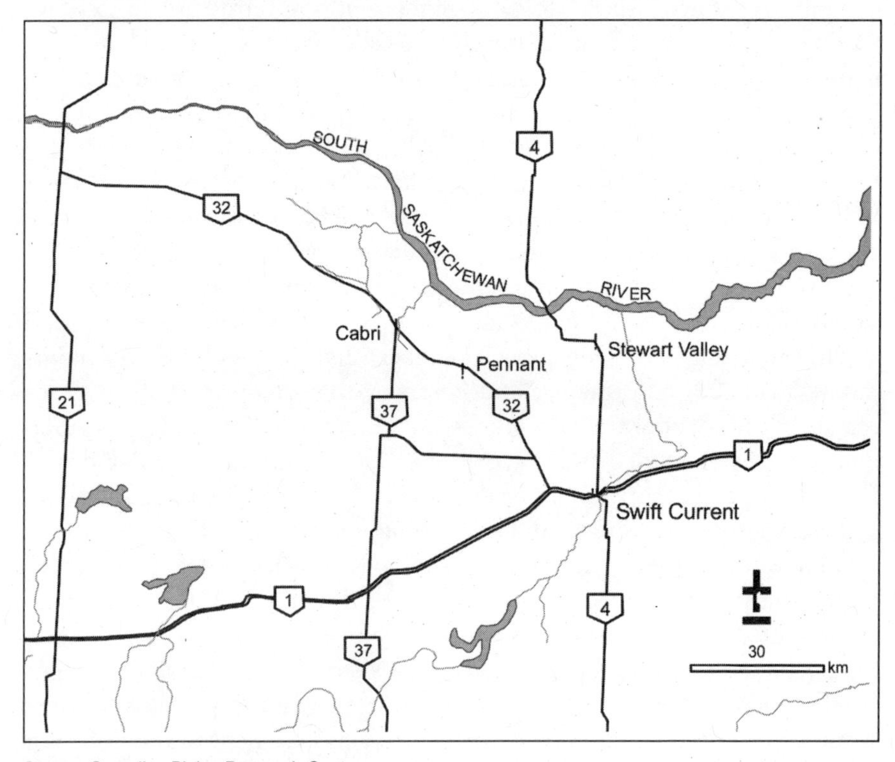

Source: Canadian Plains Research Center

Figure 1. Study Area Map.

The two communities are located in the Palliser Triangle, a semi-arid region in the southwestern portion of the province characterized by a large temperature range, extreme rain and snowfall events, and severe floods and droughts. The coldest month is generally January and the warmest is usually July. While temperature is generally similar in the two locations, precipitation is different: Stewart Valley receives, on average, 94 mm of precipitation annually more than Abbey,[4] the nearest climate station to Cabri.[5] Moisture is a limiting factor for crop production in this area and therefore, because of the connection between the entire community and agriculture, Cabri and Stewart Valley are more exposed to the uncertainties of climate than some of the other Saskatchewan rural communities.

Water resources are important for any agricultural community. The South Saskatchewan River is an important water source for both of the RMs in this paper, providing for irrigation, recreation, and potable water. Both communities are also close to Lake Diefenbaker, an important reservoir for Saskatchewan. The water in this reservoir is used for hydro-electric generation, water level control, irrigation, and recreation. With the exception of a small irrigation project located along the Miry Creek, most of the agricultural production in the study area is under dryland production systems. Groundwater is also used by residents of the study area for several purposes including rural residents' household potable water, water for livestock, and water for crop spraying. Finally, the study area contains a large number of dugouts, which also provide local farmers with water for households, livestock, and crop spraying. Dugouts are generally designed to withstand two years of drought conditions.[6]

Since the town of Cabri is dependent on the river for potable water, particularly in drier than average years, river flow and the level of Lake Diefenbaker are very important. The community's potable water supply is a reservoir northeast of the town, which, in turn, gets its water from Antelope Creek and when required from the South Saskatchewan River via a pipeline. The reservoir water is pumped to a water treatment plant, where it is treated for consumption. Cabri consumes on average (1976 to 2005) 99,664 cubic metres of water per year (m^3/y). In the last ten years the annual consumption has risen to 109,458 m^3/y. As expected, the greatest consumption period is from May to September,[7] translating into an average per capita consumption of 249 m^3/y.

Stewart Valley's potable water source is groundwater from three shallow wells located within town limits and the water is treated for consumption. This community's average consumption is 11,150 m^3/y (1981 to 2005, missing 1996 to 2000), but the consumption decreased to 8,994 m^3/y during 2001 to 2005. The greatest consumption period is from May to August[8] with a rate

YEAR	RURAL MUNICIPALITY OF RIVERSIDE (NO. 168)		RURAL MUNICIPALITY OF SASKATCHEWAN LANDING (NO. 167)		COMBINED POPULATION FOR RM 167 AND 168
	CABRI	TOTAL RM NO. 168	STEWART VALLEY	TOTAL RM NO. 167	
1986	628	677	128	551	1,228
1991	561	619	115	509	1,128
1996	529	552	101	529	1,081
2001	483	495	101	506	1,001
2006	439	511	100	480	991

Table 1. Population Trend in the Study's Rural Municipalities, 1986–2006.

of only 90 m^3/y/person, which is much lower than that for Cabri. The reason for the lower per capita consumption was not determined because of lack of information. This rate is another contrasting factor in the communities' vulnerability to drought.

As in other rural communities in Saskatchewan, the population in the study region is declining. Table 1 shows a decline of almost 19% between 1986 and 2006, a decline related to a large rural out-migration trend that has resulted in an average population decrease in most provincial RMs.[9] Most of the people migrating out of the area are younger,[10] which results in diminishing numbers and an aging local population. If the region's population is an accurate indicator of the economic viability of a community, its declining trend would forecast an uncertain economic future for both communities and the surrounding region.

The impacts of this declining population trend is exacerbated by the existence of a much older population living in both of these communities relative to the province of Saskatchewan (which incidentally has an older population compared to some of the neighbouring provinces). In Cabri, the proportion of people age 65 or higher was 30%, while Stewart Valley's was 25%; both are higher than the Saskatchewan average of 15%. This combination of overall declining population and an increasing elderly population indicates that these communities in the short- to long-term future could face tough economic and social times, thereby making them more vulnerable to many external stressors and reducing their adaptive capacity.

Within the Cabri area, agriculture is the driving industry, supporting primary producers and the majority of local businesses. Most local businesses offer services to agricultural producers or carry out retail trade providing producers with many farm inputs, household supplies and needs such as groceries. Farm service businesses are also present providing either retail goods such as fertilizer, chemicals, livestock supplements, seed, machinery and equipment parts, and tools etc., or services, such as marketing grain, cleaning grain, soil testing, and equipment rental.

The oil and gas industry is the second most valuable industry in Cabri. In recent years, this industry has had greater monetary value than agriculture but has not involved as many businesses, directly or indirectly. Four oil and gas companies plus local mechanic, trucking, construction, and trades businesses support the oil and gas industry. Many local services are required by both oil and gas and agricultural industries, including building construction, land excavation, welding needs, electrical work or automotive repair, to name a few. The oil and gas industry is very valuable to this community because it provides high paying jobs and is less sensitive than agriculture to climate and weather changes. This industry provides off-farm employment to local residents, especially during periods of low farm income.

A third important industry in Cabri is tourism, which is based upon Cabri Regional Park and seasonal game hunters. The Park attracts families from across the province as well as a few Canada-wide travelers, who provide economic spin-offs for local businesses. Game hunters, mainly from the United States, arrive in autumn typically to hunt waterfowl. These hunters are usually in the community for a week and require many services, including accommodations, meals, fuel, supplies, and gifts for their families. Although hunting is seasonal, it still contributes to the economy in a major way.

Two service sectors that are significant to this community's future are the Cabri School and Health Centre. The school has combined elementary and high school programs, and creates employment for professional teachers, teaching assistants, and maintenance personnel, with each position supporting a family in the community. The Health Centre provides emergency and long-term care for local residents as well as citizens from the area around Cabri. The Centre cares for the aging population and reassures people's decisions to continue to reside in Cabri. It employs 45–55 individuals[11] that, in turn, have families that support local businesses. A solid medical presence in the community and surrounding region results from the Health Centre, Cabri Medical Clinic, and the pharmacy, which provide further support of local businesses.

Accommodation services are met by three local businesses that are utilized by many workers in the oil and gas industry, tourists, travelers, and hunters. Most importantly, they enable travelers to stay within the community for a longer period of time, which further supports other local businesses. Restaurant services in Cabri also provide food services in-house and also for catering.

A measure of a community's status may be based on the services it has available, and these services rely on a threshold level of population within the community and would not be available if the community was not sustainable. A few more important services that provide employment are the Royal Canadian Mounted Police, Canada Post, bus service, fire, ambulance, and daycare. Also, several home-based businesses are in areas such as hair dressing, accounting, and bottled water.

Many community, sporting, and cultural organizations, ranging from the Rotary Club to hockey teams are non-profit organizations, but they still add social value to the community since they act as a source of social capital.

In Cabri, major employment areas are in health care and social assistance, accommodation and food services, and the agricultural industry.[12] These three

NDUSTRY GROUP	CABRI				RIVERSIDE NO. 168			
	TOTAL	MALE	FEMALE	PERCENT	TOTAL	MALE	FEMALE	PERCENT
Total Industry	245	120	125	100.0%	260	140	120	100.0%
Agriculture, forestry, fishing and hunting	40	35	0	16.3%	195	120	80	75.0%
Mining and oil and gas extraction	0	0	0	0.0%	15	15	0	5.8%
Wholesale trade	20	15	0	8.2%	10	10	0	3.9%
Retail trade	20	10	10	8.2%	0	0	0	0.0%
Transportation and warehousing	15	15	0	6.1%	10	0	0	3.9%
Information and cultural industries	10	0	10	4.1%	0	0	0	0.0%
Finance and insurance	15	0	10	6.1%	0	0	0	0.0%
Professional, scientific and technical services	10	0	10	4.1%	15	0	15	5.8%
Educational services	20	0	15	8.2%	15	0	15	5.78%
Health care and social assistance	55	0	55	22.4%	10	0	10	3.9%
Accommodation and food services	40	20	25	16.3%	10	0	10	3.9%
Other services (except public administration)	15	10	0	6.1%	0	0	0	0.0%

Note: In order to protect the confidentiality of Canadians, no count of less than 10 is displayed and individual counts have been subjected to random rounding with the exception of total population figures. The total shown in this row would not be an actual sum of the data for individual categories shown below. However, the total employment for these communities is accurate.
Source: Statistics Canada (2001).

Table 2. Employment by Industry in Cabri and Region, 2001.

industries provide for more than half (55%) of the total employment (Table 2). One of the features in the employment pattern is that more female workers are employed in the community because of their work in the health care and social assistance industry. When the employment pattern for the surrounding region is examined, agriculture employs three-quarters of the workers, making the RM predominantly dependent on agricultural activities.

Stewart Valley is primarily an agricultural community with one central business, which is a combination of a grocery store, a farm centre, and post office. This store services the farm sector and local residents and is a vital service for the community. The oil and gas industry is not prominent in this area; therefore, the success of the village may be a reflection of recent developments affecting the agriculture industry. The reason for the lack of services in the village is the vicinity and amenities of the city of Swift Current, which is easily accessible to the residents of Stewart Valley and the surrounding area.

The patterns of employment for Stewart Valley and its surrounding area indicate a clear predominance of the agricultural industry (Table 3); however, there is a presence of workers in industries related to retail trade, and transportation and warehousing. The RM of Saskatchewan Landing is predominantly agricultural, with some people employed in accommodation and food industries, and in administrative services.

INDUSTRY GROUP	STEWART VALLEY				SASKATCHEWAN LANDING NO. 167			
	TOTAL	MALE	FEMALE	PERCENT	TOTAL	MALE	FEMALE	PERCENT
Total Industry	50	30	15	100.0%	165	100	70	100.0%
Agriculture, forestry, fishing and hunting	20	15	10	40.0%	120	90	35	72.7%
Retail trade	15	0	10	30.0%	0	0	0	0.0%
Transportation and warehousing	10	15	0	20.0%	0	0	0	0.0%
Administrative and support, waste management and remediation services	0	0	0	0.0%	10	0	10	6.1%
Accommodation and food services	0	0	0	0.0%	30	0	25	18.2%

Note: In order to protect the confidentiality of Canadians, no count of less than 10 is displayed and individual counts have been subjected to random rounding with the exception of total population figures.
Source: Statistics Canada (2001).

Table 3. Employment by Industry in Stewart Valley and Region, 2001.

These community descriptions demonstrate the dependency of Cabri and Stewart Valley on the condition of the agriculture industry. The sustainability of both communities is affected positively or negatively by the state of the agricultural industry and by the impacts of climate upon that industry. A significant difference is that Cabri, in comparison to Stewart Valley, is somewhat more diversified, having a variety of businesses, and the oil and gas industry, which is less directly affected by climate. This last industry provides much needed employment and is a significant economic driver along with agriculture, which decreases the community's vulnerability and helps local individuals adapt in years when agriculture faces poor economic conditions.[13]

The Community Vulnerability Assessment

The vulnerability of a community is shaped by its exposure-sensitivity to the various economic, social, political, and physical forces (including climate) that affect its daily life, and by its adaptive capacity which is its ability to reduce its exposure to stressors. Both exposure-sensitivity and adaptive capacity are affected not only by climatic and hydrologic conditions but also by internal and external economic, social, cultural, and institutional conditions.[14] Accordingly, the community vulnerability assessments in this study involved the identification of the communities' exposures to these conditions and the adaptive capacities developed in dealing with those conditions, including the role of external institutions in strengthening these capacities. Thus, the assessments sought a comprehensive understanding of the communities' vulnerabilities that went beyond climate and water related problems, and intended to understand how a diversity of stressors—climate, changes in government policies, global markets, and others—affected each other and the viability of the community.

Current Exposures

The economic exposures affecting the communities of Cabri and Stewart Valley mirror those experienced throughout much of rural Saskatchewan.[15] Changes in the farm economy and markets and the absence of employment alternatives which allow for economic fulfillment and sustainability have been identified by most respondents as some of the fundamental stressors that contribute to the increased vulnerability of rural households. The prevalence of a single economic industry, as most respondents made very clear, is problematic since the prevailing market and policy conditions framing agricultural activities reduce the viability of many farms and ranches.

A decline in crop prices and an increase in input costs were identified as specific stressors increasing the economic vulnerability of local agricultural producers. One respondent stated that grain prices were at an all-time low for his farming career. He noted that when he started farming in 1974 the return on durum wheat was roughly $6 a bushel, but was reduced to approximately $4 a bushel by 2005 (in current prices). Respondents also noted that the prices of machinery, fertilizers, and pesticides have increased in a geometrical progression.

Those community members not directly dependent on agriculture for their subsistence are indirectly affected by the adverse economic conditions that affect farms and ranches. Several respondents in Cabri and Stewart Valley noted that the reduced income of farmers is one of the predominant factors that affects their livelihoods. Many of them provide a variety of services to local farmers so their income is highly tied to the uncertainties of agriculture.

Economic upheavals are accompanied by exposure to social processes that impact the sustainability of communities and households. Lack of local economic opportunities has resulted in a declining representation of youth in the communities due in part to the out-migration of youth in search of secure employment and economic sustainability. As noted by several respondents, the process of centralization of services, which has reduced infrastructure and the number of retail and service outlets in rural communities (for example, Stewart Valley), has also contributed to this process of depopulation. They noted that the regional trend is for a concentration of services and business in Swift Current, which is a cause for alarm to local community members, especially those in Stewart Valley. Cabri is a larger centre, farther way from Swift Current; therefore, the preoccupation for the future of local business is less pronounced, but several respondents expressed concern that the local school may be amalgamating with other school boards and perhaps moving to Swift Current in the near future.

This concentration of services in larger centres and the process of out-migration have implications not only for the economic sustainability of the communities, but also for their social capital, one of the determinants of adaptive capacity.[16] As noted by many respondents, the aging population reduces the level of participation in local organizations and recreational activities, and, hence, networks for mutual support slowly disintegrate.[17] If one uses this measure, Stewart Valley was once very prosperous, but now there are no longer enough people or finances to sustain hockey, figure skating, baseball, curling teams, or community events. These were a source of community identity that has faded, and losses such as these have made communities more loosely knit and almost transparent.

Social and economic vulnerabilities coexist and are exacerbated by exposure to climate-related events. The overwhelming majority of the interviewees noted that the recent winters in the two communities have not been nearly as harsh as, and have had much less snow than, previous years. Many respondents noted that snowfall in the past was much greater and regularly drifted as high as the tops of the fences in their yards; one respondent recalls snow as high as the tops of telephone poles. This lack of snow accumulation results in decreased surface runoff, which, in turn, results in water scarcities.

The history of both communities has been intertwined with drought. Most notable for their members were the droughts of 1988, 2001, 2002, and 2003. The drought of 1988 was referred to as the worst of those mentioned. One farmer stated that 1988 was a disaster; he harvested only 400 bushels of wheat from the entire farm compared to the average yield of 40 bushels per acre for that crop type. The drought from 2001 to 2003, although not identified to be quite as devastating as the drought of 1988, still had significant economic impacts on agricultural producers, as will be discussed in the next section.

As indicated earlier, Cabri is dependent on the South Saskatchewan River for potable water. Residents reported water shortage as the predominant water-related exposure during drier years. Many respondents have noticed over the years a decreasing amount of precipitation in the winter months, which affects the snow pack in Alberta's Rocky Mountains, a primary source of water for the South Saskatchewan River. In the mountains the moist air masses from the Pacific Ocean drop a large quantity of orographic precipitation.[18] A lack of orographic precipitation results in the river having low water levels, which impacts Cabri's potable water supply.

As noted earlier, the river water is pumped through a pipeline to a reservoir just beyond the town, and then pumped to the town water treatment plant. This situation caused problems with the water supply of the community in the past (1981 and 2001) because of low water levels in the river. According to respondents, the relative lack of capacity of the water treatment plant and its treatment capacity also caused water shortages in the community. Historically, farmers used water from the water treatment plant to spray crops, sometimes filling up large containers, but the plant was unable to keep up with the quantity of water demanded by both the farmers and town residents. This resulted in farmers no longer being allowed to use town water for agricultural purposes.

Stewart Valley's residents reported no issues concerning water shortages from their wells. Raw water is pumped directly from the wells to a 10,000 gallon cistern in the water treatment plant, where it is treated by continuous chlorination. The fact that the community has a consistent source of water less dependant on surface runoff is the contributing factor that makes the community residents less vulnerable to droughts than the neighbouring community of Cabri.

With respect to temperatures, the single greatest problem noted by respondents was the extreme unpredictable variability of temperatures. For farmers, the extreme variability of temperature and its timing has myriads of possible consequences. For example, a cool spring with late frosts can delay seeding, which may result in lower crop yields. This was the case in the 2005 growing season as one respondent recalled that July was the only month in the entire year without frost. Early fall frost is also devastating to crops because it decreases the quality of crops, which reduces the income potential of the harvest. Both high and low temperatures can result in damaged crops, drying soil, and burning crops, and can allow for an environment for various crop-damaging insects such as flea beetles, saw flies, and grasshoppers to flourish. Moreover, the increasing fluctuation of temperatures and weather patterns is said to cause increased sickness in cattle. Respondents noted that chinooks can cause fluctuations of 30° C in a matter of days or even hours. As chinooks recede and the temperature decreases, "bugs" die off; when chinooks advance and temperature increases, "bugs" resurface, resulting in decreased adaptive capacity of cattle since their immune systems may be compromised, making them sick with pneumonia and they may die.

Other climate-related exposures mentioned by respondents were wind and hail. In dry years, topsoil has blown off summerfallow fields, severely damaging agricultural lands. As well, trees have been blown down, rendering shelterbelts ineffective and leaving fields more vulnerable to wind damage. Hail storms were singled out as causing the worst crop damage. Although the respondents noted no specific instances, hail storms were said to flatten crops, reduce yields significantly, and to be one of the dominant forces that increase household and community vulnerability. Floods and strong rains, while somewhat uncommon in this region, have proven to be an issue in recent years. Several respondents reported that in the last couple of years their basements flooded following a heavy rain event.

Stricter water quality regulations were seen as problematic by some of the respondents. Although stricter regulations promote high-quality safe potable water, they also hold negative connotations in that many communities cannot currently meet the standards. If this occurs, the provincial government would assume control of the community water supplies and likely charge more than if the community were managing the water treatment. Many respondents reported that this is a very likely scenario for both communities since they do not have the appropriate technological or demographic resources to meet the enforced standards.

Several respondents recognized inter- and intra-organizational communication of public agencies as an impediment because they have found it difficult to deal with government organizations, noting that the answers they receive from these organizations are dependant on the government agency and representative providing the response. This makes it difficult for communities to plan accordingly as they have no concrete guidance.

Current Adaptive Strategies

As in any other rural communities, adaptation to the environment, the economy, and other stressors in Cabri and Stewart Valley is a must. Rural people have learned, and need to continue to learn, how to deal not only with climate events but also with other stressors that impact upon their livelihoods.

As noted previously, farming under existing practices in the current economic environment is becoming less viable and is often no longer a feasible means of livelihood. Local farmers have responded by adopting a variety of measures to reduce the vulnerability of their farms and households, a coping behavior that has been documented in rural and development studies.[19] The specific measures adopted and the way in which they are combined vary from household to household, but community networks provide for sharing of experience and information that facilitates the adoption of the best possible strategies by the household. These practices involve mainly access to external institutionalized mechanisms of farm support, changes in farming practices, and/or a diversification of income sources.

Crop and hail insurance was a common coping institutional mechanism used by farmers and ranchers to ensure a degree of financial stability. Many respondents noted that in years when conditions are harsh and crop insurance is required, insurance provided by institutions covered the cost of farm operations only. With droughts and extreme weather occurring more often, hail and crop insurance is imperative to ensure a marginal source of revenue in years where conditions are harsh. However, this coping mechanism was problematic for many respondents because crop and hail insurance coverage levels are decreasing; thus leaving agricultural producers vulnerable. Many participants reported opting out of purchasing crop and hail insurance due to the high premiums, noting that adding crop insurance on top of already high input costs is not viable. They felt the gamble of not purchasing insurance and having poor yields makes them less vulnerable than if they had insurance and did not need it. Nonetheless, producers also recognized that crop insurance covers the cost of production and provides enough for farms to continue to operate the following year.

An important change in farming practices has been crop diversification. With returns diminishing on traditional crops that flourish in the southwestern Saskatchewan environment, such as durum wheat, many local farmers switched to crops with lower yields but higher value, such as canola and pulses. Most farmers have adapted to wind erosion by using continuous cropping, so that wind is less likely to blow topsoil around as plants and reduced tillage operations protect the soil. This protects investments and in effect allows for a more stable source of income. The replacement of discers by air seeders has also become a common way to maintain adequate soil moisture levels, because air seeders leave the soil less disturbed, which preserves the protective cover that holds in soil moisture.

A well-known mechanism to reduce financial vulnerabilities has been to find a secondary source of income. For the two communities, and especially Cabri, the presence of an oil and gas industry in the area has been financially beneficial, helping many households to maintain, and even increase, a secure income. Other respondents find work in the trades or try to obtain employment in larger centres, such as Swift Current. Finding employment for females was reported to be more difficult because of the lack of jobs for women in the area. For example, one of the respondents worked two part-time jobs, while at the same time studying to obtain a youth care worker certificate to obtain employment at the school—all to subsidize low income from the farm.

In terms of climate vulnerabilities, the two communities (especially Cabri) have implemented several adaptive measures to reduce problems related to water scarcity. Dealing with drought has been a learning process that has contributed to the development of these measures; for example, when water scarcity was a serious issue, both communities implemented restrictions to minimize water use. In both communities a water use scheduling policy was put into place by the local government as a way to increase their water capacity. Residents in both communities who required a more substantial supply of water hauled it from wells located in the RMs or brought it from larger centres such as Swift Current. Respondents noted that in times of water shortages, neighbours who had a water truck to haul water would water lawns or gardens of those around the community without the capability of hauling water. Those residing outside the communities, mostly farmers, often filled their cisterns from their dugouts or nearby creeks. In Cabri, as noted earlier, farmers had restrictions on using town water for spraying their crops causing some farmers to obtain water directly from the South Saskatchewan River or other water bodies. This practice created problems for farmers because of poor water quality, resulting in frequent equipment repairs and reduced efficiency of chemical applications used to control weeds and insects. The communities' water conservation measures not only controlled water demand, but also ensured an adequate supply of water in case of a fire emergency.

The majority of respondents in both communities are extremely conscious of water use and abide by the policies implemented by community governance bodies. For pre-emptive measures, some respondents still practice water conservation as if in a drought context; for instance, many respondents noted that they continue to make use of the rinse water from their laundry by recycling it for the garden or toilet, some sweep rather than wash their driveways, and others collect rain water for their gardens.

The water issues that Cabri has experienced, have resulted in the town council, the town foreman, and a group of local water experts creating a water committee. The rationale of the committee is to address any water-related issues or problems the community may encounter, and to attempt to determine solutions in the most appropriate manner.

In both communities, cooperation with other community members—'social capital'[20]—has been instrumental in developing the coping capacities of both households and the communities as a whole. Respondents maintained that rural people are "a tough bunch who stick together." If the community has needs, it will address those needs itself, without government aid and without funding. These two communities use social capital in many ways to decrease vulnerabilities and increase their adaptive capacity and coping range. For example, a few respondents reported that they could recall cases when storms resulted in crops and buildings being damaged, and, that in such cases, the community assembled crews to help deal with the crops, and to shingle and rebuild buildings. Social capital is also an effective tool in dealing with water related problems. Respondents reported that during water shortages there is community pressure to conserve water, and that there is a serious negative social stigma attached to those individuals who do not abide by water scheduling policy; because, by their failing to do so, the entire community may suffer.

Respondents were also asked about the role of external organizations in reducing the conditions of community vulnerability. According to the majority of respondents, several external organizations have played a major role in aiding communities in reducing exposures; perhaps most commonly noted was the role of the Canadian Wheat Board (CWB). Although many respondents consider the returns that the CWB provides to be low, they recognize the stability that CWB presents. Many farmers believe they may be able to find better returns elsewhere, but this is a gamble that most producers cannot afford to take. Saskatchewan Agriculture and Food[21] (SAF) was also cited as an institution that increased the coping range of communities, because in the recent drought years SAF supplied farms with irrigation pipeline and pumps to transfer water from sloughs to dugouts, allowing farmers to water cattle in the winter months. Respondents acknowledged the important role of the Prairie Farm Rehabilitation Administration (PFRA) in helping to secure access to water and in promoting sustainable irrigation practices. Historically, the PFRA constructed a reservoir outside of Cabri for farmers and also implemented a program for building dugouts. At the time of the assessment the PFRA, in cooperation with Cabri, was conducting a study to obtain ideas to secure safe reliable water for irrigation, livestock watering, and domestic and potable use. Specifically, the study focused on watershed management to protect both surface and groundwater sources to reduce water quality deterioration and scarcity during drought years.

Both communities are undoubtedly concerned about their future viability. Their main preoccupations are related to the economic and social processes that negatively affect their communities. Increased farming costs, the decline of prices for crops and livestock, loss of population, and further centralization of basic services are the most serious issues for them. In this context climate change is still a minor issue. The communities are, however, well aware of the negative impacts that future climate events, like droughts, could have on their continued viability.

The 2001–2002 Drought

Antecedent conditions are known to contribute to the severity of subsequent droughts. Abbey, the location of Cabri's closest climate station, began 2000 with above normal precipitation amounts but the year ended with seven months of below average precipitation. The year 2001 was exceptionally dry with 11 months of below average precipitation, and seven of those months were at less than 50% below average. Abbey's temperature was near normal in 2000, but 2001 turned hot with summer mean monthly temperatures as high as 1.9° C above average. The mean maximum temperatures in August and September of 2001 were 4.5 and 3.7°C above average respectively. 2002 started out dry and cold in Abbey. March had above average snowfall, but was more than 11.5° C below normal in average temperature. The cold temperature continued through April and May with below normal precipitation amounts as well. June, August, and September, 2002, had above normal precipitation, but the cool temperatures returned in August. The winter (November, December and January) was generally warm, with lower than normal temperatures returning in February and March. April 2003 was wet and warm, however, the rest of 2003 was quite dry with only September recording above normal precipitation amounts, and except for November, most of 2003 had above normal temperatures.

Stewart Valley was approximately 0.5° C cooler and wetter in 2001–2002 than Abbey. In 2000, Stewart Valley received 156.3 mm more precipitation than Abbey, most in the critical growing season. In 2001, Stewart Valley received about 76 mm more precipitation than Abbey, mostly in May and July. In 2002, July was the anomalous month, with Stewart Valley receiving 60 mm more precipitation than Abbey. In 2003, Stewart Valley received more precipitation in March, May, June, and September than Abbey. The higher precipitation and cooler conditions in the Stewart Valley area likely resulted in higher crop yields and more reliable water supplies than Cabri.

Other conditions also demonstrate the severity of the drought, for example, evapotranspiration, which is the combination of evaporation from soil and transpiration from plants. Evapotranspiration (PET) was greater than 600 mm over a larger part of the prairies in 2001 compared to 2000[22]; on average, in Saskatchewan, 600 mm PET is exceeded in a small area only in the far south.[23]

The Palmer Drought Severity Index (PDSI) is a meteorological drought index and is used to characterize the degree of wet and dry conditions.[24] PDSI values usually range from +4 (or more) indicating extremely wet conditions, to -4 (or less) indicating extreme drought conditions. Environment Canada's PDSI information for the Canadian Prairies shows the evolution of the drought on a province-wide scale. The fall of 2000 showed dry conditions in southeastern Alberta beginning to encroach into southwestern Saskatchewan. This trend continued during the winter of 2000–2001. The spring of 2001 had PDSI values in southwestern Saskatchewan ranging from -3 to -5, indicating severe to extreme drought conditions. By August 2001, a large region of western Saskatchewan had PDSI values of between -5 to -7 indicating extreme drought conditions. These values continued through the fall of 2001 and winter of 2001–2002. The drought weakened only slightly in the spring of 2002 with values of -4 to -5. The summer of 2002 continued to improve in southwestern Saskatchewan with values of -1 to -4, indicating mild to severe drought. By the fall of 2002, the PDSI values were classified as normal to slightly above normal indicating the end of the drought and this trend continued into the winter and spring of 2003.

The Impacts of the Drought on Regional Natural Resources

Water supply issues are undoubtedly critical in drought periods. Cabri's water consumption was the highest in 2001 even though water usage restrictions were in place. This contributed to increasing vulnerability. The water consumption in Cabri during 1976 to 2005 was the largest in 1998, followed closely by 2001. The water consumption records for Cabri indicate a 25% increase in 2001 and a 15% decrease in 2002 based on a 30-year average. Stewart Valley has a dependable water supply and there have been no known problems with water sources or water quality during drought conditions.[25] The water consumption records for Stewart Valley indicate a 17% decrease in 2001 and a 27% decrease in 2002 based on a 20-year average. The reasons for the decreases in water use are not known, but they may be due to a lower population (approximately 100 people) plus possible water conservation measures implemented by the village. It is projected that the populations of the studied communities will continue to decrease; however, if the populations were to increase, the communities may not be able to adapt to more severe droughts (that is, of greater intensity, area and duration), and a limit to current adaptation may be reached.

The drier average conditions of the early 21st century led to lower than average stream flows and reservoir levels. The South Saskatchewan River generally had decreasing mean annual stream flows at both Medicine Hat and Saskatoon over the 1911–2004 recording period with the higher flows in the earlier part of the recording period. At Medicine Hat, the lowest flow year in recorded history was in 2001 and the preceding year was below normal as well. At Medicine Hat, 27 consecutive months recorded below average stream flows on the South Saskatchewan River, extending from March 2000 to May 2002 inclusive. At Saskatoon, 2001 was the third lowest flow year recorded since 1912, despite flows being highly regulated by the Gardiner Dam. The South Saskatchewan River at Saskatoon had 19 months with below average stream flows for the March 2001 to September 2002 period. In 2001, Lake Diefenbaker was 1.7 metres below its average water level—the reservoir's lowest level since 1984. Reservoir levels influence the ability of downstream water users to access the water. For example, the water level was extremely low in 2001 resulting in Cabri extending its water intake line farther into the river. The Miry Creek Irrigation District in the study area had sediment problems in 2001 because of the low water levels.

As described earlier, groundwater is widely used in the study area. Saskatchewan has a network of observation wells used to measure natural fluctuations in groundwater levels. The Tyner and Verlo wells are the groundwater observation wells located closest to the Stewart Valley/Cabri region. The Verlo observation well has had a general decrease in water levels of approximately 2.5 metres since 1972. Most years had a recharge during spring, but the recharge in 2000 to 2002 was minimal. The Tyner observation well level has generally increased by over 0.5 metres since observations began in 1965. This general increase of the Tyner well level plateaued during mid-2000 to mid-2002. The Tyner water levels continued the increasing trend from mid-2002 to a record high level in mid-2005.

Data collected by the Prairie Farm Rehabilitation Administration (PFRA) show that the majority of the dugouts in the region were half full during the spring of 2000. By July 2000, the dugouts in the northern halves of the two RMs were only one quarter full while the dugouts in the southern halves of the two RMs were still half full. During the 2000–2001 winters, the dugouts in the entire area of the two RMs were only one quarter full. By November 2001, only the dugouts just south of the South Saskatchewan River remained one quarter full and the survey indicated dry dugouts in the southern half of the RMs. By May 2002, the dugouts were classified as being one quarter full, and by early June the dugouts just south of the South Saskatchewan River were dry.

Weather conditions have a major role in determining the severity of grasshopper outbreaks. Dry areas—those that receive less than 750 mm of precipitation annually—are most susceptible to infestations. Temperature is also a contributing factor in determining spring grasshopper population numbers (a warm spring results in early hatching and quick development and a cool spring delays the onset of grasshoppers).[26] The grasshopper forecast for 2001 in the Stewart Valley/Cabri area was very light to moderate.[27] To the west of the RMs, the infestation forecast was from moderate to very severe. The forecast for the 2002 season ranged from severe to light, indicating that the grasshopper numbers were a possible problem in the southern portion of the RMs. The unusually cold spring of 2002 resulted in the delay of grasshopper hatching so the grasshoppers attaining adulthood late in the summer laid fewer eggs than under normal climatic conditions.[28] The delay in grasshopper maturation likely decreased the impact of grasshoppers on the 2002 crop.

Dry conditions increase the possibility of wind erosion, resulting in loss of topsoil, plant damage, air pollution, traffic accidents, and clean-up costs.[29] The respondents gave no indication that blowing dust was a factor in the two RMs in 2001. Personal observation indicates severe blowing dust occurred in the Swift Current area.[30]

Socio-Economic Impacts of the Droughts

Social-economic impacts of droughts emanate from various bio-physical changes brought forth by the droughts. The analysis of the impacts of the 2001–2002 drought on the two RMs where the communities are located focused on the agricultural impacts, including crop and livestock enterprises. No estimates of the spill-over effects on other industries were undertaken.

In order to assess the impact of drought on crop yields, a long-term perspective was taken. Yield data for major crops were collected for the 1954–2004 period. For other crops, data were collected for the more recent, but shorter period. The 2001–2002 drought had a net economic loss to the producers in the two RMs, primarily a result of decreased crop yields. There was a loss of $14.3 million in 2001 and another $7.3 million in 2002 relative to the 1998–2000 average net returns. The 2002 drought year was not as severe—however, producers' net returns were still negative when all production inputs were accounted for. The economic situation improved in 2003, but was still grim as it resulted in a loss of $4 million (relative to a base period). Total impact of the drought in the two RMs during 2001 to 2003 was estimated to be $25.8 million. From this economic analysis, simply stated, the crop production sector is very vulnerable to drought. The trend towards diminishing returns, due to the increasing cost of production and bearish markets (1998–2004), requires consistent yields and results in significant vulnerability to drought.

The livestock sector within the RMs of Riverside and Saskatchewan Landing consists primarily of primary production, involving cow-calf and on-farm feeder operations. This sector of the local agriculture industry is supported by forage production on farms. It is reported that 22.98% of Riverside's and 19.89% of Saskatchewan Landing's total farm area was in pasture for grazing.[31] Estimation of drought impact on livestock production concentrated mainly on cattle production because of its high dependence on availability of good quality forage and range conditions, which are both affected by drought.[32]

One of the most immediate decisions producers must make during a drought is whether to sell part of their cattle herd in response to feed and water availability. This is recognized usually by the culling (sale or disposition) of cows, decreased retention of replacement heifers, and selling feeder cattle at lighter weights. Of these actions, the culling of cows and the reduced retention of replacement heifers results in the most negative drought impacts in the short and long term. The drought of 2001–2002 created a need to sell cattle and calves, resulting in a decreasing trend in the number of cattle in 2002–2003, and increased marketing of cows during 2001–2002. In 2003, cattle and calf numbers continued to decrease until May 20, 2003, when an export ban was confirmed after the discovery of Bovine Spongiform Encephalopathy (BSE), resulting in an explosion of total cattle and calves numbers, which reached record highs in 2004. Although cattle numbers were not as affected in the long term, the situation became worse especially if debt was incurred because of the drought years. This is because cattle prices plummeted from the initial drought year in 2001 from $112.91 to $79.35 per hundred weight by 2004. If producers did not recover immediately then they were injected into a fragile economic situation.

The Responses of the Communities

An important finding of this work is that the communities of Stewart Valley and Cabri and their surrounding regions provided a valuable contrast for an examination of adaptation and vulnerability to drought. In spite of being relatively close to each other, the communities had different exposures as Cabri had more severe drought intensities. Also, Stewart Valley has a lower sensitivity as it has a more reliable water supply leading to decreased vulnerability. The community vulnerability assessments provided significant insights into how community members perceived the drought impacts and the adaptation strategies implemented by the different parties who were being impacted.

Many of the respondents in the Cabri region indicated that low water levels and lack of precipitation were an issue in the 2001–2002 drought. Low water levels impacted Cabri and area residents because it affected water available for cattle, irrigation, personal, and town usage. The low precipitation levels reduced crop production with crop quality being variable, resulting in crop producers having a decrease of income directly in relation to yield loss. Producers reported that their input costs, which are relatively beyond their control, have narrowed their profit margin so significantly that there is no room for yield loss. The result is their break-even yield has increased and the impacts of the low 2001–2002 crop yield left them with very low to negative income returns. Farmers offset this by accessing government programs such as crop insurance and the National Income Stabilization Account (NISA) program. However, as mentioned before, many of the farmers found that crop insurance barely covered operating costs. Grain farmers also seeded earlier in the growing season to try to take advantage of the available soil moisture. Producers located along the Cabri pipeline reduced their exposure to drought by having the option to access water to fill their dugouts.

In the Stewart Valley region, the 2001–2002 drought were not considered to be as problematic as in the Cabri region. The Stewart Valley region had average to slightly below average crop yields. Many of the grain producers had crop insurance but were unable to collect because their yields were above the cut-off collection line.

As stated earlier, producers found that grasshoppers were a problem during the drought years. Also, increasing costs of insecticide and spraying in general increased as water was limited in quality and quantity. No effect on spraying operations or re-spraying was noted because of water quality, but such impacts may have been present.

Livestock producers had reduced income because of the increase in cost of management of resources. Feed and water became sparse and therefore the associated procurement costs decreased their income. Producers did not indicate any dispositions as a result of resource shortages; therefore, long term impacts as a result of a decrease in herd stock cannot be assessed. To reduce their exposure to distress, producers hauled water from RM wells for livestock consumption and, where possible, producers irrigated so they could attain a hay crop.

Cabri adapted to the crisis by imposing strict water conservation measures. Residents adapted to water shortages in various ways including hauling water from municipal wells to water trees, installing rain barrels to catch water from eaves, collecting residential grey water and using it on the gardens, and washing their vehicles in a different community. Most residents complied with the water restriction order. The town also attempted to find groundwater but was unsuccessful.

Since these communities are dependent on agriculture, the impacts of the droughts were also felt by local businesses. Producers had less money and therefore spent less. The restaurants experienced less business as community members ate out less frequently and retail outlets experienced a decrease in sales as less was spent on non-essential items. For every business that was affected, they in turn made the same decisions as agricultural producers by cutting their costs and spending less.

The depression in the economic state of agriculture was accompanied by a renewed prosperity in oil and gas. The oil and gas industry helped support most of the community businesses throughout the drought period and also provided an option of secondary employment for producers and other people directly dependent on agriculture.

The tourism industry in the Cabri area proved to be exceptionally vulnerable during the drought because the low water levels and unattractiveness of the landscape caused by the drought resulted in Cabri Regional Park experiencing less business. Also, tourism declined with fewer sportsmen coming to hunt waterfowl as fewer waterfowl were in the area because of low water levels. The retail, food, and accommodation businesses were again affected the most by these conditions.

Several examples of adaptation strategies used to deal with drought affecting Cabri and Stewart Valley and their RMs are in Tables 4 and 5. The source of this information was the discussion notes compiled from the interviews mentioned in the methods section. The strategies are organized by the group undertaking the measure (e.g., business, water utility, livestock producers), the level of success of the adaptation, and the reason for the success or failure. This information is valuable for estimating the ability to adapt to future droughts, as well as the level of success of and barriers to adaptation. A comparison of the Tables also emphasizes the dearth of impacts and adaptations needed for Stewart Valley and area, as compared with Cabri.

A major conclusion is the effects of and responses to drought were more prominent in Cabri and surrounding area than at Stewart Valley because of several factors. Individuals and families simply spent less and people involved in the agriculture or tourism industries and water sector were affected the most. Community businesses achieved less success than in other years, although no receiverships were reported. Most people stated that the economic state of their communities had been complicated by uncontrollable external factors, especially with regard to agriculture. However, the 2001–2002 years are remembered by all as especially difficult as the result of the drought.

TYPE OF RESPONDENTS	ADAPTATIONS		
	SUCCESSFUL	NOT SUCCESSFUL	REASON
The Community	• Used water conservation strategies in households • Hauled water from municipal wells to water trees • Installed rain barrels under eaves • Collect grey water and used on garden • Washed cars in Swift Current • Purchased bottled water • Worked together to conserve water when it was most required	• Examples were not available	• Developed a personal water reserve to assure a water supply during drought • Attempted to prevent further depletion of a limited water supply and manage weather extremes
Local Businesses	• Utilized available water sparingly • Recycled bath water for garden • Purchased bottled water • Waited until the water levels returned to normal before reopening the car wash	• Managed water used by other people staying at the hotel • Increased sales volume	• Conserved limited water resources to maintain businesses
Water Utility	• Pumped water into reservoir from the river • Town foreman took weekly water samples and sent them to the provincial lab • Water users were advised to not put bleach into their water due to the high manganese content	• Drilled test holes outside the community to tap into the "Judith Aquifer" but were not successful in attaining suitable water	• Attempts to alleviate water shortages caused by drought
Tourism Industry	• Water was transported from municipal well to the local pool	• Sustaining the tourism levels was difficult in the drought	• To keep hunters in the area and tourists visiting the park
Livestock Producers	• Transported hay from Northern Alberta • Cattle were relocated to pastures with water •Hauled potable water from town • Irrigation	• Examples were not available	• Attempts to keep their herd healthy
Grain Producers	• Scavenged for water in dugouts and secondary sources such as municipal wells • Used crop insurance (hail insurance (Net Income Stabilization Account, Canadian Agriculture Income Stabilization) • Southwest Conservation Association put on field days to give farmers ideas to conserve water and use crops requiring less water • Farmers along the South Saskatchewan River/Cabri pipeline filled their dugouts with water from the pipeline • Hauled potable water from town • Sprayed for grasshoppers • Continuous cropping appeared to have decreased the potential impact of soil erosion • Seeded earlier to use the advantage of available moisture	• Crop insurance barely covered the cost of production which left them with no income	• Solved water shortage • Crop insurance did not cover all the costs of farming and pay back is decreasing • Net Income Stabilization Account and Canadian Agriculture Income Stabilization takes a long times to get payment
Mixed Farms (same as livestock and grain)	• Irrigated less but at optimal times • Hauled potable water from town • Communicated the issues of concern within the household • Reduced fertilizer input costs	• Control of the grasshoppers	• Utilized water resource available such as municipal wells • Prevented further crop and forage loss • To try to sustain the farm business economically
Local Governments	• Implemented water conservation measures • Restricted lawn watering • Farmers were not allowed to use town water for spraying—only household use • Town rigged a pump to extend into the South Saskatchewan River to access more water to fill town reservoi • More policing of water usage • Shut down car wash • Sent out informational packages on water conservation strategies	• Tried to find groundwater but was unsuccessful (Even if groundwater was found the water treatment plant is set up for surface water. Costs for conversion are great.)	Lessons learned: • Need to manage water better • When the river is full and storage is available in the town reservoir, water should be stored

Source: Adapted from Wittrock et al. 2006.

Table 4. Drought Adaptation Strategies for Cabri and Area.

TYPE OF RESPONDENTS	ADAPTATIONS		
	SUCCESSFUL	NOT SUCCESSFUL	REASON
Local Businesses	• Crop Insurance • Hail Insurance	• Examples were not available	• Examples were not available
Grain Producers	• Crop Insurance • Hail Insurance • Zero till helps manage moisture in fields	• Couldn't collect crop insurance because yield was too high	• Examples were not available

Source: Adapted from Wittrock et al. 2006.

Table 5. Drought Adaptation Strategies for Stewart Valley and Area.

Conclusion

The vulnerability of a community is shaped by its exposure-sensitivity to various physical, economic, social, and political stressors, and by its adaptive capacity. The communities of Cabri and Stewart Valley are affected by a large number of social and economic stressors. These stressors are further exacerbated by extreme climatic events such as the drought of 2001–2002. Changes in agriculture at the level of production, and additionally in the markets, negatively impacted the communities. Other stressors were the water supply and water quality issues, as well as the increased water consumption documented in the region. No less critical were the social issues of an aging population, depopulation, and the centralization of services. All of these stressors threaten the long-term sustainability of the communities and local livelihoods.

As in many rural communities, adaptation to adversity in Cabri and Stewart Valley is a necessity. Rural people have learned how to deal with a multitude of stressors that impact their lives. Agriculture has become less sustainable because of economic, environmental, and political issues among others, resulting in farmers having to respond by adopting a variety of measures suited to their individual needs. The communities have also adapted to their own stressors, such as water scarcity, by implementing various coping strategies. The contrasting exposures and responses of the communities provided a valuable insight into adaptation and vulnerability.

The future viability of these communities is a concern, particularly in the context of the economic and social processes that negatively affect them. The communities are generally aware of climate change, but their current adaptation strategies are largely reactive. In the perspective of the expected impacts of global warming for the region, which indicate an increasing rate of recurrence and intensity of drought, there is a clear need to strengthen the proactive adaptive capacities of these communities. Climate change could also, potentially, create new opportunities. Expanding our knowledge and fostering community resiliency beyond *ad hoc* approaches regarding climate change impacts and adaptation is essential for effective management of both threats and opportunities. This could be achieved by planning and acting across traditional sectors, issues, and political boundaries, and integrating ecosystem-management, disaster reduction, and social and economic development measures.[33]

Endnotes

1. This article is a product of the Institutional Adaptation to Climate Change Project. Funding for this project was provided by the Social Sciences and Humanities Research Council of Canada's (SSHRC) Major Collaborative Research Initiatives (MCRI).
2. A graduate student of the University of Regina, Stephanie Jeanes, lived in Cabri from June to August of 2005. During this period of time, and with the support of another graduate student, Erin Knuttila, she contacted and interviewed people in the two communities and participated in their activities. Without their support, hard work, and dedication this article would not have been possible.
3. For example, E. Wheaton, S. Kulshreshtha, V. Wittrock, and G. Koshida, "Dry Times: Lessons from the Canadian Drought of 2001 and 2002," *The Canadian Geographer*, Vol. 52, No, 2 (2008): 241-62.
4. Abbey is approximately 25 km west of Cabri.
5. Environment Canada 2006a, Canadian Climate Data [on line]. Available at: http://www.climate.weatheroffice.ec.gc.ca/advanceSearch/searchHistoricDataStations_e.html (accessed January 2006).
6. G. Bell, personal communication, 2002. Mr. Bell is a hydrologist with Agriculture and Agri-Food Canada, Prairie Farm Rehabilitation Administration, Saskatoon, Saskatchewan, in V. Wittrock, "Canadian Droughts of 2001 and 2002: Water Resource Conditions in Western Canada: Impacts and Adaptation," in E. Wheaton, S. Kulshreshtha and V. Wittrock, (eds.), *Climatology, Impacts and Adaptations, Volume II* (Saskatoon: Saskatchewan Research Council, 2005).
7. D.F. Anderson, personal communication, 2006. Mr. Anderson is a Senior Technologist with the Regulatory Coordination, Operations Division of the Saskatchewan Watershed Authority, Moose Jaw, Saskatchewan.
8. Anderson, Saskatchewan Watershed Authority.
9. Statistics Canada. *Statistical Profile: Saskatchewan Community Profiles*, 2001. Available at: http://www12.statcan.ca/english/profil01/CP01/Index.cfm?Lang=E (accessed June 2007).

10. Recent discussions with the community members have suggested that since 2006, some young people are starting to return. However, exact data on this in-migration are not available.
11. Statistics Canada, *Census of Agriculture,* 2001. Available at: http://library.usask.ca/data (accessed March 2006).
12. Employment data were obtained from Statistics Canada (special tabulations). However, the data released for public use is not always accurate. To preserve confidentiality, some of the employment data categories are rounded up to a higher round number (such as a 10 or a 20). On account of this rounding, the sum of reported data does not add up to the total number of workers in the region (or community).
13. These poor economic conditions generally result in low farm income, and could be a result of low market prices for products, poor weather, the outbreak of plant and animal diseases, among other factors.
14. See, for example, W. Adger, "Social Aspects of Adaptive Capacity" and B. Smit and O. Pilifosova, "From Adaptation to Adaptive Capacity and Vulnerability Reduction." These two books chapters are found in J. Smith, R. Klein, and S. Huq (eds.), *Climate Change, Adaptive Capacity and Development* (London: Imperial College Press, 2003), 29–49 and 9–28; see also B. Smit and J. Wandel, "Adaptation, Adaptive Capacity and Vulnerability," *Global Environmental Change* 16 (2006): 282–292.
15. M. Jones and M. Schmeiser, "Community Economic Viability in Rural Saskatchewan: The Culture of Survival," *Prairie Forum* Vol. 29, No. 2 (2004): 281–300.
16. See Adger, "Social Aspects of Adaptive Capacity," 24; IPCC (Intergovernmental Panel on Climate Change), *Climate Change 2001: Impacts, Adaptation, and Vulnerability Technical Summary.* A Report of Working Group II of the Intergovernmental Panel on Climate Change (New York: WMO and UNEP, 2001), 73
17. Similar findings have been documented in other studies, See H. Diaz and M. Nelson, "Social Capital and Adaptation to Climate Change," *Prairie Forum* Vol. 30, No. 2 (2005): 289–312
18. Masaki Hayashi and Garth van der Kamp, "Snowmelt Contribution to Groundwater Recharge," in D. Sauchyn, M. Khandekar, and E.R. Garnett (eds.), *The Science, Impacts and Monitoring of Drought in Western Canada: Proceedings of the 2004 Prairie Drought Workshop* (Regina: Canadian Plains Research Centre, 2005), 21–26.
19. See, as an example, M. Redclift, "Survival Strategies in Rural Europe. Continuity and Change. Introduction," *Sociologia Ruralis* 26, 3/4 (1986).
20. Adger, "Social Aspects of Adaptive Capacity."
21. In December 2007, the Department was renamed Ministry of Agriculture.
22. E. Wheaton and V. Wittrock, "Canadian Droughts of 2001 and 2002: Climatological Description of the Droughts in Western Canada," in E. Wheaton, S. Kulshreshtha, and V. Wittrock (eds.), *Climatology.*

23. O. Lundqvist, "Agroclimate," in K. Fung (Director and Editor), *Atlas of Saskatchewan,* (Saskatoon: University of Saskatchewan, 1999).
24. W.C. Palmer, "Meteorological Drought," Research Paper 45. (Washington, District of Columbia: United States Department of Commerce, 1965).
25. G. Hagen, personal communication, 2006. G. Hagen is a Senior Technologist of Regional Operation, Saskatchewan Watershed Authority, Swift Current, Saskatchewan.
26. J. Calpasand and D. Johnson, *Grasshopper Management. Agri-Facts: Practical Information for Alberta's Agriculture Industry* (Alberta Agriculture, Food and Rural Development and Agriculture and Agri-Food Canada, 2003). Available at: http://www1.agric.gov.ab.ca/$department/deptdocs.nsf/all/agdex6463/$file/622-27.pdf?OpenElement (accessed January 2006).
27. D. Johnson, personal communication, 2006. D. Johnson is a Professor of Environmental Science, Canada Research Chair in Sustainable Grassland Ecosystems, Department of Geography, University of Lethbridge, Lethbridge, Alberta.
28. D. Johnson and J. Calpas, "2003 Grasshopper Forecast for Alberta and adjoining regions of Saskatchewan." Available at: http://www.sidney.ars.usda.gov/grasshopper/Extras/cafrst03.htm (accessed January 2006).
29. E. Wheaton, "Prairie Dust Storms—A Neglected Hazard," *Natural Hazards* 5 (1992): 53–63.
30. Bell, personal communication, 2002; Data received from L. Campbell, Operations and Planning Engineer, Department of Highways and Transportation, Saskatoon, Saskatchewan; E. Wheaton, "Canadian Droughts of 2001 and 2002: Wind Erosion in Western Canada: Impacts and Adaptations," in E. Wheaton, S. Kulshreshtha and V. Wittrock (eds.), *Climatology.*
31. V. Wittrock, D. Dery, S. Kulshreshtha, and E. Wheaton, *Vulnerability of Prairie Communities' Water Supply During the 2001 and 2002 Droughts: A Case Study of Cabri and Stewart Valley, Saskatchewan* (Saskatoon: Saskatchewan Research Council, 2006).
32. Data availability is a major constraint to estimation of exact drought impacts at the Rural Municipality level. This is because yearly inventories of cattle are not available. No farm surveys were exercised on account of resource availability. As a crude substitute, data for the Census Agricultural Region Level (CAR) 3BN were used.
33. See International Institute for Sustainable Development (IISD), *Designing Policies in a World of Uncertainty, Change, and Surprise* (Winnipeg: IISD, 2006)

THE BLOOD TRIBE: ADAPTING TO CLIMATE CHANGE

LORENZO MAGZUL

Abstract

A community vulnerability assessment to water related impacts of climate change was conducted in the Blood Tribe in the South Saskatchewan River Basin (SSRB). The vulnerability of the Blood Tribe to climate change impacts such as floods is compounded by pressing socio-economic concerns such as health, housing shortages, and unemployment. This array of socio-economic challenges limits their capacity to adapt to present and future climate change impacts. For communities like the Blood Tribe, addressing issues such as housing shortages and unemployment can also serve as measures that minimize their vulnerability to climate change impacts.

Sommaire

Une évaluation de la vulnérabilité de la collectivité, par rapport aux impacts liés à l'eau qui sont dus aux changements climatiques, fut effectuée chez la tribu des Blood dans le bassin de la rivière Saskatchewan-Sud (BRSS). À cette vulnérabilité de la tribu des Blood aux impacts liés aux changements climatiques, tels que les inondations, se superposent des préoccupations pressantes de nature socio-économique telles que : la santé, la pénurie de logements et le chômage. Cet éventail de défis socio-économiques auquel fait face la tribu réduit sa capacité à s'adapter aux impacts présents et futurs liés aux changements climatiques. Pour les communautés telles que la tribu des Blood, s'attaquer aux questions telles que la pénurie de logements et le chômage peuvent aussi être des mesures qui contribuent à minimiser leur vulnérabilité à ces impacts.

> One of the main issues in the community is the lack of employment; we are basically living in *third world* conditions here . . . There is only a certain number of dollars that come into the reserve on a yearly basis . . . I don't know what the current financial transfer agreement is with the federal government . . . You see, our treaty did not have a cap on it. We had under the treaty, under the true spirit of the treaty, that the Queen, in exchange for sharing some of the land—we did not surrender our land—she was going to share the land with us; because of that she was going to care for us as long as the sun shone, the grass grew and the rivers flowed. That meant medication [health care], education, basic living essentials (housing, water, whatever); but that has since stopped, the federal government has put a cap on it, a cap on the services.[1]

Introduction

The Blood Tribe, a Blackfoot community in the South Saskatchewan River Basin (SSRB), is one of Canada's largest reserves and has significant natural assets that include agricultural land, oil, gas, minerals, and water. However, like other First Nations communities in Canada, most of the people of the Blood Tribe live in Third World conditions. Unemployment figures in 2005—according to some members of the community—was as high as 80%. Housing conditions are poor and overcrowded, alcoholism and drug addiction are rampant, and the prospects for economic development opportunities are almost non-existent. Over the past decade, impacts of floods and droughts have had adverse consequences on the community's farming operations, homes, roads, and drinking water supplies.

Climate change and variability in the form of droughts and floods has a long history in what is now the SSRB.[2] There is evidence however that prior to the arrival of Europeans indigenous people in the SSRB area, including ancestors of the Blackfoot people, had developed successful adaptation strategies that allowed them to cope with the effects of climate variability. For example, in times of droughts, indigenous people were able to hunt buffalo through traditional pedestrian techniques because they were able to predict and to a certain extent control the movement of buffalo herds. They predicted the movement of buffalo herds because they learned the location of water sources that would not dry up during drought spells and hunted the buffalo that sought out the green grass and drinking water available from these water sources. They also recognized the importance of ponds and reservoirs created by beavers in times of droughts. Water trapped in these ponds and reservoirs provided a water supply for the buffalo herds. By not exploiting the beaver population they were able to control and manage some of the water sources and hence predict the movement of the buffalo herds.[3]

With the arrival of Europeans in the 18th century, the type and means of livelihood for indigenous people underwent profound changes. With the use of horses and guns to hunt buffalo, indigenous people no longer needed to know where the water sources were located nor did they need to maintain beaver populations and thus preserve water sources in order to predict the movement of buffalo. Therefore, as the type and means of their livelihood changed, their vulnerability to climatic variability such as droughts increased.[4]

For indigenous people in the SSRB the incremental loss of their traditional livelihood and a decreased dependence on their natural resources for their survival is today almost complete. At present the well-being of most of the Blood Tribe people is directly related to the annual transfer of funds by the federal government—funds promised upon the signing of Treaty 7 in 1877[5]—for housing, health, education, social welfare, economic development and various other needs,[6] rather than on the health of their environment and an in-depth knowledge of their surroundings.

The disruption of the Blood Tribe people's traditional livelihood, combined with increased dependence on federal government funds and the impacts of government policies such as the Residential Schools, have generated high levels of poverty, unemployment, poor housing and other hardships, resulting in great social problems including violence, addictions, and poor health. These social problems and issues are today the primary concerns of the Blood Tribe people because they impact the great majority of them on a daily basis. Although floods and droughts have impacted the community, only some of the people have been directly affected. Those affected would have been people whose homes were flooded, or who had a road they rely on washed out, or who had crops they grew lost due to water shortages or flooding. Consequently, for the Blood Tribe finding solutions to poverty, unemployment, housing, and so on are of greater priority than adaptations to climate change.

In light of the social, economic, and political context of the Blood Tribe community, this paper discusses some of the key vulnerabilities to climate change impacts, as well as some of the challenges confronted in developing adaptation strategies to those impacts. In this sense, highlighted is the importance of considering policies such as the Indian Act and the Residential Schools, which have produced institutional arrangements, reshaped people's relationships with their environment and with each other, all of which facilitates or constrains the adaptive capacities of the community to climate change impacts.

The data collection for this study involved community-level fieldwork through in-depth personal interviews with key informants, focus groups, and

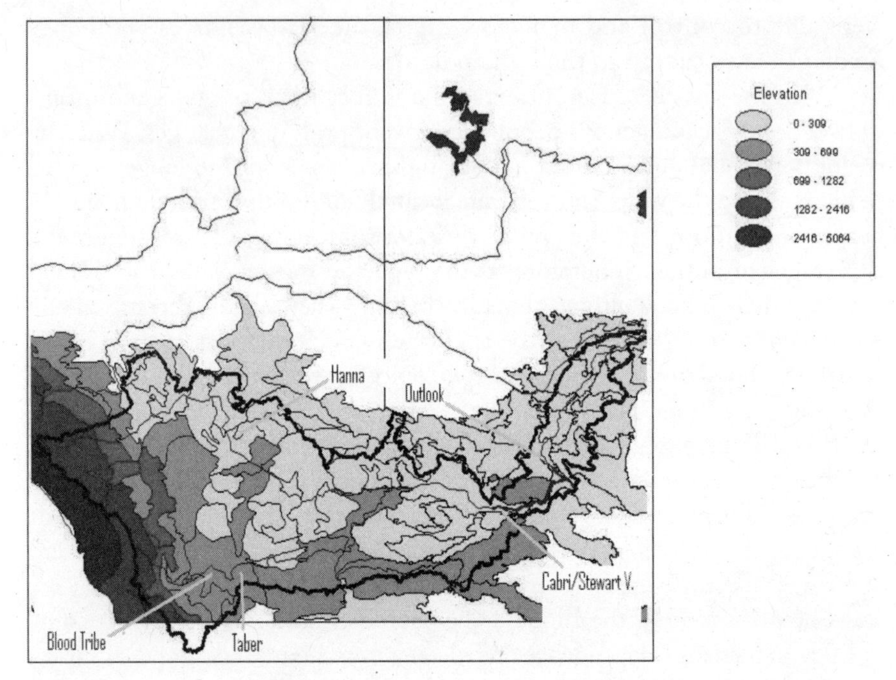

Figure 1. The Blood Tribe and other selected rural communities in the South Saskatchewan River Basin.

participant observation. Community members interviewed included male and female, youth, elderly, business people, and Band employees. Due to the sensitive nature of some of the information or opinions provided by the respondents, the anonymity of each respondent is maintained by citing them by an assigned code (e.g., BT1 for Blood Tribe respondent number 1, BT2 for Blood Tribe respondent number 2, etc.). All interviews were transcribed with the results coded and interpreted using a NVivo software package.

Blood Tribe: Background

The Blood Tribe reserve, located in southern Alberta (Figure 1), is one of the largest in Canada. A member of the Blackfoot Confederacy and the Treaty 7 Tribal Council, the Blood Tribe people lived in this region for hundreds of years before European contact. The Blood Tribe territory consists of two parcels of land: Reserve No. 148 and Timber Limits No. 148A.[7] Reserve No. 148 is bounded by three rivers: the Belly River to the west, the St. Mary River to the east, and the Oldman River to the north. Reserve No. 148 consists of 351,961 acres and it is the land base where the various communities that make up the Blood Tribe are scattered. These communities are Standoff, Moses Lake, Levern, Old Agency, Fish Creek, Fort Whoop-Up, and Bullhorn.

Timber Limits No. 148A, consisting of 4,795 acres, is located in the foothills of the Rocky Mountains and has been allocated for public use including recreation and hunting and gathering. These two parcels of land make up 356,756 acres, which is the total territorial domain of the Blood Tribe people.[8]

The economic base for the Blood Tribe community is agriculture. Of the total land base, 180,140.8 acres are cultivated and 55,601.4 acres are grassland, all of which is leased on three-year leases to about 61 non-native farmers. Title to the Blood Tribe reserve lands is held by Her Majesty in right of Canada and set apart for the use and benefit of the community members. However, within the Blood Tribe about 80% of the land is privately held through occupancy rights by about 10–12% of the people; the remaining 20% of the land is held communally and under the control of the Band Council. Individuals with occupancy rights to the land—although they do not have fee simple title to the land—can transfer, exchange or sell their lands to other members of the community but not to an outsider. Most to the land held through occupancy rights is leased out to non-aboriginal farmers. Proceeds from the leasing of the land benefits only the 10–12% with occupancy rights, and this has generated frictions between those who own land and those who do not.[9] At the same time, community members without occupancy rights cannot own a home because they do not have the land to build it on, even if they had the means to own a home. Therefore, most of the people without occupancy rights move to townsites such as Standoff, Moses Lake, and Levern, where they live in rental housing managed by the Band Council. Others move out of the reserve to nearby communities such as Cardston, Lethbridge, and Fort McLeod.

Collective lands under the control of the Band Council are used by community agricultural corporations such as the Kanai Agribusiness Corporation (KAC) and the Blood Tribe Agricultural Project (BTAP). These corporations function at arms length from Chief and Council and cultivate 68,191.8 acres. BTAP operates one of the largest irrigation projects in Canada, consisting of 20,000 acres cultivated with timothy hay destined for Japan; the operation also has its own hay processing plant. About another 1,385 acres are also public lands, and are set aside for recreational, residential, commercial, and industrial development.[10]

According to Statistics Canada,[11] in 2001 the population of the Blood Tribe living on reserve was of 3,850 people; however, Blood Tribe administration employees interviewed for this study estimated the 2005 population to have been about 9,000 to 10,000 people, with 6,000 to 7,000 people living on reserve. One possible explanation for the discrepancy between Statistics Canada's figures and the estimates by the Blood Tribe administration—according to the Blood Tribe administration employees—is that there is a

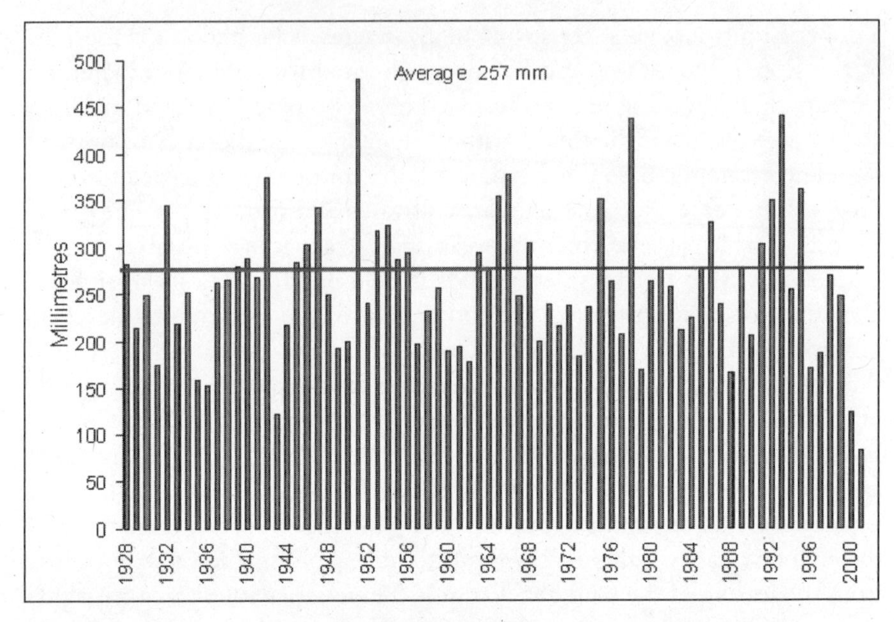

Source: Alberta Agriculture, Food and Rural Development, 2002.

Figure 2. Average growing season precipitation (April 15-October 15) of 257 mm, obtained from Bow Island, Lethbridge, Medicine Hat and Taber precipitation data (Environment Canada's data, 2001).

constant outflow of the population to nearby towns in search of employment or housing, and that many people return to the community when they fail to find jobs or accommodation.

With respect to environmental conditions, the southern region of Alberta, where the Blood Tribe community is located, has historically been affected by periodic floods and droughts.[12] As illustrated in Figure 2, since 1928 there were severe droughts in 1936 and 1943, but the worst drought occurred in 2001.

In the 2001 drought, only 83 mm of rain fell for the growing season (April 15 to October 15), compared to the historical average of 257 mm. To make matters worse, for the previous growing season only 123 mm of rain fell, which was also well below the historical average.[13] Even though the southern Alberta region has water reservoirs, the two consecutive drought years of 2000 and 2001 had severe impacts on the irrigation industry. Figure 2 also shows that there have been years when precipitation was well above the historical average of 257 mm. The Blood Tribe was impacted by floods in 1995, 2002, and 2005.

With regard to temperature, trends show that in 1990–98 the mean annual air temperature increased by an average of 0.9° c over southern

Canada, especially in western Canada, and the greatest warming occurred during the winter and spring.[14] A 2002 study on behalf of Alberta Environment concluded that one of the most dramatic impacts of climate change in the prairies has been the increase in mean temperature over the last 50 years, primarily due to the increase in the minimum temperature in the winter season.[15] For Alberta, the temperature in the winter has increased by 1° C to 3° C, but it has cooled slightly in the summertime.[16]

In the Prairie Provinces the agricultural community has already experienced substantial economic losses due to moisture constraints; moreover, climate projections forecast greater moisture limitations by 2040–2069. Although precipitation is expected to increase, moisture availability will decrease due to warmer temperatures and higher rates of evapotranspiration.[17]

Do First Nations in Canada live in Third World conditions?

The difficult living conditions of First Nations people in Canada is often compared to the living conditions of the poor in Third World countries. This comparison is not an exaggeration because First Nations communities in Canada and some poor people in Third World countries share a similar colonial European conquered-conqueror relation that has resulted in dispossession, marginalization, discrimination, and assimilation. This relation and its consequences persist, and is perhaps somewhat renewed, due to the increasing economic, social, and political restructuring of societies worldwide due to globalization. The legacies of colonial policies and their persistent modern forms represent probably the greatest challenges to the adaptive capacities of indigenous peoples, like the Blood Tribe, and the poor in Third World countries, to impacts of climate change.

According to the Assembly of First Nations, based on the United Nation's Human Development Index, in 1998 First Nations in Canada ranked 63rd in the world in terms of quality of life indicators.[18] In the case of the Blood Tribe, according to the Community Well-Being Index (CWI)[19] developed by Indian and Northern Affairs Canada's Research and Analysis Directorate, in 2001 the community had a score of 61 out of 100, compared to 84, the average for non-First Nations communities in Alberta.

With respect to education, the level of education in the Blood Tribe population is much lower relative to the province of Alberta. For example, in 2001, 40% of the Blood Tribe population aged 20–34 had less than a high school graduation certificate, against 18% of the same age group for Alberta. More recent figures by Statistics Canada (2008) show that only 8% of the Blood Tribe's labour force has a university degree, compared to 21% for the province of Alberta.

With respect to income, according to figures by Statistics Canada, in 2001 the median family income for the Blood Tribe was $20,768, compared to the median family income for Alberta which was $60,142.[20] Statistics Canada's unemployment rate for the Blood Tribe for 1996 and 2001, are 29% and 45% respectively.[21] Statistics Canada does not have more recent unemployment rates for the community; however, several of the Blood Tribe people interviewed in 2005–2006 estimated the rate of unemployment to range between 60% and 80%.

According to the Blood Tribe Housing Department,[22] there is a considerable shortage of houses to satisfy the community's housing demand. Due to the shortage of houses it is common to find 2–3 families living in a single home. This shortage of houses forces families to move to the nearby towns as the Housing Department cannot build homes fast enough to fulfill the demand.[23] In March 2006 there were 439 applicants on the Housing Department waiting list, while the average number of homes built annually has only been 20.[24]

Of 915 adults (15 years old and above) surveyed by Statistics Canada in 2001, only 60 reported working in agriculture and other resource-based industries, and, ironically, despite the assertions by the Blood Tribe Lands Department[25] that the economic base for the Blood Tribe community is agriculture, none reported agriculture as an occupation.

These key socio-economic characteristics indicate that Blood Tribe people have low educational levels, a high unemployment rate, lower levels of income, a shortage of housing, and only a few of the people depend on the land for their livelihood. These characteristics constitute serious challenges and constraints for the Blood Tribe in adapting to climate change impacts.

The Poor: Vulnerabilities and Adaptive Capacities

Increasingly, there is an acknowledgment that impacts of climate change will be greater for the poor because they are more frequently exposed to climate related risks. Poor people are more exposed to these risks because they live in marginal lands and poor housing conditions, and have a higher sensitivity to those impacts due to limited economic, human, and institutional capacity to respond to the impacts.[26] It is also argued that for the poor the impacts of climate change will exacerbate existing vulnerabilities such as limited access to drinking water, poor health, and food security. For example, in developing countries the recent years of droughts, floods and storms have had disastrous consequences on agricultural resources and therefore on the food security of the poor, thus rolling back decades of food security progress from development efforts.[27]

The vulnerability of the poor to climate change impacts is best understood by taking into account other stresses to which they are frequently exposed:

> Vulnerability is both a condition and a determinant of poverty, and refers to the (in)ability of people to avoid, cope with or recover from the harmful impacts of factors that disrupt their lives and that are beyond their immediate control. This includes the impacts of shocks (sudden changes such as natural hazards, war or collapsing market prices) and trends (for example, gradual environmental degradation, oppressive political systems or deteriorating terms of trade).[28]

Similarly, it is generally acknowledged that the capacity to adapt to climate change impacts is uneven within and between societies because some individuals and groups have less capacity to adapt than others. The level of adaptive capacity of these individuals or groups is positively influenced by their "economic and natural resources, social networks, entitlements, institutions and governance, human resources, and technology."[29] On the other hand, the adaptive capacity of individuals and groups is negatively affected by "technological, financial, cognitive and behavioural, and social and cultural constraints."[30]

Given that populations that are most vulnerable to climate change impacts are those that experience other stresses such as poverty, unequal access to resources, food insecurity, environmental degradation and risks from natural hazards, it is logical that by addressing these other stresses the vulnerabilities to climate change impacts can also reduced.[31] Therefore, some of the mechanisms that can be used to enhance the coping ability of vulnerable populations to climate change include increased access to information and technology, equity in the distribution of resources, and enhanced human and social capital.[32]

Unfortunately, as observed in many instances worldwide, development and economic activities continue to exacerbate climate-related vulnerabilities of the poor rather than reduce them. These activities exacerbate the vulnerabilities of the poor because they fail to address the non-climate stresses such as poverty, food insecurity, and so on. In fact, despite the seemingly obvious benefits of promoting sustainable development as an adaptation strategy to climate change impacts, only a handful of efforts explicitly recognize and acknowledge this fact.[33] For the poor, the promotion of development that addresses existing non-climate related vulnerabilities should be viewed as a key mechanism for reducing climate change impacts.[34]

Blood Tribe Exposures to Climate Change

Based on the interviews conducted, there are recent examples of significant climate change impacts that the Blood Tribe people have experienced. Some of these experiences can be discerned from comments such as "kids do not go sleighing anymore because of the lack of snow; some birds no longer migrate; there are more grass fires during the early spring; there are more mosquitoes in the summer; and, sometimes there are not enough units of summer heat for growing a second crop."

However, extreme climate related events such as floods have had the most impacts on the Blood Tribe community. Over the last ten years the Blood Tribe experienced severe floods in 1995, 2002 and 2005, and according to the Housing Department the cost for the 2002 and 2005 floods were $8 and $6.5 million, respectively. In 2005, out of the total of 1,280 homes in the community 397 were affected through flooded basements and/or water damage because of roof and wall leaks or poor drainage.[35]

Due to the 2005 floods a new well had to be dug for one of the communities because flood waters damaged the existing well through siltation. The cost for the new well was $800,000. Flood events have also added extra costs to the maintenance of roads because these were often damaged by washouts.[36] Also, for the 2002 floods, for two months Kanai Resources, which manages the Blood Tribe's oil and gas resources, closed down oil wells located near the river bottom. The two month closure of the oil wells caused a loss of $20,000.[37]

Recent droughts have also affected the operations of Kainai Agribusiness Corporation (KABC) and the Blood Tribe Agricultural Project (BTAP). The 2001 drought, which affected most of the Canadian prairies, caused the loss of feed crops for the 700 heads of cattle that KABC had at the time. As a result, KABC reduced the size of the herd from 700 to 200.[38] Water shortages during the 2001 drought also affected the operations of the BTAP because, like other farmers in southern Alberta that depend on the St. Mary's river, it received 125 mm less water than required for the 2001 growing season.[39]

Given the history of periodic floods and droughts in the area, the likelihood of future droughts and floods affecting the Blood Tribe is fairly high. If floods were to increase in frequency, the Blood Tribe Fire and Emergency Services foresees that in order to be better prepared they require more equipment: a small all-terrain vehicle that could be used to access more remote areas, and a boat in order to gain access to flooded homes in the low lying areas.[40] With regard to long term planning, there is a need for moving about 500 homes away from the flood plains that are at risk of future flooding.[41]

According to the Lands Department, to date there have been no cases whereby non-native farmers have cancelled land leases because of continuous years of drought. However, if in the future the intensity and length of the droughts increase there is the potential risk that non-native farmers may stop leasing the land from members of the Blood Tribe.[42] The loss of the income from land leases will add further stress on Blood Tribe members who depend on that income.

Blood Tribe Adaptive Capacity

The Blood Tribe is governed by a Head Chief and a Council of 12 people who are elected by the community for a four-year term of office. Over the past two decades Indian and Northern Affairs Canada has transferred decision making powers to the Blood Tribe government over matters such as education, health, policing, and social services. The belief is that as the Blood Tribe community increasingly regains control over managing its own affairs, there will be an increase in opportunities for the people to regain the capacity to develop their human and natural resources and create a more stable economy in the community. The various departments that make up the administrative body of the Blood Tribe include health, education, economic development, police, public works, housing and others.[43] Blood Tribe departments that normally deal with the impacts of droughts and floods include health, public works, housing, and police.

Since floods have affected the community most, some strategies to avoid and/or recover from flood impacts have been developed, but these strategies have been mostly reactive; that is, these strategies have been developed or implemented after the occurrence of a flood.

Examples of reactive strategies for the 2005 flood include the quick reaction by the Housing Department in covering leaking doors and roofs and using sump pumps to drain water out of flooded basements.[44] The Housing and Public Works Departments joined efforts and coordinated the availability of sump pumps for flooded basements. The Health Department provided bottled drinking water for residents because of the contamination of water wells with surface water; and, in the absence of bottled water, it had an advisory campaign for people to boil water before drinking it. Since the flood of 2002, Public Works has improved the quality of roads through more compaction and better drainage, which will prevent future road washouts during heavy precipitation events.[45] Also, due to the increase of flood events over the last ten years, the Fire and Emergency Services department has acquired hip gaiters, flood lights, and self-containing generators in preparation for future floods.[46]

For preventative or anticipatory strategies, the Housing Department now conducts soil tests on lands where a home is to be built, making sure that the soil has the proper drainage capacity; thus, minimizing the risks of future flooding. Also, the Housing Department is negotiating with Indian Affairs for funding to move homes away from the flood plains, but this option is limited by issues of land occupancy rights. Homeowners that do not have rights of occupancy to lands other than in the flood plains, cannot easily move to another location; therefore, some of these homeowners are asking the housing department to put their homes on stilts, lifting their homes above flood levels—this could be the only option for some homeowners.[47]

Blood Tribe Constraints to Adaptation

The Blood Tribe's departments of Housing and Public Works have been the main institutional bodies that have implemented adaptation strategies to address flood impacts. Although some of the strategies by these departments aim to prevent future impacts, most of the strategies implemented so far have been to cope with damages that have already occurred. The community's limited ability to develop long term strategies to address impacts of floods is very much mediated by some factors outside their control. In addition to the risks of being physically more exposed to climatic hazards, the Blood Tribe community faces challenges in developing adaptation strategies to climate change impacts because of the limitations posed by the imposition of a political system. This political system has greatly influenced the social, economic, and environmental conditions in the community as well as its own political and institutional arrangements.

The imposition of this political system has been done through policies such as the Indian Act, a piece of legislation that some First Nations people consider greatly discriminatory: the Indian Act as a piece of legislation is "discriminatory from start to finish . . . [that no] just society, and no society with even pretensions of being just can long tolerate."[48] Moreover, through the Indian Act First Nations defer the control of their resources such as land, oil, gas, and water to the Canadian government. In the case of the Blood Tribe, the Department of Indian Affairs and Northern Development (DIAND) administers "royalties and other revenues from natural gas and oil production on reserve lands."[49] Yet, at the same time, the Blood Tribe people "receive more than 80% of their annual revenue from the federal government in the form of funds allocated for specific purposes."[50] For some, the Indian Act and the historical legacy of residential schools have articulated the Canadian government's efforts to "civilize, that is, to assimilate, Indian Tribes."[51]

Interviews conducted for this study reflect that the control over the political structure, resources, land tenure system, and economic development of the Blood Tribe have been severely constrained by the Indian Act, and for some of the people the ramifications and consequences are fairly clear:

> Economic development, it's probably the biggest [challenge in the community]. I don't know how many millions of dollars are spent on social services and other programs like that . . . from an economic standpoint, we are nothing if we don't have an economy, that's why you see so much poverty, suicide, drug abuse . . . If we could get our own economic engine started here, people would be a lot more prosperous and happier, and cut down on a lot of the social ills that are upon us. This bad economy which was forced upon us by Indian Affairs . . . in

my opinion, that's the biggest problem . . . if we could get our economy started somehow that would alleviate a lot of our problems.[52]

Our Indian Reserve Land [is] held collectively [by] the Chief and Council for the members of the tribe; we don't have individual legal title. There's no real estate value; we can't sell it . . . around the turn of the century bison disappeared, which was our livelihood, we were forced to adopt agriculture . . . the Bloods adapted to agriculture very well and we were out-producing the non-natives across the river in grain, hay, and livestock production. After we started doing that, they started complaining and they were lobbying the government. They implemented restrictions on how much we could produce after that. When the playing field was level, so to speak, we were out-producing the non-natives. Then agriculture started to improve with technology and getting mechanized with farm machinery. At that time, because we couldn't take our land to the bank and borrow on it to buy equipment, we couldn't afford the new equipment and that's when we started falling behind. We never used to rent to non-natives. In 1960 there was a five year lease agreement to allow a number of non-natives to farm our land . . . It's been 45 years and 200,000 acres of our crop land is being farmed for the most part by non-natives . . . That is something that has been a challenge that we're trying to address. How do we get back out there?[53]

With respect to the forceful internship of First Nations children into residential schools (run by the Roman Catholic, Anglican, and Presbyterian churches), people in the Blood Tribe believe that the consequences of separation from parents and community, the prohibition of their language and traditional practices, continue to reverberate in the Blood Tribe community and manifest themselves in the form of many social ills including alcohol and drug abuse, crime, family violence, and a general disintegration of social cohesion:

They established residential schools on the reserves throughout Canada, mostly Anglican and Catholic . . . They established those schools and forced the children to go; took them out of their homes . . . Obviously, at that time they didn't have any idea of the social implications. They broke up families, and they [children in the schools] were persecuted if they spoke their Native tongue and were forced to speak English. They didn't allow us to practise our traditional customs and ceremonies; they were trying to assimilate us. We were considered savages.[54]

For most of the Blood Tribe people the high level of unemployment in the community is the inevitable consequence of the lack of skills, training, and business opportunities, and, more importantly, the lack of access to capital. The lack of access to capital is deemed to be one of the biggest factors for the lack of economic activity in the community. Not being able to access capital prevents community members from starting up small businesses or engaging in agricultural activities. According to some respondents, to undertake a farming operation requires large amounts of funds for machinery and equipment, and therefore individuals most likely to succeed are those that undertake a large farming operation so as to be able to recover the investments in expensive machinery and equipment. However, individual members of the Blood Tribe have difficulties accessing capital because they do not have fee simple title to their lands. Faced with these limitations, the best option for those who have occupancy rights to lands in the Blood Tribe is to lease them out:

> Most of the [land] owners lease out their land; I only know of one or two native farmers [who do not lease], but everybody else leases. We can't afford to buy this machinery these days, because, as I said, I only have 100 acres, I can't expect to make a living off of 100 acres, you need maybe 1,000 acres at least before I buy a half a million dollar tractor, half a million dollar combine. All these stuff costs big money, so they have a stranglehold over us there.[55]

Difficulty in accessing capital is not the only factor that influences the low level of economic activity in the community. For some, the lack of strong leadership by Chief and Council is also a significant factor that contributes to the lack of economic opportunities:

> We have the largest reserves in Canada, but we are poor, we are very poor. Basically, because all the employment that you see here on the reserve ... [is] in this building [the building that contains most of the band administration offices, a grocery story and a café], everything else ... for example, as far as major highways is concerned, the maintenance of the reserve, farming, even the rare rock ammonite that is being mined, all of these are being done by non-natives. All the major, major employment is done by non-natives; that's not our doing; that's the doing of our leaders, and that's what Indian Affairs is doing now.[56]

The type of land tenure also influences the type of economic activity in the community. Although the Indian Act stipulates that the land belongs to the Band collectively, not all members have access to or benefit from the land.

Economic revenues from leasing the land only benefits the 10–12% of the Blood Tribe people that have occupancy rights to the land, but the vast majority do not obtain any benefits. This inequity with regards to benefits from the land has created divisions and conflicts in the community. Several respondents believe that resolving the issue of land tenure is critical for addressing the many land disputes in the community and can potentially lead to economic activities that are more locally controlled. A study conducted in the community regarding the flow of money in and out of the community showed that the leasing of land to non-natives results in a net annual outflow of $48 million.[57] One non-native farmer alone rents 70,000 acres from the Blood Tribe.[58] There is a feeling that the issue of land tenure would not easily be resolved by Chief and Council because of its sensitivity and the influence and power of landowning families.

> It would probably be fairly difficult [to change the land tenure system], it would have to be phased in . . . There has been talk that the majority that don't occupy land [are] frustrated . . . it's very clear in the legislation that governs this place that it's [the land] for the use and benefit of everyone. So it'll take political will . . . Why it hasn't changed is because the ones that occupy land have more money. They are maybe more upstanding citizens in the community and maybe have more influence and their families have elected them to leadership over the years. The majority of the Chief and Council occupy land. I think there's only one or two that don't occupy land. It's a . . . change that I think needs to happen sooner than later.[59]

The combination of social and economic hardships, land disputes, and mistrust of the Chief and Council, and managers and directors of entities such as the Blood Tribe Agricultural Project, the Kainai Agribusiness Corporation and Kainai Resources, appear to be contributing to feelings of internal marginalization and the loss of social cohesion and networks in the community:

> We don't know where that money [from resources] goes. They say that it goes to cap off operations for the reserve, but yet when you look at it, there is only a handful of people here who are benefiting off this land. Now there is a motion where they are saying, "well, we are going to control all our oil resources, oil and gas resources" . . . this is going to happen, but how is this going to benefit the people? It will not. It's just going to disappear that much faster [the money], so everything that the people on this reserve feel that—or they

should—by nature, by law, should be benefiting them, is not. So people are left . . . well, you see them walking around, they don't have anything, while you see others driving brand new vehicles.[60]

Chief and Council need to put their feet down and [see] where [the] money is going. When the dollar becomes your almighty then you start having problems. They [Chief and Council] have to start sitting down with the people. It goes to their heads and they forget about us. There's no communication with the people. They don't come down to our level.[61]

We just had an election seven months ago. During the election everyone [politicians] promised transparency: "there is going to be a huge change, we are going to work for the people, blah, blah, blah.' . . . [But] they have no new ideas to introduce to the people to help them start their own economy. There is no economy on this reserve, there is nothing here. If you look at it, there are gas stations, but even then those gas stations are struggling.[62]

Another thing that really destroyed our members . . . it used to be with the clan system that families lived together; extended families helped one another, they never saw anyone starve or to be without the basic: food, clothing and shelter; that's how we were. In the 60s we were forced to go under the election system and that's when we started creating the town sites and so many of our people were forced to live in the town sites; they were just put there, and their families were far away. So we lost contact with who our families are; many of our members now don't know which clan they are from.[63]

The adaptability of indigenous people like the Blood Tribe to coping with climate variability prior to the arrival of Europeans has been eroded. With the change in their type of livelihood, the strategies that they had developed from past experiences and accumulated knowledge have largely been forgotten. Although the Blood Tribe is one of the largest reserves in Canada, this reserve land is only a fraction of the territory that they had access to prior to signing a treaty with the Canadian government. Therefore, for the Blood Tribe their access to natural resources, which is one of the key factors for coping with climate variability, has been profoundly changed and has compromised their adaptive capacity. Moreover, effects of the Indian Act and Residential Schools have generated personal trauma as well as a break down in interpersonal relations, and have negatively affected the level of trust, social networks, and support in the community, and therefore further reduce the adaptive capacity of the community.

Given the many and more immediate social and economic challenges that the Blood Tribe community faces, it is not surprising that for most of the people climate change variability and their impacts are not, for now, a major concern:

> Cancer is very high; sugar diabetes is very high. So . . . people here because they have had nothing for a long time they feel that they don't have a sense of belonging to the tribe anymore. They just don't care what happens out there with the farmland; unless you are a farmer or landowner you don't care about the farmland. Unless you have land or cattle you'll care about the moisture; other than that, people here in the community live on a day to day basis. They are wondering where they are going to get their next meal. You go to social services [and] you only get a cheque once a month, which is maybe about $330 and you have to live on that for a month. So, it's pretty bad.[64]

> So there are a lot of issues I could bring up, but what we are talking about today . . . the people don't really care about climate change here. They live on a day to day basis. Morale right here is very low. Our average age of death I think is 50 years.[65]

Conclusion

To recapitulate, the imposition of a political system and policies of assimilation by colonial and current Canadian governments on the Blood Tribe community have significantly and negatively altered their access to natural resources, governance and institutional arrangements, social networks, and human and technological resources, thus limiting the adaptive capacity of the community to climate change impacts.

Through their displacement and relocation onto reserves, the Blood Tribe, like other First Nations, lost access to once vast traditional territories and became dependent on the Canadian state for housing, education, health care, employment, and other living essentials. The combination of an imposed governing structure and the loss of their land base and traditional forms of livelihood has resulted in the loss of economic, human, and social capital. Without these, the community will continue to face great challenges in their adaptation to climate change impacts.

Climate change is affecting societies worldwide and, as is increasingly acknowledged, those affected most are societies that face other social and economic stressors. For the Blood Tribe, understanding their contextual socio-economic and political conditions informs one of their particular vulnerability and adaptive capacity to climate change impacts. Given the trends and predicted impacts of climate change in the South Saskatchewan River Basin, it is important that the Blood Tribe develop anticipatory strategies to cope with floods and droughts. The Blood Tribe community is a clear example of situation where other stresses affect the community more directly and immediately; therefore, in order to enhance the coping ability of the community to climate change impacts it is critical to address these stresses. Strategies that increase access to resources, reduce poverty, unemployment, and the lingering effects of residential schools, can greatly increase the financial, human and social capital of the community. The alleviation of these non-climate related stresses will strengthen the community's adaptive capacity to climate related stresses.

Clearly, past and present Canadian government policies have contributed to the Third World conditions in First Nations communities like the Blood Tribe. Improving living conditions in First Nations communities will rectify some of the injustices they have been subjected to, while at the same time strengthening their adaptive capacity to cope with climate change impacts.

Endnotes

1. Blood Tribe community member (BT1), personal communication, 2005.
2. D. Sauchyn and W. R. Skinner, "A Proxy Record for Drought Severity for the Southwestern Canadian Plains," *Canadian Water Resources Journal,* Vol. 26, No.1 (2001):253–272.
3. J. Daschuk and G. Marchildon, "Climate and Aboriginal-Newcomer Adaptation in the South Saskatchewan River Basin, 1700–1800," IACC Project Working Paper No. 8, 2005.
4. Ibid.
5. Treaty 7 was signed in 1877 between several Blackfoot First Nations communities—settled in what today is southern Alberta—and the Government of Canada. It is one of the eleven numbered treaties signed between First Nations and the Government of Canada. First Nation communities signatories of Treaty 7 include the Piikani, Blood Tribe, Siksika, Stoney, and Tsuu Tina. Available from Indian and Northern Affairs Canada at: http://www.ainc-inac.gc.ca/pr/trts/hti/guid/tr7_e.html (accessed December 2006).
6. The annual funds transferred to the Blood Tribe by the Canadian federal government amounts to $120 million. Of this, $117 million is destined for the social needs in the community and $3 million is allocated to economic development (Blood Tribe Council member, personal communication, 2008).

7. Lands Management Department, Blood Tribe Special Report June 2006, Published by Blood Tribe Administration Public Relations Department, 1–24.
8. Ibid.
9. Ibid.
10. Ibid.
11. Government of Canada, Statistics Canada, Community Profiles, Aboriginal Community Profiles 2001. Available at: http://www.statscan.ca (accessed October 2007).
12. Sauchyn and Skinner, "A Proxy Record," 253–272.
13. Alberta Agriculture, Food and Rural Development, "Southern Alberta Drought," 2002. Available at: http://www1.agric.gov.ab.ca/$department/deptdocs.nsf/all/irr4416 (accessed November 2006).
14. B. Harker, J. Lebedin, M. J. Gross, C. Madramootoo, D. Neilsen, B. Paterson, and T. van der Gulik, "Land practices and changes—Agriculture, " in *Threats to Water Availability in Canada 2004*, National Water Research Institute, Burlington, Ontario, WRI Scientific Assessment Report Series No. 3 and ACSD Science Assessment Series No. 1, 21.
15. M. Khandekar, "Trends and Changes in Extreme Weather Events: An assessment with focus on Alberta and Canadian Prairies." Report prepared for Science & Technology Branch, Alberta Environment, (2002). ISBN: 0–7785–2429–9 (On-line Edition): ww.gov.ab.ca/env/info/infocentre/publist.cfm.
16. Ibid.
17. Natural Resources Canada, Climate Change Impacts and Adaptation Directorate, "Climate Change Impacts and Adaptation: A Canada Perspective," 2004. Available at: http://adaptation.nrcan.gc.ca/perspective_e.asp (accessed October 2006).
18. Assembly of First Nations, *Fact sheet: the Reality of First Nations in Canada.* Available at: http://www.afn.ca/article.asp?id=764 (accessed November 2006).
19. The Community Well Being index (CWB) is a product of Indian and Northern Affairs Canada's Research and Analysis Directorate. It was derived from the 2001 Census and is a means of measuring well-being in Canadian communities. It combines indicators of income, education, labour force activity, and housing conditions into a single number or "CWB" score. CWB scores may fall anywhere between zero (0) and one hundred (100), with one hundred being the highest. A score was generated for each community that participated in the 2001 Census, allowing an "at-a-glance" look at the relative well-being of those communities. Note that the scores are not reported for communities with fewer than 65 inhabitants or those with data quality issues (INAC, First Nations Profile webpage).
20. Government of Canada, Statistics Canada, Community Profiles, Aboriginal Community Profiles 2001, online.
21. Ibid.
22. Flood Supervisor (BT2), personal communication, 2005.

23. Ibid.
24. Blood Tribe Housing Department, "Special Report," March 2006, Blood Tribe Administration Public Relations Department.
25. Ibid.
26. Poverty-Environment Partnership, "Poverty and Climate Change: Reducing the Vulnerability of the Poor through Adaptation," 2003. UNDP, UNEP, World Bank, ADB, AfDB, GTZ, DFID, OECD, EC on behalf of the Poverty-Environment Partner. Available at: http://www.dfid.gov.uk/pubs/default.asp?move=10&TypeID=0&TopicID=5&CountryID=0&LanguageID=0&YearID=0&FreetextID=0&Listby=date (accessed August 2007). Poverty-Environment Partnership–a partnership of various high profile organizations including the United Nations Development and Environmental programmes, the World Bank and the Organization for Economic Co-operation and Development Programme.
27. Ibid.
28. Task Force on Climate Change, Vulnerable Communities and Adaptation, *Livelihoods and Climate Change: Combining disaster risk reduction, natural resource management and climate change adaptation in a new approach to the reduction of vulnerability and poverty—A Conceptual Framework* (Winnipeg, Manitoba: ISSD, IUCN WCU, SEI, SACD, ISSD, 2003), 6. The Task Force on Climate Change, Vulnerable Communities and Adaptation is also a partnership of key environmental organizations including the International Institute for Sustainable Development and the World Conservation Union.
29. W. N. Adger, S. Agrawala, M. M. Qader Mirza, C. Conde, K. O'Brien, J. Pulhin, R. Pulwarty, B. Smit, and K. Takahashi, "Assessment of adaptation practices, options, constraints and capacity," in M. Parry, O. Canziani, J. Palutikof, P. van der Linden, and C. Hanson (eds.), *Climate Change 2007: Impacts, Adaptation and Vulnerability.* Contribution of Working Group II to the Fourth Assessment Report of the Intergovernmental Panel on Climate Change (Cambridge, UK: Cambridge University Press, 2007), 717–743.
30. Ibid.
31. Ibid.
32. Ibid.
33. Ibid.
34. G.W. Yohe, R.D. Lasco, Q.K. Ahmad, N.W. Arnell, S.J. Cohen, C. Hope, A.C. Janetos, and R.T. Perez, "Perspectives on climate change and sustainability," in Climate Change 2007: *Impacts, Adaptation and Vulnerability,* 813.
35. Flood Supervisor (BT2), personal communication, 2005; and, Blood Tribe Housing Department, "Special Report," March 2006.
36. Director, Blood Tribe Department of Public Works (BT3), personal communication, 2005.
37. Manager, Kainai resources (BT4), personal communication, 2005.

38. Director, Kainai Agribusiness Corporation (BT5), personal communication, 2005.
39. BTAP Operations manager (BT6), personal communication, 2005.
40. Blood Tribe Director Emergency Services and Fire department (BT7), personal communication, 2005.
41. Flood Supervisor (BT2), 2005.
42. Director Blood Tribe Lands Department (BT8), personal communication, 2005.
43. Blood Tribe (online), available at: http://www.bloodtribe.org/community/ community.html.
44. Flood Supervisor (BT2), 2005.
45. Director, Blood Tribe Department of Public Works (BT3), 2005.
46. Director Emergency Services and Fire department (BT7), 2005.
47. Flood Supervisor (BT2), 2005.
48. Report of the Royal Commission on Aboriginal Peoples, 1991, Volume 1—Looking Forward Looking Back, Part Two: False Assumptions and a Failed Relationship, Chapter 9—The Indian Act, 3.
49. A. Long, "Political Revitalization in Canadian Native Indian Societies," *Canadian Journal of Political Science / Revue canadienne de science politique,* Vol. 23, No. 4 (December 1990): 752–3.
50. Ibid.
51. Ibid.
52. Blood Tribe community member (BT8), personal communication, 2005.
53. Director, Blood Tribe Department of Public Works (BT3), 2005.
54. Blood Tribe community member (BT9), personal communication, 2005.
55. Blood Tribe community member (BT10), personal communication, 2005.
56. Blood Tribe community member (BT1), 2005.
57. Director Public Relations (BT11), personal communication, 2005.
58. Ibid.
59. Blood Tribe community member (BT12), personal communication, 2005.
60. Blood Tribe community member (BT1), 2005.
61. Blood Tribe community member (BT13), personal communication 2005.
62. Blood Tribe community member (BT1), 2005.
63. Blood Tribe community member (BT10), 2005.
64. Blood Tribe community member (BT1), 2005.
65. Ibid.

CONTRIBUTORS

Joel Bruneau is an assistant professor and undergraduate chair in the Department of Economics at the University of Saskatchewan.

Darrell R. Corkal is a senior water quality engineer for Agriculture and Agri-Food Canada.

James Daschuk is an assistant professor in the Faculty of Kinesiology and Health Studies at the University of Regina.

Harry Diaz is the Institutional Adaptation to Climate Change project director, the Executive Director of the Canadian Plains Research Center, and a professor of Sociology and Social Studies at the University of Regina.

David A. Gauthier is the Vice President (Research and International) at the University of Regina.

Margot Hurlbert is an assistant professor in both the Department of Sociology and Social Studies and the Department of Justice Studies at the University of Regina.

Suren Kulshreshtha is a professor in the Department of Bioresource Policy, Business and Economics at the University of Saskatchewan.

Suzan Lapp is completing her Ph.D. thesis on future climate drought indices at the University of Regina.

Lorenzo Magzul is completing his Ph.D. in the Faculty of Land and Food Systems at the University of British Columbia.

Gregory P. Marchildon holds the Canada Research Chair in Public Policy and Economic History at the Johnson-Shoyama Graduate School of Public Policy at the University of Regina campus.

Brett Matlock is a graduate student in Sociology at the University of Regina, and has a Cultural Resource Management designation from the University of Oregon.

Lorena Patiño is a Ph.D. candidate in the Department of Geography at the University of Regina.

Elise Pietroniro is a geographical areas analyst with Statistics Canada in Ottawa.

Jeremy Pittman completed his graduate studies in the Department of Geography at the University of Regina and now works for Agriculture and Agri-Food Canada.

Bernardo Reyes is a prominent Chilean environmentalist who works for the Institute of Political Ecology and the international organization ForestEthics.

Alejandro Rojas is an assistant professor of agroecology in the Faculty of Land and Food Systems at the University of British Columbia.

David J. Sauchyn is a professor of geography at the University of Regina and the Research Coordinator at the Prairie Adaptation Research Collaborative (PARC).

Barry Smit is Canada Research Chair in Global and Environmental Change and a professor in the Department of Geography at the University of Guelph.

Brenda Toth is a hydrologist with Environment Canada.

Garth van der Kamp is a research scientist with Environment Canada.

Johanna Wandel is an assistant professor in the Department of Geography and Environmental Management at the University of Waterloo.

Elaine Wheaton is a senior research scientist for the Saskatchewan Research Council's climatology group and an adjunct professor at the University of Saskatchewan.

Virginia Wittrock is a research scientist in climatology with the Saskatchewan Research Council.

Gwen Young completed her graduate research in community vulnerabilities to climate change in the Department of Geography at the University of Guelph.

INDEX

The body of this book is set in ADOBE CASLON. The Englishman William Caslon punchcut many roman, italic, and non-Latin typefaces from 1720 until his death in 1766. At that time most types were being imported to England from Dutch sources, so Caslon was influenced by the characteristics of Dutch types. He did, however, achieve a level of craft that enabled his recognition as the first great English punchcutter. Caslon's roman became so popular that it was known as the script of kings, although on the other side of the political spectrum (and the ocean), the Americans used it for their Declaration of Independence in 1776. The original Caslon specimen sheets and punches have long provided a fertile source for the range of types bearing his name.

Carol Twombly designed this Caslon revival for Adobe in 1990, after studying Caslon's own specimen sheets from the mid-eighteenth century. This elegant version is quite true to the source, and has been optimized for the demands of digital design and printing. *Adobe Caslon™* makes an excellent text font and includes just about everything needed by the discriminating typographer: small caps, oldstyle figures, swash letters, alternates, ligatures, expert characters, central European characters, and a plethora of period ornaments.

The titles and accents in the book are set in SWISS 721. It was designed in 1982 by Bitstream Inc., and is a variation of Helvetica.